LIVESTOCK AND MEAT MARKETING
Second Edition

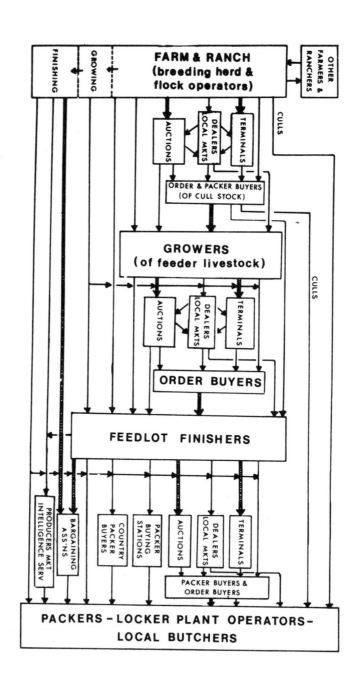

LIVESTOCK AND MEAT MARKETING

SECOND EDITION

John H. McCoy, Ph.D.

Professor of Agricultural Economics,
Kansas State University,
Manhattan, Kansas

AVI PUBLISHING COMPANY, INC.
Westport, Connecticut

© Copyright 1979 by
THE AVI PUBLISHING COMPANY, INC.
Westport, Connecticut

Library of Congress Cataloging in Publication Data

McCoy, John Henry, 1912–
    Livestock and meat marketing.

    Includes bibliographies and index.
    1. Meat industry and trade—United States. I. Title.
HD9415.M25 1979   381'.41'600973   79-12205
ISBN 0-87055-321-6

Printed in the United States of America

Dedication

Dedicated to my students—past, present, and future.

Preface to the Second Edition

In this edition I have attempted to incorporate the numerous changes which have occurred since the original publication. In addition, I have attempted to improve presentation of the material. In both regards I have been favored by many suggestions from students and colleagues, and there is no better place to test one's material than the classroom.

Interest in marketing livestock and meat has increased measurably in recent years. This reflects a growing recognition that efficiency in production alone is not sufficient to assure success in farming and ranching. There is also a growing recognition that, while many improvements in marketing are within the control of individual operators, some of the most vexing problems require group action. That does not come easily among the many independent minded livestock operators. Of particular interest is the quest for greater understanding of the futures market and its place in reducing the inherent risk of adverse price movements—a risk which has increased with spiraling operational costs and volatile product prices. Of equal interest is the desire for information on alternative methods of marketing livestock. This has stimulated innovations and experimentation in several variations of electronic markets. A persistent decline in the number of farmers and ranchers, and related increase in size of operations—together with an increase in direct marketing has put many relatively small producers in an unfavorable marketing position. The development of alternative marketing methods is imperative if the small producer is to remain in livestock production.

Changes have occurred in virtually every aspect of marketing. Livestock and meat production and marketing is a truly dynamic industry. Interesting and challenging developments are on the horizon. My major objective in preparing this text is to lend some assistance to our young men and women in understanding the system and in improving it. They are an intelligent, resourceful lot and I am confident they can meet the challenge.

JOHN H. McCOY

January, 1979

Preface to the First Edition

The task of directing production and of moving some 38 billion pounds of highly perishable products into the hands of consumers requires a vast and complex market system. In the years immediately following World War II much criticism was levelled at the system for its lack of change and adaptation in a rapidly changing world. Not all of the criticism was warranted for the historical record shows change. Nevertheless, the pace of change has quickened in recent years and, like many other aspects of current economic and social life, possible developments in livestock and meat marketing challenge the imagination.

This book is designed to present a balance of descriptive analysis of the marketing system together with the basic economic principles underlying its operation. The text is intended for undergraduate students. It is assumed that readers have an acquaintance with elementary economic theory, but some of the more relevant principles are reviewed in applying them to this particular field. Emphasis is placed on practical aspects rather than academic discussion. This text is the product of a number of years of classroom give and take with students from all agricultural curricula, but the predominant student influence has been from majors in agricultural economics and animal science. The book is intended to provide the chief reference for courses dealing exclusively with livestock and meat marketing. Relevant chapters may be utilized as supplementary reference material for appropriate sections of general agricultural marketing courses. Readers who wish to delve deeper in the various subject matter areas will find additional references at the end of each chapter.

The author acknowledges the contribution of former students and colleagues of Kansas State University and the University of Arizona in their numerous suggestions for improvements in content and organization. Indebtedness also is acknowledged to the many authors of publications cited throughout the text.

JOHN H. McCOY

January 10, 1972

Contents

Introduction

LIVESTOCK-MEAT, AN IMPORTANT INDUSTRY

The production and marketing of livestock and related products is one of the largest and most important industries in the world. Millions of producers depend upon livestock raising for a livelihood in both developed and developing countries. The degree of processing and sophistication of distribution systems varies directly with economic development, but on a global basis vast amounts of resources—manpower, land and capital—are devoted to the industry.

Marketing systems differ from country to country, and variations occur within country, but all have one primary purpose—to move products from producer to consumer. That is a staggering responsibility. The quantity is enormous, distances often are great, and most of the products are highly perishable.

The importance of an industry can be viewed in many different ways. In an economic sense, importance is attached to such factors as the number of people employed (directly and indirectly), amount of investment, quantities of physical resources used, and quantities of products produced. From the standpoint of physical health of the nation, concern is attached to wholesomeness and nutritive aspects of the products of that industry. Governmental agencies are concerned with the contributions an industry may make to tax revenues; and in recent years concern has become pronounced with social costs in such aspects as environmental pollution. Importance also may be attached to an industry's contribution to national defense. Some aspects of importance are capable of quantitative measurement, but some cannot be measured objectively. Some could be measured, but data have not been assembled. These limitations apply to the livestock-meat industry. However, enough information is available to show that, by any reasonable standards, livestock and meat are big business in the United States.

Product Output

Products of this industry are essential to the comfort, health, and well-being of countless millions the world over. Meat, of course, is the primary product. Lean meat is prized as a basic source of high quality protein. As strange as it may seem to a majority of Americans, the fat component of meat is equally desired by many peoples.

In the U.S. alone approximately 17.8 billion kilograms[1] (39.2 billion pounds) of red meat were produced and an additional 1.0 billion kilograms (2.2 billion pounds) were imported and moved to consumers during 1977. That was an average of 87.6 kilograms (193.2 pounds) for every man, woman and child in the country. A greater total tonnage of meat is consumed in the U.S. than in any other country, but on a per capita basis U.S. is outranked by Argentina, New Zealand, Uruguay and several others.

An often overlooked and sometimes unappreciated output of the industry is the wide array of products that are derived from livestock in addition to carcass meat. These products may be classified as edible by-products, inedible by-products, pharmaceuticals, and variety meats. The following is only a partial listing of such products. Among the better known variety meats are liver, heart, tongue, kidneys, brains, and sweetbreads. Edible by-products include oleo stock, oleo oil gelatin, suet, and sausage casings. Among the more commonly recognized inedible by-products are leather, wool, mohair, inedible fats, fatty acids, glues and adhesives, animal feeds, fertilizers, combs, buttons, paper, bone charcoal, and surgical sutures. Livestock are walking factories of more than 100 pharmaceutical products. Insulin is perhaps the best known. Not so well known are heparin, epinephrine, thrombin, fibrinolysin, chymotrypsin, glucogon, trypsin, parathyroid hormone, corticotropin (ACTH), thyrotropin (TSH), and vasopressen. The importance to mankind of this type of product is far beyond their monetary value.

Assets

Total assets of U.S. farmers and ranchers were $729.6 billion at the beginning of 1978 (Table 1.1). That was the current value of all equipment, livestock, land, buildings, and savings of farmers and ranchers.

TABLE 1.1. U.S. FARM AND RANCH ASSETS OF THE 48 CONTINENTAL STATES, 1976, 1977, 1978

Item	1976	1977 (Billion $)	1978[1]
Physical assets			
Real Estate	429.1	497.2	546.9
Nonreal estate	132.0	140.6	148.0
Financial assets	31.7	33.1	34.7
Total	592.8	670.9	729.6

Source: USDA (1977A)
[1]Preliminary

[1]Carcass weight.

Farm and ranch assets were twice as great as the combined assets of the 35 largest U.S. industrial corporations. Information is not available to indicate the proportion of farm and ranch assets applicable to livestock. Any such determination would show that, in addition to assets used directly in livestock production, the bulk of those assets devoted to forage and feed grain production indirectly are an integral part of the industry.

In addition to farm and ranch assets, the total industry includes the packing, processing, and distribution sectors. Beyond that are many associated, or supplementary sectors which are entirely or partially devoted to functions of the industry. Among the latter are substantial assets in facilities for transportation (truck and railroad), marketing (auctions, terminals, concentration yards, futures markets, etc.), banking and credit, veterinary services, manufacturing of veterinary equipment and supplies, feed manufacturing, livestock and meat processing equipment manufacturing, insurance companies, publication of numerous magazines and newspapers, etc. No one has ever calculated the combined value of assets devoted to livestock and meat, but there is no doubt that this industry, if not the leader, is near the top of all U.S. industries.

Cash Receipts.—Cash receipts from farm and ranch marketings totaled $96.1 billion (exclusive of government payments) in 1977 (Table 1.2). Of this total $28 billion was from meat animals, wool, and mohair. The share of cash receipts from meat animals increased up to the early 1970's, then decreased as stepped-up world-wide demand for grain coincided with a cyclical slump in cattle prices. However, in recent years cash receipts from meat animals, wool, and mohair again increased, indicating a return to the pre-1970 trend.

The importance of meat animals and wool varies considerably among

TABLE 1.2. U.S. CASH FARM AND RANCH RECEIPTS, 1975, 1976, 1977

Product	1975	Year 1976 (Million $)	1977
Cattle and calves	17,524	19,302	20,230
Hogs	7,883	7,261	7,327
Sheep and lambs	385	392	389
Total meat animals	25,792	26,955	27,946
Wool and mohair	70	98	101
Dairy products	9,923	11,428	11,776
Poultry and eggs	5,879	7,020	7,065
Other livestock items	603	651	677
Total livestock and products	43,059	46,152	47,565
Crops	45,150	48,349	48,519
All commodities	88,209	94,501	96,084

Source: USDA (1978).

states—both in absolute amount and as a percentage of cash farm and ranch receipts (see Table 1.3). Iowa was high in total dollar receipts in 1976 with slightly over $3.5 billion, while Wyoming was tops as a percentage of receipts with a combined meat animal and wool total of 73%.

TABLE 1.3.   CASH FARM AND RANCH RECEIPTS FROM MEAT ANIMALS AND WOOL FOR SELECTED STATES, 1976

State	Meat Animals ($1,000)	Wool ($1,000)	All Commodities ($1,000)	From Meat Animals (%)	From Wool (%)
Illinois	1,554,031	700	6,100,893	25.4	1
Iowa	3,540,478	1,621	7,009,696	50.5	1
Missouri	1,150,584	571	2,630,755	43.7	1
Nebraska	2,014,495	916	3,867,626	52.0	1
Kansas	1,738,711	905	3,531,560	49.2	1
N. Carolina	327,546	40	2,821,612	11.6	1
Georgia	298,873	11	2,269,021	13.1	1
Florida	235,104	11	2,532,786	9.2	1
Alabama	406,589	11	1,617,948	25.1	1
Mississippi	234,961	14	1,671,808	14.0	1
Arkansas	239,751	17	2,296,027	10.4	1
Oklahoma	1,073,801	299	1,911,932	56.1	1
Texas	2,391,864	15,879	6,298,417	37.9	.2
Montana	407,640	3,724	996,975	40.8	.4
Wyoming	280,519	7,617	386,101	72.7	1.9
Colorado	1,291,404	5,806	1,976,608	65.3	.3
N. Mexico	453,684	3,231	712,325	63.6	.5
Arizona	437,601	1,524	1,240,120	35.3	.1
California	1,151,447	6,369	9,101,860	12.6	.1

Source: USDA (1977 B)
[1]Less than ½ of 1%

A large share of the income received by farmers and ranchers is pumped back into the economy in payment for production expenses, and most of the remainder is spent for household and personal items. As meat and wool moves through marketing channels, value is added by slaughtering, processing, and distribution. Sales of red meat at retail were estimated at $50.9 billion in 1976. Thus, the value added in slaughtering, processing, and distribution pumped an additional $23.7 billion ($50.9 − $27.2 = $23.7) and this does not take into consideration the impact of wool, mohair, and exports on the economy. At these magnitudes it is apparent that the livestock-meat industry is an important component—not only in U.S. agriculture, but in the national economy as well.

Interrelationships with Other Industries

Livestock is important, not only in absolute terms, that is from the standpoint of the number of dollars turned over by the industry, but even more so in additional business activity generated in other economic sectors in servicing livestock and meat producers. This commonly is referred to as the "multiplier effect." It applies to income generated in other sectors, as well as output of goods and services in other sectors. No national studies have been made comparing the multiplier effect of meat animal sectors with other sectors, but this was done by Emerson *et al* (1973) in a study of the Kansas economy. There is reason to believe the Kansas results would apply to other important U.S. livestock areas.

Emerson and his co-workers made a comprehensive input-output analysis in which they separated the Kansas economy into 69 major sectors, or components, to study interrelationships among economic sectors. Cattle and hog production were treated separately, but sheep production, unfortunately, was not. Meat packing and processing were handled as one sector. Other agricultural sectors consisted of the major crops, dairying, poultry, grain milling, etc. Nonagricultural activity was broken down into a number of industrial and service sectors, government, and households.

Income Multiplier.—This study showed that the Kansas beef cattle industry had an income multiplier of almost 6 (Table 1.4). This means that every additional $1.00 worth of output in the beef cattle industry generated 6 times as much income in other industries as it did in the cattle industry itself. In hog production the income multiplier was 4.23, and in the meat packing and processing industry it was almost 9. Of all Kansas industries, including nonagricultural industries such as petroleum, coal, gas, building construction, aerospace and other manufacturing, only

TABLE 1.4.   INCOME AND OUTPUT MULTIPLIERS FOR LEADING KANSAS ECONOMIC SECTORS, 1973

Sector	Multiplier
Income Multiplier	
Grain mill products manufacturing	10.971000
Meat packing and processing	8.997794
Cattle production	5.997510
Dairy products manufacturing	5.551970
Crude oil and natural gas mining	4.387497
Hog production	4.230286
Dairy farming	4.010548
Food and kindred product mfg.—other than meat, dairy and grain	3.981318
Petroleum and coal products manufacturing	3.792320
Nonmetalic mining	3.624680
Output Multiplier	
Meat packing and processing	2.794279
Dairy products manufacturing	2.325515
Maintenance and repairs	2.209400
Cattle production	2.174308
Hog production	2.057495
Dairy farming	2.013863
Heavy construction	1.927127
Grain mill products manufacturing	1.839749
Building construction	1.739112
Electric, gas and sanitary services	1.643615

Source: Emerson *et al.* (1973).

grain milling had a higher (i.e., slightly higher) income multiplier than meat packing-processing. The grain milling multiplier was 10.97, compared to 9 for meat packing and processing. It may be noted that also is an agricultural sector. By comparison, the income multiplier effect of several other important sectors was: petroleum and coal products manufacturing, 3.79; dairy products manufacturing, 5.55; crude oil and natural gas mining, 4.39; nonmetalic mining, 3.62; dairy farming, 4.01; and building construction, 1.74. The income multipliers of crops were considerably lower than those of livestock activities. None of the crop sectors are in the top ten.

Output Multiplier.—Another measure of interrelationships among economic sectors is the output multiplier. The output multiplier of the cattle sector was 2.17. This means that each additional $1.00 output in the cattle sector generated 2.17 times as much output in other sectors as it did in the cattle sector itself. In hog production the output multiplier was 2.06 and in meat packing-processing it was 2.79. As Table 1.4 shows, meat packing-processing had the highest output multiplier of any economic sector, cattle production ranked fourth from the top among all industries of the state, and hog farming was fifth.

It is readily apparent from these data that the livestock and meat sectors are highly interrelated with other industries in the state of Kansas. Logically, comparable relationships would be expected in other important livestock states or regions. The significance of these relationships lies in the indicated importance these sectors have in the economic development of a state or region. They provide the clue that economic development of the state could be enhanced by research and extension activities which would promote development of livestock and meat production.

Land Utilization

Livestock production utilizes a far greater area of land than any other activity. On a worldwide basis it is estimated that about two-thirds of the land devoted to agriculture is in range land, meadow, and permanent pasture (National Research Council 1977). In the United States 69% of the agricultural land was grazed in 1969 (Nix 1975). Even if one includes non-agricultural land in the United States, more than one-half of the total land area is devoted to "Range, consisting of grassland, shrublands, and open forests. . . ." (USDA 1974). In the western states, a much higher percentage is devoted to grazing. The production of supplementary pastures, forages, and feed grains necessitate additional vast acreages. A major share of the feed used by western livestock comes from native range, pastures, and meadows on land which, under present conditions, has virtually no other feasible alternative food production use. The same situation prevails through the Flint Hills and Osage Pasture areas of Kansas and Oklahoma, and many other localized areas in all parts of the United States.[2] Livestock (primarily cattle and sheep) utilize this native vegetation and enormous quantities of roughage, crop aftermath, etc., which otherwise would contribute nothing to the food supply. Forages, both native and cultivated, provide an economical means of growing livestock preparatory to finishing for slaughter. Requirements of the grain-finishing operation support virtually the entire feed grain industry which in itself is a major user of much of the best land in the United States. Consumers spend more for meat than any other food item, which attests to the importance attached to the end product.

Employment

Published data are not available on the number of people employed in the production, processing, and marketing of livestock and meat. Directly or indirectly, nearly all farmers and ranchers have a hand in it. Most of those who do not produce livestock are involved in feed production to a greater or lesser degree. The relationship is obvious for those in feed grain and forage production. It is less obvious in such crops as cotton and sugar beets, but many crops do have feed by-products such as cottonseed meal and beet pulp. Even fruit and vegetable producers are involved in a minor way because limited quantities of pulp and refuse are used for livestock feed. In 1976 there were 4.4 million persons employed in farming (USDA 1977C). Assuming 85% of that number had some input in either direct livestock production or feed production there were approximately 3.7 million engaged in production phases. In all of agriculture there are about 8 people in marketing for every 6 engaged in production. If that same ratio holds true for the livestock sector, there would be some 5 million people engaged in the marketing of livestock and meat. In total, then, there would be 8.7 million people employed in production and marketing. And if all service and complementary employment were included, such as veterinarians, insurance agents, publishers, equipment manufacturers, etc., it is probable that the total amounts to 10 to 11 million.

MARKETS AND MARKETING

Marketing—the Concept

Marketing is that area of economics concerned with the exchange and valuation of goods and services.[3] This definition encompasses (1) ac-

[2]Some of the lands, of course, have other uses such as recreation, wildlife habitat, mining, and watersheds for hydroelectric, irrigation and urban water supplies.
[3]Agreement is not universal on an acceptable definition of marketing (or of the term market). The usefulness and validity of a definition is associated with its application. A different definition may be perfectly proper and correct depending upon the use made of it.

tivities associated with the physical movement and transformation of goods and (2) the pricing of goods and services. In some aspects the physical functions of marketing are related to production. Production, in economic jargon, is defined as the creation of "utility" or usefulness. Utility may be created by changing the form, location, availability over time, or possession (ownership) of a product. In a narrow sense, marketing sometimes is construed to apply only to exchange of title (Bakken 1953). This would limit it to problems of possession utility including pricing and activities associated with buying and selling. However, by long tradition marketing has been considered to be broader than this. By the so-called functional approach to the study of marketing it would include, in addition to buying and selling, such functions as transporting, storing, processing, packaging, advertising, collecting and disseminating market news, standardizing and grading, inspecting, financing, and risk bearing. This would appear to include the creation of all types of utility listed under the definition of production; and prompts the question of whether there is a difference between production and marketing. There is little to be gained in belaboring this point. For administrative purposes, the USDA considers marketing to be limited to those activities which take place from the time products leave the farm gate. Activities performed prior to that are considered to be production. But clearly, there can be a direct relationship between the two. A marketing program may be influenced by prior decisions that determine the quantity, quality, or timing of production. Even prior to that, a decision which determines the quality of feeder livestock purchased may directly affect the feeding program. Marketing and production are closely intertwined. There is a growing awareness that efficiencies may be gained by considering production and marketing as one integrated or coordinated system. This is one application of the so-called systems approach to economic as well as noneconomic problems, which has received increasing attention in recent years. We will not attempt to draw a firm distinction between marketing and production.

What Is a Market?

At first thought, the concept of a market may appear rather straightforward. One can see the livestock auction building and pens on the outskirts of his home town, or the big office building and yards of public terminal livestock markets at such places as Kansas City or Omaha. But it also is rather common to hear references to the "cattle" market, the "corn" market, and the "automobile" market. On other occasions the reference may be to the "futures" market and the "cash" or "spot" market. In the first instance, a market was associated with a particular place; in the second, the association was with a commodity; and in the third, with an element of time. In other cases a market has been defined as: a particular group of people, an institution, a mechanism for facilitating exchange, the perfect market, and the imperfect market. The market concept also has been linked to the degree of communication among buyers and sellers and the degree of substitutability among goods. While there are differences among these concepts, any one could be correct for the purpose for which it might be used. The concept of a perfect market is an abstraction used by economists as a benchmark for evaluating performance of market situations that deviate from its specifications. Further reference will be made to the perfect market in later sections.

Our application here of the term market does not necessitate a single choice of definitions. Depending upon the particular context, reference will be to certain geographically located markets; on other occasions to markets for particular class, grade, and weight of livestock. Various aspects of the cash and futures markets, as well as the agencies and institutions that make up these markets will be analyzed at some length. In other words, a comprehensive concept of the market will be used. It is generally agreed that a viable, competitive market requires more than physical facilities. The assembly and concentration of salable livestock may tend to promote competition. But the essential ingredients are people and a communication system—people who are adequately informed with respect to (1) the quality of livestock being offered (2) the current or prospective value of that quality of animals, and (3) bargaining techniques. If all parties interested in transacting business are adequately informed and have an efficient communication system, it is apparent that the concept of a market would need little reference to geographical location or space. Some recent developments in livestock marketing are making this more and more obvious.

Role of Markets and Marketing in the Economy

It was implied in previous paragraphs that we are concerned with marketing under competitive conditions. This normally is taken for granted for we live in a market economy—also sometimes called an exchange economy, a competitive economy, or capitalism—where competitive forces are "relatively" free to exert their influence in the formation of prices and in direction of the economy. This is not the case throughout the world. While all countries have some degree of government regulation or control over marketing activities, the situation is one of degree. The United States is at one end of the spectrum with a relatively low degree of government control. Such countries as the Soviet Union, Peoples Republic of China, Albania, and Cuba are at the other end with a relatively high degree of central planning.

Regardless of the economic system, certain functions must be performed. Goods must be produced and distributed. Income must be distributed among the participants. In a market economy, competitively-determined prices are the guiding force which gives direction to what is produced, what technologies are used in production, where production takes place, when production is carried out, when and where consumption takes place and who gets the proceeds from the whole process. In a completely centralized economy such decisions as these would be dictated by the government through administrators, committees, boards, etc., responsible for operation of the economy. The attainment of an optimum which meets or approaches economic, social, and political objectives is an extremely complex problem. The problem is no less complex in a market economy, but the approach is vastly different. Here, chief dependence is placed upon impersonal, competitive market forces which generate prices that give direction to the economy—again with the expectation that economic, social, and political objectives will be met to a satisfactory degree. This places a tremendous burden on the markets. If markets do not operate efficiently, resources used in production may be misallocated; consumers may not have goods available in the form, quantity, quality, place, and time desired; and inequalities may occur in the distribution of income among individuals. Departure from generally-desired objectives or goals prompts government intervention in the name of public interest, or general welfare. Intervention may take many forms, e.g., inspection, licenses, regulations designed to curb monopoly or enhance competition.

Marketing Problems

Problems which have arisen in the system stem directly from the functions markets are expected to perform in directing the economy. These may be grouped under three general classifications—determination of consumer demands, reflection of these demands back through market channels to processors and producers, equitable distribution of income generated in the system, and physical movement of goods through market channels to consumers. Problems arise with respect to the efficiency with which these functions are performed. From an analytical standpoint, the first three are encompassed in the study of problems of "pricing efficiency" and the latter under problems of "operational efficiency." In actual operation the two are often interrelated.

Operational Efficiency.—Operational efficiency in marketing is analogous to the engineer's concept of physical efficiency, i.e., it is concerned with measuring input-output relationships. In marketing, the relevant relationships are in the physical movement of products from point of production to point of consumption. An improvement in marketing technology that permits an increase in quantity of goods marketed without a proportional increase in resources devoted to marketing or, what amounts to the same thing, a decrease in resources used in marketing without a proportional decrease in quantity of goods handled, would represent an increase in operational efficiency. In either case, it is assumed that unit costs of marketing would be lower with improvement in operational efficiency. There are many examples of this. Studies have revealed that labor requirements, per unit of livestock handled, can be reduced at many livestock auctions by improvements in pens, alleys, gates, scales, and sale ring layouts which facilitate the movement and sale of livestock (McNeeley *et al.* 1953). Modernized methods and office equipment can reduce bookkeeping costs per livestock unit handled in larger marketing operations. Another example is the use of multiple-deck and drop-center trucks which lead to a reduction in unit costs of transportation.

But not all types of operational efficiency lend themselves to such straightforward measurement. Suppose the output were marketing services instead of physical quantities. Service cannot be measured in physical units. Yet, services most certainly are an output of the marketing system. The commission agent engaged to sell livestock on a terminal market is expected to perform a service, as is the order buyer engaged to purchase feeder cattle. One approach would be to consider that the value of the commission agent's services is the additional value he obtains for livestock over that which could have been obtained without his services. Likewise, the value of the order buyer's services may be presumed to be the difference between the value (cost to you) of feeder cattle purchased by the order buyer as compared to the value (cost to you) if you had not used his services. The principle is clear enough, but quantification can be difficult in some cases.

In a broader sense, value added in manufacturing can be used as an imputed value of services performed by a firm engaged in manufacturing. Calculated marketing margins of farm products, defined as the margin between farm prices and retail prices, is an approximation of the value of services performed in processing and moving goods from farm to consumer. Both value added and marketing margin are crude measures of operational efficiency. The general assumption involved is that competition is high enough to prevent inclusion of excess profits in value added and/or in marketing margins. This may be a valid assumption but conclusive evidence is hard to come by. In spite of recent investigations which failed to uncover existence of unreasonable profits in food industries,[4]

---

[4]Investigations carried out by the National Commission on Food Marketing, an investigating agency formed by Congressional action in 1964. Relevant reports by this commission are discussed in later chapters.

many farmers and ranchers feel they are disadvantaged in bargaining for the sale of their products and suspect excess profits in the processing and distributing sectors.

Farmers and ranchers are interested in marketing livestock at the lowest cost consistent with price received (i.e., as long as marketing economies enhance the net price). Over the years, producers generally have questioned whether gains in operational marketing efficiency accrue to them, to the market agencies, or to consumers. In a competitive economy, any reduction in marketing costs, which results in above normal profits, will attract additional competitors. With the possibility of increased profits, competition for products to handle tends to enhance prices paid to producers. Competition by buyers, farther down the marketing channels, to get products as cheaply as possible will also tend to lower selling price. Lack of adequate competition anywhere in the system may permit above normal profits for some time; but if competition is keen, an improvement in operational efficiency will benefit all parties in the long run.

Nevertheless, farmers and ranchers on occasion have been dissatisfied with what they considered exorbitant marketing costs. Probably the most extensive counteraction has been the organization of farmer-owned marketing agencies, such as cooperatives.[5] There are examples of cooperative livestock commission agencies, cooperatively-operated auctions, various types of cooperative livestock and wool pooling arrangements, and cooperative packing plants. Early history is replete with examples of farmer agitation for the lowering of freight rates. And, largely as a result of farmer dissatisfaction, the Packers and Stockyards Division of the USDA exerts a degree of control over the level of commission and yardage and feed charges at public livestock markets.

Pricing Efficiency.—In the traditional sense (Phillips 1961), "Pricing efficiency . . . is concerned with the price-making role of the market system. It concerns how accurately, how effectively, how rapidly, and how freely the marketing system makes prices which measure product values to the ultimate consumer and reflects these values through the various stages of the marketing system to the producer . . .

"Economic theory suggests that prices which reflect more accurately the preferences of consumers will do a more efficient job in allocating productive resources to maximize consumer satisfaction and producer incomes."

Thus, pricing efficiency is concerned with such questions as how well

---

[5]Not all cooperatives were organized solely in the interest of operational efficiency. Cooperatives may also be designed to improve pricing efficiency or to act in dual capacity for both operational and pricing objectives.

the price system interprets changes in consumer demands, how well prices transmit changing demands back to producers and induce a proper allocation of resources among alternative productive uses, and how well the price system distributes income among producers and marketers.

If consumers have a preference for meat with certain quality specifications it is presumed that competitive market forces, acting through the pricing mechanism, will transmit this message through market channels back to producers and induce an increase in production of meat with those specifications—and vice versa with meat of less desirable qualities. A much-discussed example of this is consumer reaction to overfat pork. An efficient pricing system would be expected to record this situation in consumer willingness to pay more for leaner pork. Thus, a price differential would be transmitted from the retail level back through distribution channels—processors, packers, and market agencies to the producers. The differential, if it were of significant magnitude, would be expected to induce swine producers to increase production of lean, meat-type hogs relative to the production of lard-type hogs. This assumes the increase in price equals or exceeds any associated increase in production costs. This is an example of prices directing a reallocation of resources. Presumably, producers and consumers both would benefit from such a change. Whether market agencies or packers and processors would benefit would require an analysis of their costs and returns. In spite of considerable criticism over the speed and effectiveness of pricing efficiency in the hog market, records show that over a period of years a substantial increase has occurred in the production of leaner hogs (Agnew 1969). This is not to say that the market exhibited perfect pricing efficiency, but it serves as an example of the principle. Further improvement undoubtedly can be made, and if consumer preferences change, the market will be expected to transmit those changes.

Classical economic theory, built around a model of perfect competition, was able to show that self-generating competitive forces would optimize the allocation of resources, the production and distribution of goods, and returns to factors of production, i.e., wages, rent, and interest. There would be no pure profits, but management would receive a minimal, normal profit. From the beginning, the prevailing economic philosophy in the United States has been based on relatively free and competitive markets. However, it also was apparent from the beginning that unbridled competition produced some undesirable social results, and regulations of many sorts have been incorporated into the system. The attributes of perfect competition never existed and no one has argued that they did. It always has been obvious that from the private, individual standpoint, advantage could be gained by introducing or developing imperfection in the system, e.g., developing inequality in bargaining power.

In other words, private interest is not necessarily compatible with the general interest. As a member of society, each person is presumed to be cognizant of social responsibility. At the same time experience has shown that individuals or firms, by and large, will go about as far as the law allows (or as far as personal moral convictions allow) in the enhancement of personal or private interest, which generally means enhancement of profit.

The recent increase in concern about farmer bargaining power is, in reality, concern about pricing efficiency. A term used in this connection is "market performance." Farmers have shown dissatisfaction with the distribution of returns. Some would resort to greater governmental intervention, directly or indirectly. Others recommend voluntary farmer organization with the exercise of self-discipline in marketing. This area will be discussed l.ter in relevant chapters, but it should be noted here that market organization (the structure which gives rise to the behavior or conduct of firms and their performance in terms of pricing efficiency) is not a static phenomenon. As noted by Farris (1965), producers of most farm products ". . . traditionally have had readily available to them markets in which prices, though subject to various kinds and degrees of imperfections, were generated in a relatively impersonal manner. This is changing; such markets are fading from the scene."

BIBLIOGRAPHY

AMERICAN NATIONAL CATTLEMEN'S ASSOCIATION. 1965. Proc. 3rd Coordinated Beef Improvement Conf., July, Texas A & M Univ.

BAKKEN, H. 1953. Theory of Markets and Marketing. Mimir Press, Madison, Wisc.

EMERSON, J. et al. 1973. The interindustry structure of Kansas. Dept. Econ. Analysis, State of Kansas, Topeka.

FARRIS, P. L. 1965. Talk given at 3rd Coordinated Beef Improvement Conf., July. Texas A & M Univ.

FOWLER, S. H. 1961. An introduction to livestock marketing. In The Marketing of Livestock and Meat, 2nd Edition, Interstate Printers & Publishers, Danville, Ill.

McNEELEY, J. et al. 1953. Texas livestock auction markets—methods and facilities. Texas Agr. Expt. Sta. Misc. Publ. 93.

NATIONAL SCIENCE COUNCIL. 1977. World food and nutrition study, the potential contributions of research. National Academy of Sciences, Washington, D.C., June.

NIX, J. E. 1975. Grain-fed versus grass-fed beef production. USDA Econ. Res. Serv. Circular 602, April.

PHILLIPS, V. B. 1961. Price formation and pricing efficiency in marketing agricultural products: The role of market news and grade standards. Proc. 26th Ann. Conf. Natl. Assoc. Social Sci. Teachers, Econ. Sect., April 20, Howard Univ., Washington D.C.

USDA. 1955. Guide to agriculture, USA. USDA Econ. Res. Serv., Agr. Econ. Rept. 95.

USDA. 1974. Opportunities to increase red meat production from ranges of the United States. Six Agencies: Econ. Res. Serv., Ext. Serv., Forest Serv., Soil Cons. Serv., Agr. Res. Serv., Co-op. State Res. Serv. (unnumbered) June.

USDA. 1977A. Agricultural finance outlook. USDA Econ. Res. Serv. AFO-18.

USDA. 1977B. State farm income statistics. USDA Econ. Res. Serv. Supp. to Stat. Bull. 576.

USDA. 1977C. Agricultural statistics 1977, U.S. Govt. Printing Office.

USDA. 1978. Farm income statistics. USDA Econ., Stat., and Co-op. Serv., Statistical Bull. No. 609.

2

Historical Perspective

Many changes have occurred in methods of marketing livestock and meat in the United States, and the dynamic nature of the industry practically assures that changes will continue. The following thumbnail sketch of historical development is presented, not so much in the interests of history per se, but rather to point out factors which were responsible for changes in the marketing system. In many respects the marketing system developed in conjunction with and as an integral part of the production process. For both it has been a gradual, but steady evolution.[1] Among the more important factors influencing market development were shifting geographical centers of livestock production and population; shifting import-export opportunities; changes in technology of transportation and refrigeration; changes in structural characteristics of the producing, packing, wholesaling, and retailing components of the industry; developments in grading, and in the collection and dissemination of market intelligence.

Marketing was of little significance in the self-sufficient, subsistence economy of the early settlers. Meager surpluses of livestock and meat were readily bartered for other necessities. In time, a continuous influx of immigrants gave rise to population concentrations in cities in the present Central Atlantic states. Simultaneously, livestock production expanded in the eastern coastal region and slowly pushed westward and southward. Cattle were introduced in the southwest—through Spanish influence and largely by missionaries—as early, if not earlier, than on the East Coast, but the pattern of a marketing system which was to emerge as the prevailing model had its origins in East Coast livestock development. As specialization of labor increased, there quickly developed a need for trade. Barter between producers and consumers soon was replaced by merchants (middlemen) who, for a fee, acted as the go-between for producers and consumers. As long as distances were not too great, most slaughtering was done by producers who delivered dressed carcasses to local retail merchants. As both distances and demand for meat increased, the feasibility arose for another intermediary—a merchant who would buy live animals. This person might slaughter the animals and retail the

---

[1]The introduction and subsequent development of domestic meat animals is an extremely important and interesting chapter in U.S. annals. For those who may be interested, a considerable body of literature is available on the subject. Examples are: Clemen (1923), Duddy and Revzan (1938), Thompson (1942), Wentworth (1948), and Williams and Stout (1964).

meat himself, or slaughter and sell dressed carcasses to a retail merchant. Thus in our earliest period, direct marketing—producer to slaughterer —was the rule. As will be pointed out later, the trend for some years has been back to direct marketing of slaughter livestock.

At first, all movement of livestock was on foot. No other means was available. In time, limited river and coastal water transport was used. Slaughtering was limited to winter months. Meat preservation was crude, being limited to smoking, salting, and pickling. The meat was packed in barrels for storage and shipment. It was these techniques which gave rise to the term meat "packer," a term which has remained and is now used synonymously with slaughterer. Commercial trade in "packed" meats got under way around the mid-1600's with shipments to the British West Indies and the provisioning of sailing vessels (Clemen 1923). The West Indies trade was initiated by New England colonies, but in time the geographical advantage of Virginia and the Carolinas gave the latter the competitive edge—an early U.S. example of regional competitive advantage. (Probably the latest analogous example is the development and introduction of hybrid grain sorghum in the Southern Plains states, which gave that region a competitive advantage over others in cattle feeding.)

As a rule, curing and packing was done by individuals, farmers or merchants, until about 1662. At that time William Pynchon launched what is purported to be the first commercial venture in meat packing (Clemen 1923). The plant was located at Springfield, Mass. Prior to that, Pynchon had been a cattle drover. According to Webster's New World Dictionary, the word "drover" has two meanings: (1) a person who takes a drove of cattle to market, and (2) a cattle dealer. This occupation is worthy of mention because at that time, and for some time later, it was an important element in the marketing system. Many farmers drove their own livestock; but as distances lengthened this became more of a problem. Additionally, some preferred to sell at home for a known price and let the professional drover carry the risk from there on. As the definition indicates, drovers would drive farmers' livestock to market for a fee; or, acting as dealers, they would buy livestock from farmers and market them on their own account. The definition given above implies that droving applied only to cattle. While used most extensively for cattle, it also was used to some extent for hogs and sheep and, in some cases, even for turkeys.

From a dealer's standpoint, droving was a risky profession. Market information was almost nonexistent; there were physical hazards on the trail; and marketing consisted basically of locating a slaughterer who wanted livestock at that particular time—at best an uncertain proposition. Nevertheless, drovers continued to be a significant link in the marketing system until the advent of railroads and emergence of terminal markets.

Marketing apparently continued on an informal, unorganized basis until the mid-1700's although there is no question that the pace of activities and the volume of trading increased significantly with the passing years. While a number of cities emerged as market centers, Boston was particularly active and it was there, or nearby, that the formalities of organized marketing were first reported. This happened for meat a few years prior to live animals. In 1742 in Boston, a meat market known as Faneuil Hall was erected. Here ". . . the old-fashioned ways came to an end . . . Regulations were . . . voted regarding the quality of meat sold and as a result both the producer and consumer were greatly aided" (Clemen 1923). The thriving meat market fostered the meat packing business and this, in turn, fostered livestock marketing in the vicinity—or possibly the order of causation was reversed. In any event, the Brighton, Mass., Market is cited as a leader. Brighton is now a part of the Greater Boston area. This market was established by Jonathan Winship, who ". . . established a large slaughter house there, and producers of beef cattle soon began to find their way to this place to effect a sale. Other butchers came here to compete with him in making purchases, and in a short time the business became centered here, Mr. Winship having wisely encouraged and fostered it. This is believed to have occurred at the time of the old French War (1756) or very soon after. From that time this was the cattle market of New England . . .

"Brighton Market was the model for many others in the East and later in the West. [It was] . . . an example of the old-fashioned market institution which was the forerunner of the great, modern centralized livestock market . . ." (Clemen 1923).

Following the War of Independence, settlement rapidly spread westward over the Allegheny Mountains and into the rich Ohio Valley. Shortly after the beginning of the 19th Century the center of livestock production shifted to that region. Droving continued as a means of transporting cattle to eastern markets. The distances, however, necessitated development of the packing industry, particularly for hogs. The activity originally centered at Cincinnati, which became known as "Porkopolis." Not only was Cincinnati in the heart of a rich production area, but it also was on the Ohio River. River transport down the Ohio and Mississippi provided an outlet for livestock and meat to southern cities and for export. As livestock production spread westward, the packing industry also spread and the packing of beef increased with the westward movement, although at this stage beef packing was limited. As production spread into Illinois and beyond, the distances to eastern markets became too great for the droving of fat cattle. This necessitated the movement of stock cattle eastward for finishing in Ohio or beyond. But New England no longer could successfully compete with the west in fattening cattle and was rapidly being industrialized. Simultaneously the south was specializing in cotton.

The region now known as the Corn Belt virtually took over livestock production. But the southern meat market soon proved to be the most lucrative, and the movement of meat and livestock down the Ohio and Mississippi rivers increased to the detriment of eastern markets. This provided a stimulus for eastern interests to promote improved transportation to the east. Canal construction was greatly expanded during the first half of the 1800's and canals became important links in the transport system. Private toll roads and public roads were pushed throughout the settled area. But more importantly, the drive for trade gave a great impetus to railroad construction. By the 1850's railroads had reopened the eastern meat trade and quickened the pace of livestock-meat industry expansion.

It was about this time that livestock auctions gained recognition. The auction method, of course, was not new. It had been used in Great Britain for centuries. Early American colonists used auctions to some extent, but apparently they were not used extensively for livestock. Auction marketing was reported with development in the Ohio Valley. The first mentioned was the Ohio Company for Importing English Cattle, in 1836 (Henlein 1959). Others were organized in the following 25 years or so in Ohio and Kentucky in connection with the sale of imported purebred cattle (Henlein 1959). And Clemen (1923) reported that they were used for cattle, mules, jacks, jennies, horses, sheep and swine. The extent of auction marketing is not specifically reported. However, after a relatively short period of growth this method waned for new factors were shaping the marketing system. Among these were the influence of railroads in association with locational aspects of production and growing need for yarding facilities at points of assembly.

The railroads were pushed through the Ohio, Mississippi, and Missouri River Valleys. With an ever-growing volume of livestock, drovers and shippers were finding it increasingly inconvenient that no public stockyards were available for holding stock during its sale. Earlier, this was a minor problem as stock generally could be sold upon arrival; but with time this changed. One privately-operated yard, the Bull's Head Market, was reported in Chicago as early as 1848. By the early 1860's each of the five major railroads serving Chicago had constructed yards individually as an enticement of business. Livestock was a major source of freight revenue. A multiplicity of railroads converging in Chicago from the heavy production areas to the west and south soon made Chicago the major market center.

Another simultaneous development of major significance about that time was the emergence of the livestock commission agent. Prior to this the owner (drover or farmer) acted as his own agent in selling. This was time-consuming and owners were often ill-informed on market conditions. Drovers undoubtedly could see that use of an agent would relieve them of expenses of accompanying livestock to market and allow them more time to solicit business. Another advantage to drovers and farmers lay in financing. Commission firms were able to remit full receipts upon sale of livestock, while drovers formerly bought on credit and sold on credit with delayed payment all down the line. The first bona fide livestock commission firm began operation in Chicago in 1857 (Clemen 1923). This must be considered a major milestone in livestock marketing.

The Civil War (1861–1865) brought on unprecedented demands for meat. Many rail-oriented markets were operating by that time but Chicago was the greatest in volume. However, it was painfully apparent that the individually rail-owned stockyards, scattered about the city, were inadequate and inefficient from both an operational and pricing standpoint. Buyers could be at only one yard at a time. Shippers often had only one choice of rail line and ran the risk of arrival at a yard devoid of buyers. This was soon remedied. At the urging of virtually all interests—although a trifle belatedly on the part of railroad officials—the Illinois legislature in 1865 incorporated the Union Stockyards and Transit Company. This was a single facility to accommodate all rail lines and it served as the model of public terminal livestock markets for many other cities. Thus, the tendency toward centralization of marketing, which already was moving under individual auspices, received a shot-in-the-arm through public sponsorship. At that time Chicago had all the ingredients for a great market, a position it held for many years. Railroads converged on Chicago from the tremendously productive regions to the west and south, directing livestock into the city like an enormous funnel. The Union Stockyards provided physical facilities for handling the stock, and rapid expansion of slaughtering and processing plants provided packing capacity. Outgoing rail and water transportation facilitated shipments of meat and reshipment of live animals. And credit resources were available to finance necessary investments and operations.

Other centrally-located cities also possessed certain advantageous attributes and quickly came to prominence as terminal markets. Among these were East St. Louis, Kansas City, Wichita, St. Joseph, Omaha, Sioux City, and Indianapolis, to mention only a few. The stream of cattle flowing into these markets by this time had gained additional sources of supply. These were the vast range lands of the Southern Plains states and the Southwest. The Northern Plains states followed in a short time. Droving continued to be a vital link, for the railroads had not yet penetrated that far. Figure 2.1 shows the principal trails over which thousands of cattle were driven to reach these markets.

Prior to the Civil War, cattle production in Texas had expanded faster than available markets warranted. Southern coastal cities provided about the only outlets beyond the state itself and these outlets were severely restricted by monopoly tactics of shipping interests. With the outbreak of the Civil War these markets were virtually eliminated. The cattle industry of Texas suffered a severe depression. Wartime demands inflated the price of cattle in eastern markets, while range prices dropped. That, of course, provided a tremendous incentive for the range cattlemen to reach the lucrative eastern markets. The only alternative was on foot to a rail head. Many obstacles lay in the way, but in time they were overcome. Trails were established and thousands of cattle were delivered. As the rails were extended, the drives shortened, so by about 1880 this era

MAP BY DON BUFKIN

Courtesy of Arizona Highways

FIG. 2.1.   MAP OF THE PRINCIPAL CATTLE DRIVES

ended. In the process, however, some of the loading points (which also were market points like Abilene and Dodge City) gained fame and a sort of notoriety that seems destined to continue.[2]

The year 1857, mentioned earlier as the beginning of commission firm operation, also marked the inauguration of another innovation of incalculable significance. Summer meat packing made its debut by use of natural ice refrigeration. The use of natural ice had obvious limitations but it soon was augmented by artificially-produced ice and this, in turn, by mechanical refrigeration. The use of mechanical refrigeration was well established by 1890 (Ives 1966). The use of refrigeration revolutionized the meat packing industry. It could now be a year-round business, which added greatly to operational efficiency. Storage and shipment of fresh meat became possible and feasible. Not only was the quantity of meat available increased on an annual basis, but the quality also was enhanced which undoubtedly boosted demand. Recognition of quality, however, did not come about automatically. Eastern packers were averse to inshipments of fresh western beef. Railroads had cattle cars, but not refrigerated box cars, and they resisted change. Rumors were spread of objectionable quality and unwholesomeness of refrigerated meat. There may have been some basis for this at the beginning, but techniques were soon perfected. Year-round markets were opened for producers, although the seasonality of production was to continue—and continues to this date at an abating rate.

Centralized terminal marketing remained the dominant method until after World War I. However, even before that, other factors were at work which were to verify the proposition that, in the world of economics, "things never stay put for long." Improved varieties permitted the northwesterly expansion of corn production. Consistent with previous tendencies, the packing industry was prone to follow livestock production. A number of other factors were involved in this movement but probably the most significant was expanded use of motor trucks and improvements of roads and highways. A sharp drop in livestock prices following World War I made farmers relatively cost-conscious and more critical of marketing charges and practices at terminals. Improvements in market news services and adoption of grade standards made possible a wider dissemination of market information (Duddy and Revzan 1938). The trend toward decentralization of markets became obvious following the close of World War II and is continuing to date. As packing plants were constructed nearer to the point of production, producers tended to

---

[2]For an interesting account of this period see *Historic Sketches of the Cattle Trade of the West and Southwest* by Joseph G. McCoy published in 1874 by Ramsey, Millet and Hudson of Kansas City and reprinted in 1932 by the Rare Book Shop, Washington, D.C.

revert to direct marketing of slaughter livestock. The movement initially was much more extensive with hogs than with cattle or sheep. In later years it has spread more and more to cattle and sheep. This movement is particularly apparent with the expansion of finished cattle production in the Southwest, on the West Coast and in the Southern Plains states. It is significant that in May, 1970, Chicago, the former "hog butcher of the world," stopped accepting hogs at the Union Stockyards. Trading in cattle and sheep ceased in August, 1971.

Auction markets which were known in colonial times and came on with a flurry about the mid-1880's, only to fade into obscurity, came back in a big way following World War I. While of only limited use for slaughter livestock (there are exceptions to this), they assumed significant proportions in the marketing of feeder livestock and cull cows. The 1930's was a period of very rapid expansion, with a slowing of the growth rate during the 1940's and a slight decline in the number of auctions during the 1950's. Annual data are not available but it appears that the number of livestock marketed through auctions has tended to stabilize in recent years.

During the period since World War I structural changes at the retail level have changed materially the wholesale and retail system of meat marketing. This was associated with the emergence of large chain supermarkets and their influence on the relative competitive position of retailers versus packers. Their attempt to obtain a steady supply of uniform quality meat has been felt back through channels (through packers to the producer level). The increase in bargaining power of the retail sector, as exemplified by chain stores, is a near reversal of the situation prior to World War I.

Producers have always been concerned with their relative position in bargaining. This concern has come to the foreground with particular emphasis during the last two decades. Farm groups are attempting, by various means, to improve their bargaining position. This will be discussed in some detail in a later chapter.

Over the years the import-export balance also has reversed itself. While there was considerable variation in earlier periods, exports were an important element in the livestock-meat industry during the latter half of the 1800's. From that time on, however, exports dwindled and, with the exception of war periods, were a minor fraction of total meat production, but did remain of major proportions for certain items such as lard, variety meats, and hides. In recent years the United States has become a major importer of meat and has continued to be a substantial exporter of lard, fats (both edible and inedible), hides, and variety meats. Exports of live animals are growing in importance. Livestock interests have been instrumental in the passage of national legislation designed to limit imports. Imports continue to be a matter of national concern.

BIBLIOGRAPHY

CLEMEN, R. A. 1923. The American Livestock and Meat Industry. Ronald Press, New York.

DUDDY, E. E. and REVZAN, D. A. 1938. The Changing Relative Importance of the Central Livestock Market. University of Chicago Press, Chicago.

HENLEIN, P. C. 1959. Cattle Kingdom in the Ohio Valley, 1783–1860. University of Kentucky Press, Lexington.

IVES, J. R. 1966. The Livestock and Meat Economy of the United States. American Meat Institute, Chicago.

THOMPSON, J. W. 1942. A history of livestock raising in the United States, 1607–1860. USDA Agr. History Ser. 5, 1, 14–15, 37, 54, 108.

USDA. 1966. Agricultural markets in change. USDA, Econ. Res. Serv., Agr. Econ. Rept. 95. July.

WENTWORTH, E. N. 1948. America's Sheep Trails. Iowa State College Press, Ames.

WILLIAMS, W. F., and STOUT, T. T. 1964. Economics of the Livestock-Meat Industry. Macmillan Co., New York.

3

Economic Theory and Principles

Use of the word "theory" has a tendency to disturb many people who have only a layman's concept of its meaning. One way to avoid an issue on the point would be to avoid use of the word. But it is a perfectly legitimate word and in an academic setting, at least, there is justification for common agreement on its meaning. In the vernacular, theory has come to be synonymous with "impractical" or "nonfactual." That is a totally distorted view from the standpoint of a scientist, and this would hold for scientists of any discipline, be it economics, chemistry, animal science, physics, etc. In briefest terms "theory" to a scientist is a "systematic explanation." It is an explanation ". . . which describes the workings and interrelationships of the various aspects of some phenomenon" (Baumol 1961). Used in this sense it would be contradictory to label theory as impractical, per se. If a theory gives an accurate explanation, it by definition cannot be classed as nonfactual. A determination of whether it is practical or feasible may require additional analysis.

There is no guarantee, of course, that every theory is a satisfactory or adequate explanation. In some instances it simply can be bad theory—an inaccurate explanation. But that is not equivalent to being impractical or nonfactual. In other cases a theory may appear to be an adequate explanation of a phenomenon, given the state of knowledge at that point in time, but turn out to be inaccurate or inadequate at a later time when additional knowledge is available. Many such examples exist in the physical and biological sciences. Before man landed on the moon various "theories" had been advanced regarding the nature of its matter and origin. The moon landings produced evidence which showed that some of these theories were wrong.

In economics it is recognized that some of the theoretical models of various market situations do not precisely correspond to the real world. The widely-used, perfectly competitive model is an example.[1] It is an abstraction of the real world—a simplified model used as an illustration of certain principles and as a guideline or benchmark with which other market situations may be compared.

In a previous chapter a quote was presented which indicated that the traditional, relatively impersonal market is fading away. By implication this means we are drifting toward a situation where more prices and

---

[1]It is assumed that readers have a knowledge of elementary economics. Readers who have not had recent work in economics should review a modern elementary text on price theory.

terms of trade are being set by negotiation, formula, or some other institutional arrangement. The markets for a number of agricultural commodities already have faded considerably from the traditional, impersonal, free, and open market. Examples are milk, eggs, and a number of fruits, vegetables, and specialty crops. These are commodities marketed under federal or state marketing orders and agreements. To a degree, farm price support programs also have altered the working of the traditional market. This is an example of an attempt to set up an institutional arrangement (i.e., the support program) within the traditional market—to supplement without eliminating or replacing it.

Livestock marketing is influenced to a lesser degree by personal or institutional arrangements than are other major agricultural commodities. However, even here contracting is being done by farm organizations and individuals under negotiated conditions. Vertically integrated operations are becoming more commonplace in the production, processing, and marketing of all species—cattle, swine, and sheep. Some farm groups continue to lobby for government guaranteed prices. In the wholesale meat trade substantial quantities are sold by formula. Arrangements such as these deviate from the open competitive market system, but they do not lessen the need for a knowledge of economic principles. When negotiators sit down around the bargaining table or individuals take it upon themselves to do their own buying and selling, the need for an understanding of economic theory and principles is as great as it is in the free and open market, if not more so. When legislators and their staffs formulate price policy, and when administrative agencies implement the associated programs, a thorough knowledge of theory and principles is essential. Foremost among marketing theories are those which explain the relationships between demand and supply.

THEORY OF DEMAND

Demand—the Concept

"Demand" is another word that means different things to different people. During periods when prices of, say slaughter steers, are increasing or are at a relatively high level, one often hears the comment that "demand is high." Conversely when prices are low the comment may be that "demand is low" or "demand is dropping." Under certain circumstances these comments may be partially correct but they are loose, nontechnical concepts of demand. For effective communication among market analysts and for consistent use in economic analysis we need a precise, rigorous meaning. Fortunately, there is an accepted definition: Demand is the functional relationship between prices and quantities of a product that buyers will purchase in a specified market, ceteris paribus. Note that "prices" and "quantities" are used in the plural. "Ceteris paribus" refers to the assumption that all other factors—in addition to price—that can influence the quantity purchased, remain unchanged. Under this definition demand is a series of prices and quantities, not just the quantity at a particular price or the amount of money spent for some particular quantity. One of the earliest axioms established in economics was that an inverse relationship exists between quantities taken by consumers and prices of a product. Or to put it another way, consumers will buy more at lower prices than at higher prices. This is the law of demand.

The "functional relationship" may be expressed (1) in table form often referred to as a schedule, (2) as a mathematical function or (3) in graphic form. For example, the prices at which consumers will purchase various quantities of a hypothetical product may be expressed in schedule form as follows: *demand schedule.*

Price per unit $	Quantity
1.00	180
2.00	160
3.00	140
4.00	120
5.00	100
6.00	80
7.00	60
8.00	40
9.00	20

If we consider that the quantity consumed is a function of price (i.e., that quantity purchased depends upon price), by rather elementary mathematical calculations we can "fit" a function to the above data. It would be as follows:

In general form $Q = f(P)$
In specific form $Q = 200 - 20P$
where Q = quantity, and
P = price

This series of prices and quantities also can be shown in graphic form as in Fig. 3.1 where line $D_1 - D_1$ is a demand curve illustrating this functional relationship.

The above three forms of expression simply are different ways of presenting the same data—the same functional relationship. All three are used extensively in economics.

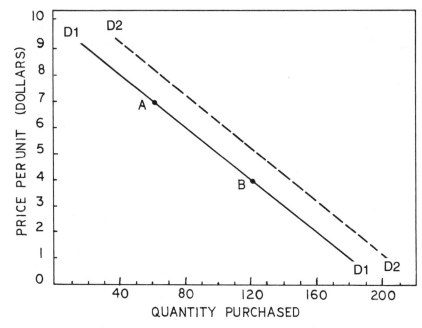

FIG. 3.1.   HYPOTHETICAL DEMAND CURVES

Conceptually we speak of individual demand and aggregate or market demand. The latter simply is the sum of the demands of individuals who make up the market.

The ceteris paribus condition, "other factors being equal or unchanged," requires some clarification. It is obvious that price is not the only factor which determines the quantity of goods consumers buy. Among the more important "other factors" are: population (the number of buyers); income levels (the purchasing power of buyers); personal preferences or tastes; the price, quality, and availability of substitute or complementary products that compete with the product under consideration for the buyer's dollar; buyer's expectations of future prices of the product under consideration and his expectation of prices for substitute or complementary products. What we mean by the ceteris paribus assumption is that at a given time in a given market these things will be at some observable level. Their effect upon quantities purchased will be incorporated in the price-quantity relationship, but at that time the only thing which is considered to vary is price. We want to know the effect on quantity purchased due solely to price variation when everything else is constant.

This does not mean that we are uninterested in the effect "other factors" have on quantity purchased. On the contrary, we are intensely interested and also try hard to determine the relationships between these factors and quantity purchased. In economic parlance they are called "demand shifters" or it may be said that a change in one (or more) of these factors results in a "change in demand" or a "shift in demand." A change in the factor implies that some time elapses for one set of ceteris paribus conditions to another.

Using the same data presented in the previous illustrations, let curve D_1-D_1 in Fig. 3.1 represent a demand curve where all the so-called demand shifters are at some given level. Points A and B do *not* represent different demands, nor do the different locations of A and B represent change in demand. They both are on the same demand curve. What these points mean is that at $7.00 per unit, buyers will purchase 60 units, but *if* the price were $4.00 they would purchase 120 units. They simply represent 2, among an infinite number of price-quantity relationships on that demand curve, only 1 of which could be realized in a given market at one time.

Now, assume that some time elapses and incomes of buyers increase. With higher incomes buyers would be expected to pay higher prices for any specified quantity (or what amounts to the same thing, purchase a greater quantity at any specified price). Buyers may now purchase 80 units at $7.00 per unit where originally they would have purchased only 60 units, or at $4.00 per unit they may now purchase 130 units where originally they would have purchased only 120—and so on at other price levels. Thus, the change in income (the increase in income) shifted the entire demand curve from D_1-D_1 to D_2-D_2. The changed demand shows that greater quantities would be taken at all possible prices (or what amounts to the same thing, higher prices now would be paid for all possible quantities). In this example it was considered that only income increased. But the same reaction could occur from changes in any one or any combination of the demand shifters. (An increase in demand is indicated by a shift to the right on the graph—a decrease would show up as a shift to the left.)

Population in the United States has increased year by year, providing one of the most stable and important shifters (increasers) in the demand for farm products. On the average, per capita personal disposable incomes have increased fairly consistently for a number of years. Studies show that consumers have a strong tendency to increase purchases of beef as incomes increase, but the same relationship has not always held for pork. These relationships will be discussed in some detail at a later point.

Derived Demand

Consumer demand for meat makes itself felt at the retail level—at the supermarket. To be more specific, a demand exists for each separate cut of meat. How is this demand transmitted to livestock producers? From the farmer's standpoint, the demand for slaughter steers, for example, is a "derived demand"—that is, the demand derived from consumers' demand for beef. The aggregate retail value of all retail cuts establishes a value for the carcass at the packer level. The carcass value, in turn, establishes the value of the live steer. The same sort of derived demand exists for hogs and lamb and, as a matter of fact, for the various classes, grades, and weights of animals within each species.

This is an oversimplified statement, but it contains the essential features of the concept of derived demand. It is obvious in this situation that the degree of pricing efficiency throughout market channels is of considerable importance in transmitting to producers the demands of consumers.

Elasticity of Demand

A knowledge of the functional relationship that exists between prices and quantities of a particular product is valuable to market analysts. But probably of even more importance is a measure of the degree of sensitivity of change in quantity purchased occasioned by a change of price. This is what is meant by elasticity. Traditionally, in the study of demand we have taken quantity as the dependent variable.[2] In the study of price elasticity we are interested in determining the percentage change in quantity purchased, occasioned by (or associated with) a specified percentage (say 1%, or 10%) change in price. Of major importance, also, is income elasticity which is the percentage change in expenditures for a product (or quantity purchased, depending upon model used by the researcher) associated with a specified change in income. And still another elasticity concept of substantial importance is cross elasticity. An excellent example of cross elasticity of direct interest among livestock producers is this: What effect would a decrease in pork prices have on the quantity of beef purchased by consumers, assuming beef prices remain unchanged? Thus, there are three concepts of elasticity to be noted (1)

---

[2]It may be noted that in graphic analysis the vertical axis traditionally is used for the dependent variable. But in graphing demand curves it is conventional to put price on the vertical axis. In a sense this is inconsistent. However, there is nothing sacred about these conventions. The graph would be equally valid one way as the other. Our major concern is with the relationship between variables—not which one is the dependent variable or which axis is used for one or the other.

price elasticity, which if written fully and more precisely would read, "price elasticity of demand," but normally is shortened to price elasticity, (2) income elasticiy, which again more precisely would be written as "income elasticity of demand," and (3) cross elasticity, a cross-price elasticity.

Price Elasticity.—The importance of price elasticity lies in its association with total revenue (i.e., total expenditures of purchasers). Total revenue is the total amount of money spent by consumers for a product. It is the total of price times quantity purchased (or sold). If, for the moment, we ignored costs of production it would be apparent that the sellers of a product would want to maximize total revenue received. It readily can be shown that under conditions of an elastic demand, total revenue can be increased by cutting price (or by increasing the quantity for sale, which amounts to the same thing viewed from a different angle). If demand is elastic, a given cut in price will result in a more than proportionate increase in quantity sold, and hence, an increase in revenue.

Before proceeding, we must present methods of calculating elasticity and clarify some terms. Price elasticity is calculated as a numerical measure which is referred to as the coefficient of price elasticity. It is calculated by dividing percentage change in quantity by the associated percentage change in price, or

$$e_{D/P} = \frac{\%\Delta Q}{\%\Delta P}$$

where $e_{D/P}$ = coefficient of price elasticity of demand
Δ = change
Q = quantity
P = price

Thus, $e_{D/P}$ is a ratio of the two percentage changes. The sign of the ratio normally is expected to be negative since price and quantity usually move in opposite directions. This was referred to earlier as an inverse relationship. Some authors omit the sign of the elasticity coefficient—it is universally understood to be negative. The magnitude of the coefficient can range from zero to infinity (negative infinity).

If price changes were associated with no change in quantity purchased, elasticity would be zero or perfectly inelastic. Graphically this would be a vertical line (curve) such as D_1 in Fig. 3.2 This might be visualized as some unique, absolutely necessary product, but a case this extreme is unlikely. If an unlimited quantity would be purchased at a given price, elasticity would be infinite or perfectly elastic. This would show up as a horizontal curve (see D_2 Fig. 3.2). Viewed from the standpoint of an individual farmer, this is the situation he faces in selling most products. No individual hog producer, for example, sells enough hogs to influence

PRICE PER UNIT

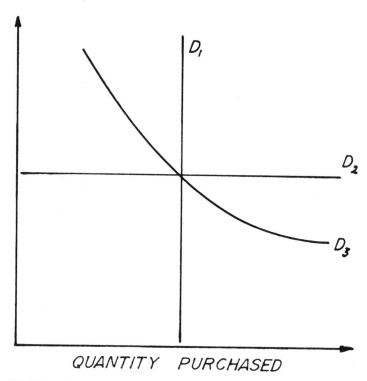

QUANTITY   PURCHASED

FIG. 3.2.   ILLUSTRATION OF DIFFERENT DEGREES OF PRICE ELASTICITY

the price. For all practical purposes, any one farmer can sell all the hogs he may have without depressing prices. In the aggregate, however, prices and quantities normally move in opposite directions.[3] Most demand curves fall somewhere between the limiting extremes—sloping downward to the right on a graph—such as D_3 in Fig. 3.2. Within this in-between area is a dividing line which separates the "relatively elastic" (often shortened simply to "elastic") from the "relatively inelastic" (often shortened to "inelastic"). The separation occurs at an elasticity of −1.0, or unit elasticity. This would occur where percentage change in quantity was exactly equal to the percentage change in price. Here demand would be unitarily elastic. For example, if a 5% decrease in quantity was associated with a 5% increase in price, price elasticity would be −1.0; that is

$$^{e}D/P = \frac{-5.0}{5.0} = -1.0$$

[3]There are exceptions to the normal situation. Some goods with high "snob" appeal sell greater quantities at higher prices than at lower prices. These cases are rare and are of no concern to this discussion.

Here the change in price is exactly offset by the change in quantity purchased. This situation can occur where a firm or family budgets a certain amount of money for a given item and sticks to the budget whether prices rise or fall.

For those with a knowledge of the calculus, determination of the coefficient of price elasticity is a simple calculation, providing the mathematical function is known. For illustration purposes we use the demand function presented earlier, i.e., $Q = 200 - 20P$. The formula for calculating elasticity is

$$e_{D/P} = \frac{dQ}{dP} \cdot \frac{P}{Q}$$

where $\frac{dQ}{dP}$ = the first derivative of Q with respect to P (from the demand function).

P = price (any relevant, specified price)

Q = quantity consumers would purchase at the specified price

For the demand function used here $\frac{dQ}{dP} = -20$. With this, the formula becomes $e_{D/P} = -20\frac{P}{Q}$ and we simply plug in applicable pairs of P and Q. A number of relevant price-quantity pairs were shown earlier. Others may be calculated by inserting any relevant price in the demand function and calculating its associated Q. We complete the illustration by calculating $e_{D/P}$ where $P = \$6.00$ and $Q = 80$ units:

$$e_{D/P} = -20\frac{6.00}{80} = -1.50$$

Thus, $e_{D/P}$ is -1.50 at the point on the demand curve where $P = \$6.00$ and $Q = 80$.

To calculate $e_{D/P}$ at another point we take $P = \$2.00$ and $Q = 160$:

$$e_{D/P} = -20\frac{2.00}{160} = -0.25$$

These two calculations should serve to show that the coefficient of elasticity may vary at different points on the same demand curve. This is further illustrated in Table 3.1 where the coefficient of elasticity is shown for nine points on the demand curve. The coefficient varies from one point

TABLE 3.1. HYPOTHETICAL PRICE, QUANTITY, ELASTICITY, AND TOTAL VALUE RELATIONSHIPS

Price ($)	Quantity	Coefficient of Elasticity	Total Revenue ($)
9.00	20	−9.00	180
8.00	40	−4.00	320
7.00	60	−2.33	420
6.00	80	−1.50	480
5.00	100	−1.00	500
4.00	120	−0.67	480
3.00	140	−0.43	420
2.00	160	−0.25	320
1.00	180	−0.11	180

to another. On only one unique type of demand curve—the rectangular hyperbola—does elasticity remain constant throughout the entire curve.

For those who do not have a knowledge of the calculus, the coefficients can be calculated by elementary arithmetic, again providing the demand schedule is known. For illustrative purposes we will use the schedule of prices and quantities in Table 3.1, which is the same schedule presented earlier. In this case we assume a movement from one point on the demand schedule to another point, calculate the percentage change in price and the percentage change in quantity, then divide percentage change in quantity by percentage change in price. This is the ratio that produces the coefficient of price elasticity. The formula is:

$$e_{D/P} = \frac{\dfrac{Q_2 - Q_1}{Q_2 + Q_1}}{\dfrac{P_2 - P_1}{P_2 + P_1}} = \frac{\%\Delta Q}{\%\Delta P}$$

where Q_1 = quantity before change
Q_2 = quantity after change
P_1 = price before change
P_2 = price after change, and
Δ = change

To complete this illustration we shall assume price and quantity were: price = \$6.00 and its associated quantity = 80 units before change; then price drops to \$5.00 which has a corresponding quantity of 100 units

$$e_{D/P} = \frac{\dfrac{100 - 80}{100 + 80}}{\dfrac{5.00 - 6.00}{5.00 + 6.00}} = \frac{\dfrac{20}{180}}{\dfrac{-1.00}{11.00}} = -1.22$$

The same procedure could be repeated at alternative places on the demand schedule to obtain measures of price elasticity throughout the schedule.

As pointed out in elementary economics, this is a measure of "arc elasticity." It represents an arc, or area, of the demand curve rather than a point as was obtained by the method involving the calculus. The method involving calculus is more precise and therefore considered to be preferable but is not necessary for an understanding of the concept.

Any coefficient of elasticity within the range 0 to −1.0 (noninclusive) is classified as relatively inelastic. Any coefficient greater than −1.0 is classified as relatively elastic. These two categories encompass virtually all goods and services that enter market transactions. Some elementary graphic explanations of elasticity purport to show that a steeply sloped demand curve is inelastic and that a gently sloped curve is elastic. This can be misleading. The slope of a curve is not equivalent to elasticity. The linear (straight line) demand curve drawn on arithmetic graph paper will have the same slope throughout its entire length, but when elasticity is calculated at various points along that demand curve it is found that the coefficient changes continuously throughout the length. Do not confuse slope with elasticity.

It was mentioned earlier that total revenue is associated with elasticity. The price-quantity schedule presented earlier is repeated in Table 3.1 along with calculations of the coefficient of elasticity and total revenue at various points throughout the demand schedule. Note that the coefficient of elasticity varies throughout the demand schedule. It ranges from −9.00 to −0.11. The schedule became less elastic with movement from higher prices (lower quantities) to lower prices (higher quantities). Total revenue also varies. It rises from $180 to a peak at $500 and then declines to $180 again. Note that total revenue rises until elasticity reaches unity. As the coefficient becomes less than unity, total revenue decreases. The lesson to be learned here is that a firm, or an industry, which has (or which can obtain) sufficient control over prices *or* production can maximize total revenue by proper manipulation of these variables. If demand is relatively elastic it can, by cutting prices, increase sales and total revenue up to the point where demand becomes unitarily elastic. If it is operating in an inelastic demand it can increase total revenue by raising the price and cutting back the quantity offered for sale, to the point where demand becomes unitarily elastic.

Maximization of total revenue does not automatically maximize profits. The firm or industry also would need a knowledge of cost behavior to maximize profits. Generally, a reduction in production would not increase per unit costs. An increase in production could lead to higher, lower, or constant cost per unit depending upon characteristics of the cost functions of that particular firm or industry.

The price elasticity faced by individual firms is greater than that faced by an entire industry. This clearly is the case for farmers. Individual farmers face a perfectly elastic demand for hogs, or cattle, or corn, etc., but the demand faced by any one of these sectors of the entire industry is substantially less. Elasticity at retail ordinarily is greater than at any earlier stage in the marketing-production process. For instance, the price elasticity for ham or pork chops is greater than for hogs at the farm level.

Income Elasticity.—The concept of income elasticity is analogous to that of price elasticity in that it refers to sensitivity of change in expenditures for a product associated with a change in income, everything else remaining constant. It is known that as incomes increase, consumers will buy more of some products, but less of others. Products that fall in the former category are classed as superior products while those in the latter are classed as inferior products. Hamburger may be considered an inferior product by people who already have relatively high incomes. That is, they probably would eat less hamburger with further increases in income. But people in low income brackets likely would increase hamburger consumption if their incomes were increased. To the latter, hamburger is a superior product.

Studies several years ago showed pork to be an inferior product on the basis of average incomes in the United States. This presumably was associated with an image (at that time) of pork as a relatively fat cut of meat. This means that consumers would spend less for pork as average income increased. With substantial improvement in quality of pork in recent years there is evidence that consumers now react differently than they did in an earlier period. Roy and Young (1977) found a positive income elasticity for pork beginning with the 1956–1965 period.

Livestock producers and meat marketers obviously have no direct control over consumer incomes. Indirectly they have some control in fostering (or opposing) national economic policy that may have an impact on employment and wage rates. On the other hand, producers and marketers do have direct control in improvement of the quality of a product which may, in turn, improve its income elasticity. In a sense, this is changing the product rather than changing the income elasticity, but the net result to producers and marketers may be the same.

The coefficient of income elasticity is calculated in a manner similar to that of price elasticity. The formula for income elasticity by the arithmetic method is:

$$e_{D/I} = \frac{\%\Delta E}{\%\Delta I}$$

where $e_{D/I}$ = coefficient of income elasticity of demand
Δ = change
E = expenditures (for a specified product)
I = income

By use of the calculus the formula is:

$$e_{D/I} = \frac{dE}{dI} \cdot \frac{I}{E}$$

The sign of the coefficient is positive for superior products and negative for inferior products.

Cross Elasticity.—Cross elasticity refers to price-quantity relationships between or among different products. As mentioned earlier, beef
producers are interested in knowing what will happen to the quantity of
beef purchased if the price of hogs declines. Beef and pork are "substitutes" for one another, or in other words, they are competing products. It
is not enough simply to know that a drop in pork prices will hurt beef
consumption and cause a lowering of beef prices. A market analyst needs
to know in quantitative terms how much a specified drop in hog prices
(say 10%) will affect beef consumption and beef prices. The general relationship between substitute products is shown by curve A, Fig. 3.3. If
products are substitutes, an increase in the price of product Y will result
in increased purchases of product X, assuming no change in the price of
product X.

While the vast majority of products are substitutes in a greater or
lesser degree, this is not true of all. Some products are "complementary."
In this case, an increase in the price of product Y will be associated with a
decrease in the purchase of product X (see curve B, Fig. 3.3). Au-

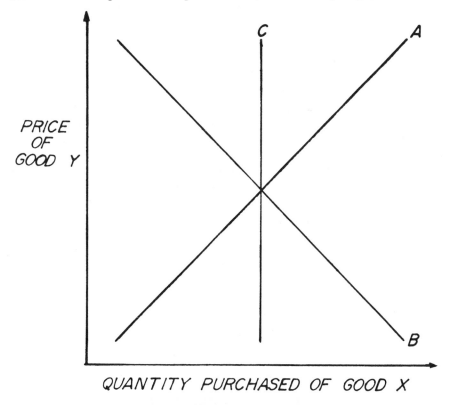

FIG. 3.3.   CROSS RELATIONSHIPS BETWEEN GOOD X AND GOOD Y

tomobiles and gasoline provide an example. An increase in the price of automobiles would be expected to result in decreased sales of gasoline, because fewer autos would be sold and hence less gasoline would be needed. A decrease in the price of sweet corn probably would lead to an increase in purchases of butter (or margarine). More sweet corn would be sold and since most people butter sweet corn, more butter would be purchased. Included in this category also are "joint" products. Beef and cattle hides are joint products. An increase in the price of beef (assuming this would result in fewer cattle being purchased) would lead to fewer hides being sold. This could, of course, bring about a longer term effect. If the increased price of beef persisted, farmers and ranchers would be induced to produce more beef cattle and thus lead to increased sales (and purchases) of hides—though prices of both might decline in the process.

In another class are "supplementary" products. Supplements are products whose demands are unrelated. This means that the quantity purchased of product X has no relation to the price of product Y (curve C in Fig. 3.3). Products that comprise a minor element in the purchases of a firm likely would fall into this category. Supplementary products are not of particular importance in this study of livestock and meat marketing.

THEORY OF SUPPLY

The concept of supply, like demand, is widely misunderstood. In a technical sense supply is the functional relationship between prices and the quantities of a product that producers are willing to place on a market. Like demand, this concept envisions a schedule of alternative prices and quantities which can be shown in table form, as a mathematical function, or in graphic form. It also envisions the situation where quantity marketed is a function of price alone—all other factors which can influence the quantity marketed are assumed to be constant. In the calculations by which supply (and demand) functions are determined, these other factors are held constant by statistical processes. We logically expect a supply curve to have positive slope, i.e., we expect producers to place more of a given product on the market at relatively high prices than at lower prices. Figure 3.4 illustrates a supply curve.

A supply function looks like a demand function except for algebraic sign. Following is a hypothetical example:

$$Q = 1.45 \, P$$

The constant is omitted in this function to indicate that at zero prices nothing would be marketed. The similarity of supply and demand functions may prompt the question of how one differentiates between the two. This is the so-called problem of "identification."

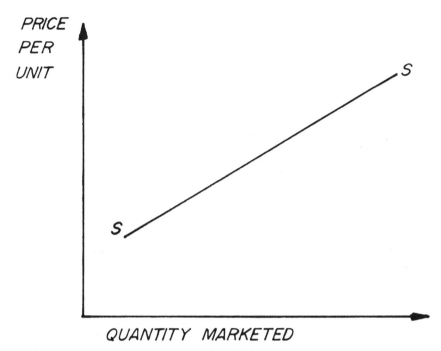

FIG. 3.4.   HYPOTHETICAL SUPPLY CURVE

"Identification (Breimyer 1961) is the term given to the principle that few observed values are of either pure supply or pure demand. Price changes in the market place do not carry any labels accommodating to analysts which explain whether they were due to demand forces or supply forces." Elmer Working's (1927) article of more than 30 yr ago remains the basic reference. Working concerned himself with how to identify supply and demand forces as represented in statistical data, and with how to separate slopes of demand and supply curves from shifts in their positions.

". . . as Working (1927) pointed out in his article on identification, it is not necessary that true demand and true supply curves be constructed if the analyst only wants a method that will help him forecast future trends in prices" (Breimyer 1961).

The concept of elasticity of supply is comparable in meaning to that of demand. Supply elasticity refers to the sensitivity of change in quantity marketed to specified changes in price. The formula for calculating elasticity of supply is the same as that for demands:

$$e_{S/P} = \frac{\%\Delta Q}{\%\Delta P}$$

where ^{e}S/P = coefficient of elasticity of supply
Δ = change
Q = quantity
P = price

By use of calculus, the formula is

$$^e\text{S/P} = \frac{dQ}{dP} \ \frac{P}{Q}$$

The coefficient of supply elasticity normally carries a positive sign while that of demand normally is negative. In some instances ^{e}S/P = O. This can be the case if a certain stock of a product is available for a market and no more can be made available at any price. Here the supply would be perfectly inelastic. When the ^{e}S/P is less than 1 but greater than zero (i.e., $0 <\ ^e$S/P< 1) supply is said to be relatively inelastic—usually referred to simply as "inelastic." When ^{e}S/P = 1 the supply is unitarily elastic, and when $\infty\ ^e$S/P> 1 the supply is said to be relatively elastic or simply "elastic." The limiting case on the upper end is ^{e}S/P = ∞ or infinitely elastic. In this limiting case the quantity offered has no relation to price.

While the notion of supply abstracts from the effects of everything except price, it is well known that "other factors" do affect the quantity offered. These other factors are "supply shifters" and are conceptualized and handled in a manner comparable to demand shifters. Among the supply shifters are (1) technical innovations (improvements in physical input-output conditions), (2) changes in factor prices (changes in prices of inputs), (3) development of new technology, and (4) changes in opportunity costs.

ORGANIZATIONAL CHARACTERISTICS

An area of market research which has received increasing attention in recent years is the study of market organization or market structure analysis. This approach holds that relationships exist between structural characteristics of an industry and the competitive behavior of firms in that industry, and beyond that, these attributes are associated with economic performance.

Performance is the critical issue. This was touched upon in Chapter 1 in the discussion of pricing efficiency and operational efficiency. Individual producers as well as the general public have a stake in this matter because the degree of efficiency attained affects producer prices and profits. It affects the costs to consumers, and thereby their real income; and it affects general resource utilization. Society as a whole has an interest in optimum utilization of resources. Performance has many other facets such as the degree of price stability in the industry, equitability of income distribution among people, degree of economic progress, and level of employment.

There is substantial evidence that, for decades, farmers have been dissatisfied with the economic performance of markets for farm products. Of recent origin are the National Farmers Organization (NFO) and the American Agricultural Movement which were organized in an attempt to improve farm income. The inauguration of the National Cattlemen's Association CATTLE-FAX program is another example of an attempt to improve prices and incomes but with a different approach. Through the years, history is replete with other examples, both private and public attempts to alleviate farm problems. To the extent that these problems are associated with structural characteristics and firm conduct or behavior, market structure analysis is a legitimate area of marketing.

Market Structure

Among the major structural characteristics of firms or an industry are: (1) Degree of concentration. This refers to the number of firms, size of firms, and their size-distribution. Is the industry composed of only a few large firms—such as automobile manufacturing? Is the industry composed of many firms, but dominated by a few large ones? Is the industry composed of many small firms? Economic theory indicates that such characteristics can influence output and profit rates. They also may influence income distribution, progressiveness of the industry, and other measures of performance.

The multitude of relatively small farmers does not have the control over output or of profits that the steel industry has, for example.

(2) Product differentiation. Do individual firms have the power, by advertising or otherwise, to convince consumers that their products are different from those of competitors? Is there a relation between the degree of concentration and the ability of firms to differentiate their products? Not only is farming characterized by a low degree of concentration, but its products are relatively homogenous. There is little opportunity for individual farmers to differentiate products. Exceptions occur such as purebred livestock breeders who have been able to show excellence in their stock. Citrus growers have been successful in differentiating "Sunkist" oranges as have walnut growers in marketing "Diamond" brand walnuts.

Processors of farm products, of course, make extensive use of brand names. This is particularly true in canned vegetables and fruits. Some large meat packers have successfully identified their brands of processed meat. Progress has been slow in differentiating fresh meat, although Swift & Co. apparently has made some headway with its Proten Beef[4] through an advertising campaign over a number of years.

While there are a few examples of differentiating agricultural products at the farm level (accomplished largely by farmer cooperatives) the vast bulk leaves the farm unbranded and undifferentiated. Outside of agriculture, differentiation is the rule.

(3) Barriers to entry into the industry. The relative ease (or difficulty) for a firm to gain entry into an industry is closely associated with degree of concentration. Traditionally, farming has been an industry with easy entry. Nearly anyone who wished could start farming. This no longer is true for commercial-sized farms. Capital requirements associated with economies of size necessitate large-scale financing that is not available to everyone.

In nonfarm sectors, successful differentiation of product by already-established firms may be an effective barrier to entry by new firms. In other instances absolute cost advantages and/or economies of scale already obtained by existing firms may prevent new firms from entering that industry.

Firms in an industry with effective barriers against new competitors exercise a different sort of behavior than firms in a highly competitive industry.

Market Conduct

Market conduct refers to the behavior of firms—the strategy they use (1) individually in competition with other firms, in purchasing inputs and selling output, and (2) in conjunction with other firms which may take the form of informal cooperation or collusion. The practices arising out of market strategy generally are related to price or product manipulation. The objective usually is to optimize profits.

Farmers acting individually, for all practical purposes, can exercise no positive market strategy. No one individual has any effective control over price, industry output, or differentiation of product. In contrast, where a few firms dominate an industry they can, by informal agreement or through collusion (explicit or tacit), exercise some degree of control over output and prices. They can differentiate products. They can divide the market. They can carry out advertising and promotional programs.

---

[4]A registered trademark of Swift & Co.

Market Performance

It has been apparent for years that, in the market arena, farmers have been operating at a competitive disadvantage relative to other economic sectors (Clodius 1959). The implication is that market structural characteristics are associated with the differences in conduct and performance. It follows that there are two logical approaches available to alleviate the situation: (1) make the nonagricultural sectors more like agriculture, i.e., make them conform more to the structure, conduct, and performance of the more highly competitive model that typifies agriculture, or (2) make agriculture more like the nonagricultural sectors, i.e., allow agriculture a degree of economic power more commensurate with that of nonagricultural sectors. There is nothing new or novel in these approaches. They have been discussed for decades. Over the years numerous actions on both approaches have been taken by both private and public (government) means. Among the public actions are freight rate regulations, antitrust laws, legislation aimed at unfair competition and price discrimination, farm price and income support legislation, specific legislation designed to regulate livestock marketing charges and to promote competition on public markets, laws providing for standardization in grading, laws that implemented market news reporting, specific restrictions on certain large meat packers under a "Consent Decree," and legislation which facilitated the organization of farmers' cooperatives. This is merely a sample of public actions which presumably were aimed at correcting or alleviating unsatisfactory conduct and performance of firms and industries. There are many more. Some are directly related to agriculture and some are not.

Private efforts on these problems probably are best exemplified by cooperatives. The organization of a farmer cooperative is recognition that a joint effort may provide farmers with some economic power where individually they have none. There are many types of cooperatives including some devoted exclusively to livestock marketing. The activities of the National Farmer's Organization are conducted as a cooperative effort.

A number of commodity-oriented organizations also illustrate private efforts. Among these are the National Pork Producer's Council with a check-off program for financing promotional programs, The American Sheep Producers Council with a similar program, and the National Cattlemen's Association with its subscription-financed CATTLE-FAX.

There appears to be a growing recognition among livestock producers (and other farm sectors) that the eventual attainment of economic power (bargaining power) necessitates a joint effort in order to gain "countervailing power" (Galbraith 1952) in the market place. If this is an indication of the trend, it will mean a further fading of the impersonal open markets. Opportunities for joint efforts increase as the number of producers becomes smaller (and operations become larger)—a trend which has been in evidence for many years.

PRICE FORMATION

Price formation is related to the degree of competition in the market. A considerable body of price theory has been developed covering perfect competition, oligopoly, duopoly, and monopoly. The available theory generally is satisfactory except for the case of oligopoly, i.e., an industry dominated by a relatively few large firms. Under these conditions it is difficult to generalize because the action one firm takes will depend upon the action his competitor takes. This can make price and output indeterminate except for special cases under rigid assumptions.

We present here only the basic concept of price formation as generated by the interaction of forces of demand and supply in a highly competitive market situation. It is noted, however, that institutional factors such as structural characteristics and market conduct discussed in the previous section may permit price modification or manipulation, particularly in the short term.

It sometimes is argued that the supply of livestock is relatively fixed for any given day or even for longer periods.[5] This is based on the assumption that for a short period there is a certain number of livestock on hand and price cannot influence that number. If this were the case, supply would show up on a graph as a perfectly inelastic, i.e., vertical curve. This is illustrated in Fig. 3.5. The aggregate demand almost always will be a negatively sloped curve (as D−D in Fig. 3.5) which will intersect the supply curve at some point. This point represents the market clearing price for that number of livestock.

The assumption of fixed supply, however, is not very realistic for livestock. Producers can withhold or rush livestock to some degree. Given some time they can make animals heavier or lighter depending on timing of marketing and variations in feeding programs. A positively sloping supply curve, even for the short period, probably is normal for livestock markets.

Normal, short-period industry supply and demand curves are shown in Fig. 3.6. Industries, of course, are made up of firms, and these industry supply-demand functions are derived from firm supply-demand functions.

---

[5]The fixed supply assumption may be appropriate for a commodity which is harvested at one particular time, thereby setting the quantity available for the subsequent year.

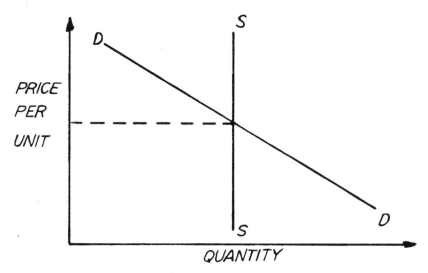

FIG. 3.5.   SHORT-RUN EQUILIBRIUM PRICE, FIXED SUPPLY

The intersection of D–D and S_1–S_1 mark the short-period equilibrium price, P_1, at which quantity Q_1 would clear the market. Demand and supply can shift as was discussed earlier. An increase in supply would shift the entire curve to the right as shown by S_2–S_2. If demand remained unchanged, the new point of intersection with D–D would be at a lower price, P_2, and an increased quantity, Q_2. A simultaneous increase in demand and supply can have offsetting effects on price, and an increase in

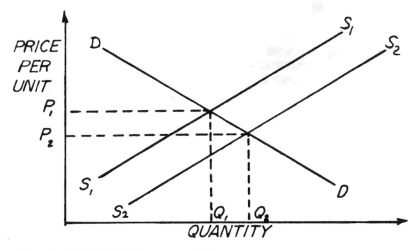

FIG. 3.6.   SHORT-RUN EQUILIBRIUM PRICE, NORMAL SUPPLY FUNCTION

demand without proportionate increase in supply can result in a higher price. It should be readily apparent that the effect on price of a change in either supply or demand will be related to the degree of elasticity associated with supply and demand.

Over the long run, prices, under highly competitive conditions, are associated with long-run costs. Costs must be covered in the long run for firms to stay in business. Anything above normal profits would attract new firms. But it takes time for adjustments to be made. There are lags between the time decisions are made and carried out, and uncertainties exist even in the most competitive markets. Imperfections in the market system result in continuous price variations about a moving long-period equilibrium. Moving equilibrium has been compared to ". . . a man riding a bicycle" (Shepherd and Futrell 1969).

BIBLIOGRAPHY

BAUMOL, W. J. 1961. Economic Theory and Operations Analysis. Prentice-Hall, Englewood Cliffs, N.J.

BREIMYER, H. F. 1961. Demand and prices for meat. USDA Econ. Res. Serv. Tech. Bull. *1253*.

CLODIUS, R. L. 1959. Opportunities and limitations of improving bargaining power of farmers. *In* Problems and Policies of American Agriculture. Iowa State University Press, Ames.

DAHL, D. C. and HAMMOND, J. W. 1977. Market and Price Analysis, the Agricultural Industries. McGraw-Hill Book Company, New York.

GALBRAITH, J. K. 1952. American Capitalism—The Concept of Countervailing Power. Houghton Mifflin Co., Boston.

ROY, S. K. and YOUNG, R. D. 1977. Demand for pork: a long-run analysis. Texas Tech University, College of Agricultural Sciences Publication No. T-1-153.

SHEPHERD, G. S. and FUTRELL, G. A. 1969. Marketing Farm Products, 5th Edition. Iowa State Univ. Press, Ames.

TOMEK, W. G. and ROBINSON, K. L. 1972. Agricultural Product Prices. Cornell Univ. Press. Ithaca, New York.

WILLIAMS, W. F. and STOUT, T. T. 1964. Economics of the Livestock-Meat Industry. Macmillan Co., New York.

WORKING, E. J. 1927. What do statistical demand curves show? Quart. J. Econ. XLI, No. 2, 212–235.

Livestock Production and Supply

4

It is readily observable that livestock and meat prices rise and fall over time. The financial fortunes of producers rise and fall with those prices. Occasionally, psychological and/or emotional disturbances trigger short-run price fluctuations, and sometimes institutional factors affect the market. But the factors which give the market its basic direction are changes in supply and demand, and the market is global. No country with trade connections can isolate itself from world conditions.

As mentioned in the last chapter, more and more imperfections are being introduced into the competitive open market system in the ever present struggle to gain economic power. Some groups have suggested price control by private means, others have suggested government programs. In essence, either approach would necessitate control of supply and/or demand. Individuals have no hope of supply or demand control, but by astute management may take advantage of favorable price movements and avoid unfavorable movements. Regardless of the approach or the objective, a thorough knowledge of supply and demand factors is essential. This chapter examines some of the more important supply characteristics. The following chapter presents a similar study of demand.

WORLD LIVESTOCK AND MEAT PRODUCTION

Leading Livestock Producers

Accurate livestock inventory records are not available for a number of Asian and African countries. A notable omission among larger countries is The Peoples Republic of China which is known to produce substantial numbers of hogs. If inventory statistics are collected in that country they are not made available to the public.

Table 4.1 shows the 10 leading producers of each specie for which data are available. The United States leads the world in number of cattle. The USSR is second, with Brazil a strong third. In world livestock statistics, water buffalo are counted with cattle in many countries. Among the top 10 leaders, however, water buffalo are not important. The USSR out-ranks all other countries in number of hogs and sheep. The United States is second in hogs, followed by Brazil which again is a strong third. Aus-

TABLE 4.1.   WORLD LIVESTOCK NUMBERS, LEADING COUNTRIES, JANUARY 1, 1977

Cattle		Hogs		Sheep	
Country	Mil. Hd.	Country	Mil. Hd.	Country	Mil. Hd.
United States	122.9	U.S.S.R.	63.0	U.S.S.R.	139.7
U.S.S.R.	110.3	United States	55.1	Australia	136.0
Brazil	96.0	Brazil	47.0	New Zealand	55.9
Argentina	58.4	Germany, West	20.7	Turkey	41.7
Australia	32.0	Poland	16.8	Argentina	37.5
Mexico	28.6	Mexico	13.2	South Africa	31.8
Colombia	23.9	France	11.6	Iran	30.0
France	12.5	Germany, East	11.3	Brazil	25.0
Turkey	15.0	Romania	10.2	United Kingdom	19.5
Germany, West	14.5	Italy	9.2	Uruguay	16.2

Source: USDA (1977).

tralia holds second place in sheep production and is not far from first place. New Zealand is third in sheep production.

World livestock numbers vary from year to year, due primarily to economic factors and weather conditions. Overriding these short-run fluctuations is an increasing trend in cattle (including water buffalo) numbers. A 13% increase was recorded from the 1968–72 average inventory to 1977. Hog numbers increased by 7% during that period, while sheep numbers declined 9%.

Leading Meat Producers

The United States is unquestionably the largest producer of meats, producing almost twice as much as the nearest competitor, the Soviet Union. The United States leads in beef and veal and in pork, but is eighth in lamb, mutton, and goat meat production. The first place ranking by the United States in pork production, while it ranked second in hog numbers, is an indication of a highly efficient swine industry. Table 4.2 shows production by continent and by leading country for the several species during recent years. World meat production is increasing in response to a growing demand. This largely reflects increases in population and gains in spendable incomes. Production gains were greatest in beef and veal, with a 25% increase from the 1967–71 average to 1976. Beef and veal constituted more than half of the total world red meat supply. Pork production increased 13%, while lamb, mutton, and goat meat decreased 3%. Over the last 14 years beef and veal production has increased approximately 4% per year, while the rate of increase for pork has been 2.6% and for lamb, mutton, and goat meat about 1% per year. It may be noted that European countries are heavy producers of pork. Japan, which appeared to be reducing pork production during the late 1960's, returned to a strong increasing trend during the 1970's.

TABLE 4.2. TOTAL RED MEAT: PRODUCTION IN SPECIFIED COUNTRIES—
AVERAGE 1967-71, ANNUAL 1972-76[1]

Region and Country	Avg 1967-71 (1,000 Metric Tons)	1972 (1,000 Metric Tons)	1973 (1,000 Metric Tons)	1974 (1,000 Metric Tons)	1975 (1,000 Metric Tons)	1976[2] (1,000 Metric Tons)
North America						
Canada	1,462.1	1,545.6	1,538.1	1,574.8	1,595.5	1,673.2
Costa Rica	46.3	62.7	55.1	59.1	66.3	68.6
Dominican Republic	41.1	53.5	55.9	57.0	56.1	63.0
El Salvador	32.6	36.4	41.8	37.8	37.5	41.6
Guatemala	66.0	74.4	70.5	71.0	80.6	86.9
Honduras	37.4	48.3	50.1	44.1	50.9	56.7
Mexico	907.6	1,055.0	1,214.0	1,270.2	1,326.6	1,452.6
Nicaragua	67.4	84.7	78.9	70.5	67.6	76.3
Panama	37.5	44.6	43.6	46.5	49.8	51.7
United States	16,230.7	16,810.8	15,830.1	17,188.6	16,675.6	17,982.4
Total	18,928.6	19,816.0	18,978.2	20,419.6	20,006.7	21,553.0
South America:						
Argentina	3,003.9	2,618.8	2,596.7	2,544.2	2,845.5	3,211.3
Brazil	2,404.1	2,760.3	3,260.5	2,901.4	2,992.4	3,079.9
Chile[3]	230.7	184.7	147.8	244.5	266.9	243.1
Colombia	488.4	544.1	507.0	526.6	572.1	648.2
Ecuador	75.0	103.1	104.2	81.0	88.7	74.9
Peru	182.0	168.2	163.8	173.6	171.0	164.4
Uruguay	428.3	357.6	351.9	398.5	434.0	493.2
Venezuela	241.2	274.5	296.3	301.9	336.0	384.2
Total	7,053.7	7,011.3	7,428.1	7,172.6	7,706.6	8,299.1
Europe:						
Western:						
EC:						
Belgium/ Luxembourg	659.7	815.5	866.0	964.5	928.1	915.6
Denmark	927.6	937.7	959.7	982.5	970.7	860.1
France	3,053.3	3,044.2	3,039.0	3,402.4	3,400.7	3,499.1
Germany, Federal Rep. of	3,473.4	3,572.0	3,533.1	3,794.0	3,777.0	3,913.0
Ireland	394.8	412.0	395.5	513.9	567.7	477.5
Italy	1,577.1	1,802.0	2,849.9	1,838.2	1,765.4	1,867.5
Netherlands	941.2	1,024.8	1,059.0	1,207.4	1,235.3	1,250.5
United Kingdom	2,072.1	2,137.3	2,118.6	2,331.0	2,320.0	2,168.0
Total EC	13,099.3	13,745.5	13,820.8	15,033.9	14,965.9	15,051.3
Austria[4]	430.8	452.7	455.2	494.9	503.0	514.6
Finland	207.3	238.8	225.9	246.4	242.4	252.6
Greece	216.0	252.8	278.5	326.0	348.0	335.0
Norway	139.9	149.5	154.3	162.7	162.3	155.5
Portugal	186.7	198.8	218.9	220.5	251.2	229.3
Spain	865.5	917.6	1,116.0	1,292.6	1,214.1	1,223.2
Sweden	406.6	408.4	397.1	429.7	429.0	446.7
Switzerland	312.7	339.9	354.1	382.5	371.9	389.5
Total	15,864.8	16,704.0	17,021.0	18,589.2	18,487.8	18,597.7
Eastern:						
Bulgaria	321.4	349.0	346.0	332.3	400.5	421.9
Czechoslovakia	670.6	743.5	763.3	803.0	830.0	812.6
Germany, Democratic Rep. of	899.0	980.9	1,027.7	1,098.6	1,189.5	1,141.9
Hungary	452.0	540.9	531.0	579.6	649.7	564.3
Poland	1,471.7	1,645.4	1,806.2	2,644.8	2,610.2	2,365.9
Yugoslavia	762.1	775.2	772.2	838.5	885.2	841.9
Total	4,576.8	5,034.9	5,246.4	6,296.8	6,565.1	6,148.4

TABLE 4.2. *(Continued)*

Region and Country	Avg 1967–71 (1,000 Metric Tons)	1972 (1,000 Metric Tons)	1973 (1,000 Metric Tons)	1974 (1,000 Metric Tons)	1975 (1,000 Metric Tons)	1976[2] (1,000 Metric Tons)
Soviet Union	9,150.0	10,026.1	9,940.1	10,739.5	10,987.7	9,759.6
Africa:						
South Africa	685.4	774.3	751.9	761.3	775.3	818.0
Morocco	86.4	105.3	174.8	169.9	167.5	166.3
Total	771.7	879.5	926.6	931.2	942.8	984.3
Asia:						
Republic of China (Taiwan)	326.5	388.3	471.5	417.6	358.7	467.4
Iran	246.8	273.4	286.6	290.1	352.9	354.1
Israel	26.0	28.2	27.5	32.4	32.0	36.3
Japan	821.0	1,071.6	1,074.4	1,254.4	1,255.3	1,360.3
Republic of Korea	118.7	135.0	135.0	144.3	168.1	189.2
Philippines	449.0	419.8	492.0	500.2	515.1	505.2
Turkey	501.7	460.7	479.4	555.8	613.5	658.6
Total	2,489.6	2,777.1	2,966.4	3,194.8	3,295.6	3,571.1
Oceania:						
Australia	1,882.6	2,397.9	2,304.0	1,920.0	2,417.0	2,642.0
New Zealand	953.3	1,029.8	987.5	949.3	1,048.6	1,150.0
Total	2,835.8	3,427.7	3,291.5	2,869.3	3,465.6	3,792.0
Total Selected Countries	61,671.1	65,676.5	65,798.3	70,213.0	71,457.9	72,705.2

Source: USDA (1977).
[1]Beef and veal; lamb, mutton and goat meat; pork, and horse meat, carcass weight basis.
[2]Preliminary.
[3]Excludes farm slaughter.
[4]Includes offal.

TABLE 4.2. BEEF AND VEAL: PRODUCTION IN SPECIFIC COUNTRIES—AVERAGE 1967–71, ANNUAL 1972–76[1]

Region and Country	Avg 1967–71 (1,000 Metric Tons)	1972 (1,000 Metric Tons)	1973 (1,000 Metric Tons)	1974 (1,000 Metric Tons)	1975 (1,000 Metric Tons)	1976 (1,000 Metric Tons)
North America:						
Canada	872.6	897.6	896.5	941.9	1,049.2	1,139.1
Costa Rica	40.6	55.8	47.7	53.1	60.3	62.6
Dominican Republic	30.3	37.1	38.8	39.0	37.1	42.0
El Salvador	21.5	25.7	30.5	26.3	26.4	30.2
Guatemala	56.7	65.5	60.6	61.9	71.5	77.7
Honduras	29.2	39.4	42.2	35.3	41.4	46.9
Mexico	535.8	592.0	744.0	844.0	889.0	986.0
Nicaragua	53.2	68.6	64.0	54.5	57.8	66.1
Panama	33.8	41.2	39.7	42.0	45.3	45.7
United States	9,904.5	10,377.4	9,813.1	10,715.8	11,271.4	12,169.6
Total	11,578.0	12,200.3	11,777.1	12,813.8	13,549.5	14,665.8

TABLE 4.2. (Continued)

Region and Country	Avg 1967–71 (1,000 Metric Tons)	1972 (1,000 Metric Tons)	1973 (1,000 Metric Tons)	1974 (1,000 Metric Tons)	1975 (1,000 Metric Tons)	1976 (1,000 Metric Tons)
South America:						
Argentina	2,518.2	2,191.1	2,146.6	2,163.0	2,438.6	2,791.5
Brazil	1,739.2	2,020.0	2,450.0	2,100.0	2,150.0	2,230.0
Chile[3]	164.9	117.8	89.2	175.2	215.5	198.1
Colombia	402.0	454.1	408.2	423.1	470.2	541.0
Ecuador	43.0	66.5	66.6	58.7	67.5	58.1
Peru	102.0	95.8	84.7	87.8	84.6	79.0
Uruguay	320.8	287.0	296.6	320.2	245.0	405.1
Venezuela	198.3	219.4	230.0	235.1	264.3	313.0
Total	5,488.3	5,451.7	5,773.8	5,563.1	6,035.5	6,616.7
Europe:						
Western:						
EC:						
Belgium/ Luxembourg	259.2	267.0	271.0	318.2	309.4	294.9
Denmark	201.2	170.5	184.1	237.9	235.3	241.8
France	1,595.4	1,456.3	1,459.4	1,791.4	1,745.5	1,798.6
Germany, Federal Rep. of	1,287.8	1,202.8	1,236.3	1,394.0	1,336.0	1,400.0
Ireland	213.2	204.5	207.3	335.5	419.5	315.0
Italy	936.7	1,027.0	1,061.7	1,041.9	935.2	985.7
Netherlands	298.6	268.5	279.2	362.5	373.2	362.0
United Kingdom	919.9	908.9	875.9	1,073.0	1,215.0	1,063.0
Total EC	5,712.0	5,505.5	5,574.9	6,554.4	6,569.1	6,461.0
Austria[4]	157.6	160.2	165.8	194.8	194.1	184.0
Finland	101.4	106.5	97.3	118.2	112.4	113.6
Greece	83.7	89.2	92.0	110.0	123.0	109.0
Norway	56.2	56.5	61.6	66.8	67.8	62.1
Portugal	69.7	73.0	81.2	84.4	96.7	78.7
Spain	268.8	302.6	371.0	416.0	453.7	418.1
Sweden	157.0	125.5	126.9	144.1	143.5	149.0
Switzerland	129.2	127.6	130.3	145.5	140.2	146.5
Total	6,735.6	6,546.5	6,701.0	7,834.6	7,900.5	7,722.0
Eastern:						
Bulgaria	96.3	96.0	108.7	99.9	96.4	98.1
Czechoslovakia	281.4	293.3	312.7	333.8	335.2	322.7
Germany, Democratic Rep. of	286.5	306.7	322.5	341.3	365.7	355.2
Hungary	118.3	105.5	120.1	115.9	158.9	141.2
Poland	530.8	499.9	539.1	699.0	752.5	802.7
Yugoslavia	292.8	260.0	318.0	307.0	321.0	335.0
Total	6,735.6	1,561.4	1,721.2	1,896.8	2,029.7	2,055.0
Soviet Union	5,039.1	5,321.5	5,461.9	5,937.1	6,019.9	5,952.0
Africa:						
South Africa	438.8	556.9	548.6	524.0	521.6	546.9
Morocco	55.6	72.1	92.0	90.0	93.0	92.0
Total	494.4	629.0	640.6	614.0	614.6	638.9

TABLE 4.2. *(Continued)*

Region and Country	Avg 1967–71 (1,000 Metric Tons)	1972 (1,000 Metric Tons)	1973 (1,000 Metric Tons)	1974 (1,000 Metric Tons)	1975 (1,000 Metric Tons)	1976 (1,000 Metric Tons)
Asia:						
Republic of China (Taiwan)	8.3	4.4	5.6	4.8	4.3	9.0
Iran	47.0	53.3	55.0	55.1	57.0	65.8
Israel	18.2	18.5	18.7	21.0	19.5	22.3
Japan	211.7	294.7	227.2	292.1	340.0	297.9
Republic of Korea	44.0	42.4	44.9	49.4	70.5	76.4
Philippines	77.9	95.7	128.7	123.6	125.8	238.0
Turkey	171.6	145.6	160.7	200.1	213.3	230.0
Total	578.6	654.5	640.7	746.0	830.3	329.4
Oceania:						
Australia	975.4	1,320.9	1,496.0	1,267.0	1,696.0	1,869.0
New Zealand	365.1	421.1	428.9	414.7	506.7	604.5
Total	1,340.5	1,742.0	1,924.9	1,681.7	2,202.7	2,473.5
Total Selected Countries	32,860.8	34,106.8	34,641.2	37,087.1	39,182.8	40,953.3

Source: USDA (1977).
[1]Carcass weight basis; excludes offal.
[2]Preliminary.
[3]Excludes farm slaughter.
[4]Includes offal.

TABLE 4.2. PORK: PRODUCTION IN SPECIFIED COUNTRIES—AVERAGE 1967-71, ANNUAL 1972-76[1]

Region and Country	Avg 1967–71 (1,000 Metric Tons)	1972 (1,000 Metric Tons)	1973 (1,000 Metric Tons)	1974 (1,000 Metric Tons)	1975 (1,000 Metric Tons)	1976[2] (1,000 Metric Tons)
North America:						
Canada	574.8	631.7	617.1	611.1	520.9	511.9
Costa Rica	5.7	7.0	7.5	6.0	6.0	6.0
Dominican Republic	10.8	16.4	17.1	18.0	19.0	21.0
El Salvador	11.1	10.7	11.2	11.5	11.1	11.4
Guatemala	9.0	8.6	9.6	8.8	8.9	8.9
Honduras	8.2	8.9	7.9	8.8	9.5	9.8
Mexico	300.8	395.0	386.0	354.0	370.0	400.0
Nicaragua	14.2	16.2	15.0	16.0	9.9	10.2
Panama	3.7	3.4	3.9	4.5	4.5	6.0
United States	6,062.8	6,187.1	5,783.8	6,261.9	5,218.2	5,644.1
Total	7,001.2	7,284.8	6,859.1	7,300.6	6,178.0	6,629.4
South America:						
Argentina	212.5	216.0	257.7	240.5	255.1	248.4
Brazil	592.4	645.2	700.6	723.0	760.0	784.5
Chile[3]	35.7	43.6	41.8	49.9	30.0	25.1
Colombia	82.8	86.2	95.1	97.8	98.0	102.4
Ecuador	25.8	28.7	29.6	18.3	16.2	13.4
Peru	46.3	41.8	45.0	54.6	54.7	52.9
Uruguay	22.0	19.9	23.2	26.3	25.0	26.0
Venezuela	39.9	51.8	63.0	63.6	68.5	67.8
Total	1,057.4	1,132.5	1,256.0	1,273.9	1,307.5	1,320.5

TABLE 4.2. *(Continued)*

Region and Country	Avg 1967–71 (1,000 Metric Tons)	1972 (1,000 Metric Tons)	1973 (1,000 Metric Tons)	1974 (1,000 Metric Tons)	1975 (1,000 Metric Tons)	1976[2] (1,000 Metric Tons)
Europe:						
Western:						
EC:						
Belgium/ Luxembourg	387.6	538.9	587.0	638.1	609.9	611.4
Denmark	723.2	765.3	773.9	743.3	733.5	715.7
France	1,262.1	1,401.0	1,402.5	1,427.5	1,470.0	1,498.7
Germany, Federal Rep. of	2,169.9	2,354.0	2,282.1	2,380.0	2,415.0	2,485.0
Ireland	134.4	160.3	142.6	131.3	98.8	120.0
Italy	548.0	666.0	689.2	708.6	734.8	781.0
Netherlands	628.5	743.3	768.3	828.1	842.7	868.0
United Kingdom	917.9	1,008.9	1,006.9	1,006.0	845.0	862.0
Total EC	6,772.5	7,637.8	7,652.5	7,862.8	7,749.7	7,941.8
Austria[4]	270.8	291.4	288.2	298.1	306.7	328.2
Finland	100.2	127.5	125.1	125.2	127.0	136.0
Greece	48.6	70.7	88.0	104.0	110.0	111.0
Norway	64.1	75.9	75.9	78.8	77.3	76.3
Portugal	92.9	103.1	112.0	113.2	132.0	127.5
Spain	447.8	461.0	587.0	710.1	601.9	648.8
Sweden	242.0	277.1	264.9	279.7	279.1	290.4
Switzerland	177.7	207.2	219.1	233.0	227.5	238.2
Total	8,216.6	9,251.7	9,412.6	9,804.9	9,611.2	9,898.2
Eastern:						
Bulgaria	138.8	165.4	150.5	145.6	226.0	237.0
Czechoslovakia	380.6	439.7	442.0	461.3	487.1	481.8
Germany, Democratic Rep. of	601.5	663.2	693.8	744.7	809.9	772.7
Hungary	324.1	426.7	402.8	456.6	483.1	415.5
Poland	897.8	1,098.8	1,224.0	1,885.5	1,792.8	1,495.2
Yugoslavia	410.0	454.8	402.4	483.0	508.0	450.0
Total	2,752.9	3,248.5	3,315.6	4,176.7	4,306.9	3,852.2
Soviet Union	3,156.3	3,827.8	3,571.9	3,877.0	4,041.5	2,952.6
Africa:						
South Africa	73.3	84.5	92.2	93.1	87.7	86.8
Morocco	1.7	1.9	1.9	1.1	.4	.3
Total	75.0	86.4	94.1	94.2	88.1	87.1
Asia:						
Republic of China (Taiwan)	317.1	382.7	464.6	411.7	353.1	457.2
Iran	1.1	1.1	1.4	1.5	1.4	1.4
Israel	4.1	5.7	4.3	7.8	8.8	8.8
Japan	596.7	769.1	841.6	957.6	909.6	1,056.1
Republic of Korea	74.7	92.6	90.1	94.9	97.6	112.8
Philippines	364.5	318.3	357.8	372.1	384.7	372.6
Turkey	.3		.3	.3	.2	.2
Total	1,358.4	1,569.5	1,760.1	1,845.8	1,755.2	2,009.0
Oceania:						
Australia	166.8	211.3	236.0	186.0	172.0	179.0
New Zealand	37.4	42.0	29.3	34.0	33.2	34.6
Total	204.2	253.4	265.3	220.0	205.2	213.6

TABLE 4.2. *(Continued)*

Region and Country	Avg 1967–71 (1,000 Metric Tons)	1972 (1,000 Metric Tons)	1973 (1,000 Metric Tons)	1974 (1,000 Metric Tons)	1975 (1,000 Metric Tons)	1976[2] (1,000 Metric Tons)
Total Selected Countries	23,522.0	26,654.6	26,534.7	28,593.2	27,493.4	26,962.6

[1]Carcass weight basis.
[2]Preliminary.
[3]Excludes farm slaughter.
[4]Includes offal.

TABLE 4.2.   LAMB, MUTTON, AND GOAT MEAT: PRODUCTION IN SPECIFIC COUNTRIES—AVERAGE 1967–71, ANNUAL 1972–70[1]

Region and Country	Avg 1967–71 (1,000 Metric Tons)	1972 (1,000 Metri Tons)	1973 (1,000 Metric Tons)	1974 (1,000 Metric Tons)	1975 (1,000 Metric Tons)	1976[2] (1,000 Metric Tons)
North America:						
Canada	8.5	9.0	9.9	8.2	8.2	7.9
Guatemala	.3	.3	.3	.3	.3	.3
Mexico	56.0	53.0	58.2	57.8	55.7	55.7
United States	263.4	246.3	233.1	210.9	186.0	168.7
Total	328.3	308.6	301.6	277.2	250.1	232.6
South America:						
Argentina	193.4	133.5	129.8	111.9	123.4	133.5
Brazil	55.3	57.6	58.4	52.4	56.4	52.4
Chile[3]	24.6	16.4	13.0	16.6	18.3	16.4
Colombia	3.7	3.8	3.7	5.8	3.9	4.0
Ecuador	6.2	7.9	8.0	4.0	5.0	3.4
Peru	33.7	30.6	34.1	31.2	31.7	32.5
Uruguay	84.7	50.4	31.0	51.8	62.1	60.2
Venezuela	3.0	3.2	3.3	3.3	3.2	3.3
Total	404.7	303.5	281.2	276.9	304.0	305.7
Europe:						
Western:						
EC:						
Belgium/ Luxembourg	3.7	2.8	2.8	2.9	3.0	3.1
Denmark	2.3	1.1	.9	.6	.7	.7
France	121.8	132.0	131.5	137.6	138.5	154.9
Germany, Fed. Rep. of	10.8	11.4	11.2	16.0	21.0	22.0
Ireland	44.1	45.0	43.3	45.0	46.3	39.0
Italy	48.4	53.0	49.0	44.0	47.2	47.5
Netherlands	9.4	10.7	9.7	15.0	16.4	15.6
United Kingdom	234.3	219.5	235.7	252.0	260.0	243.0
Total EC	474.8	475.5	484.1	513.0	533.1	525.8
Austria[4]	1.1	.6	.7	1.6	1.8	2.0
Finland	1.2	1.4	1.1	1.0	1.0	1.0
Greece	83.6	92.9	98.6	112.0	115.0	115.0
Norway	17.7	15.9	16.1	16.4	16.4	16.4
Portugal	22.5	20.9	24.2	21.5	21.5	22.3

TABLE 4.2 *(Continued)*

Region and Country	Avg 1967–71 (1,000 Metric Tons)	1972 (1,000 Metric Tons)	1973 (1,000 Metric Tons)	1974 (1,000 Metric Tons)	1975 (1,000 Metric Tons)	1976[2] (1,000 Metric Tons)
Spain	134.1	137.0	144.0	155.1	148.2	145.8
Sweden	3.2	3.1	3.4	4.2	4.3	4.8
Switzerland	3.4	3.4	3.1	2.0	2.5	2.8
Total	741.6	750.8	755.3	826.9	843.9	835.9
Eastern:						
Bulgaria	86.2	87.6	86.8	86.8	78.1	86.8
Czechoslovakia	6.5	7.6	7.5	6.8	6.7	7.0
Germany, Democratic Rep.	11.0	11.0	11.4	12.7	13.9	14.0
Hungary	7.0	5.5	7.0	6.3	6.5	6.6
Poland	25.4	24.6	24.5	23.2	21.6	21.8
Yugoslavia	58.2	60.0	51.3	48.0	55.0	56.0
Total	194.4	196.3	188.6	183.8	181.8	192.1
Soviet Union	954.6	876.9	906.3	925.3	926.3	855.0
Africa:						
South Africa	173.2	132.9	111.1	144.2	166.0	184.3
Morocco	26.5	27.0	76.0	75.0	71.0	73.0
Total	199.7	159.9	187.1	219.2	237.0	257.3
Asia:						
Republic of China (Taiwan)	1.2	1.2	1.3	1.2	1.3	1.3
Iran	198.7	219.0	230.3	233.5	294.5	286.9
Israel	3.7	4.0	4.5	3.6	3.7	5.2
Japan	1.2	.9	.8	.4	.4	.1
Philippines	5.2	4.4	5.0	3.9	4.1	4.1
Turkey	329.7	313.1	311.6	352.5	397.1	425.5
Total	539.6	542.6	553.4	595.1	701.0	723.1
Oceania:						
Australia	740.3	865.7	572.0	467.0	549.0	594.0
New Zealand	550.8	566.7	529.3	500.6	508.7	510.9
Total	1,291.1	1,432.4	1,101.3	967.6	1,057.7	1,104.9
Total Selected Countries	4,653.9	4,571.0	4,294.7	4,272.0	4,501.8	4,506.6

Source: USDA (1977).
[1]Carcass weight basis.
[2]Preliminary.
[3]Excludes farm slaughter.
[4]Includes offal.

Several European countries produce surprisingly large quantities of beef. Most of this beef comes from cattle also kept for milk, but many of the breeds are large-bodied cattle which yield a relatively large proportion of meat. Beef production, generally, is more evenly distributed among continents and countries than are other meats. But as will be seen, the exportable surplus is largely concentrated in a few countries.

GEOGRAPHIC DISTRIBUTION OF PRODUCTION IN THE UNITED STATES

Cattle, hogs, and sheep are produced in every state but the geographical distribution is not uniform. Not only is the distribution different for each of the three species, but for a given specie the density of production varies substantially. The density of human population also varies. It is not surprising that the human population density does not correspond with livestock production, but the fact that this is so complicates the marketing system. For example, production of beef in the states east of the Mississippi River comprises 20% of total production, but the population in those states accounts for approximately 70% of total consumption of fed beef. Beef must be moved in vast quantities to satisfy this demand.

The location of areas of livestock concentration follows the principle that governs industrial location. As a general rule, an area of concentrated production develops where that industry has a comparative advantage, or where it has the least comparative disadvantage. The availability and cost of inputs, transport charges on both inputs and outputs, and the location of existing and potential demand are controlling factors in industry location. Feed is by far the major input in livestock production. It almost is axiomatic that a region which possesses a comparative advantage in producing feed of a particular type also will have a comparative advantage in producing the specie and class of livestock adapted to the consumption of that feed.

Vast areas of range and pasture land, in the absence of irrigation, are adapted only to the indigenous vegetation. The native forage of the Plains and Mountain States is adapted only to cow herds, cattle growing programs, and ewe flocks. Most of this land has no other feasible commercial use.[1] The density of cattle and sheep production in these areas is relatively light. This should not obscure the fact that the areas typically are rather fully utilized. The nature of livestock production systems adapted to these areas is classified as extensive (in contrast to intensive) from the standpoint of land use.

Areas of highly concentrated red meat animal feeding (finishing) programs mirror areas of highly concentrated feed grain production. It was pointed out in Chapter 1 that livestock feeding followed feed grain (corn) production as farming spread westward.[2] Feeding activities still are fol-

---

[1]It is recognized that with increasing affluence of an increasing population, pressures are mounting for both commercial and public recreational use of land suitable for those purposes. In some particular areas, residential development is taking substantial acreages. With the passage of time, nonagricultural demands will take additional acreages.

[2]The location of broiler production is not oriented to feed grain production to the extent of red meat animals. Feed inputs are significant in broiler production, but are relatively less important. The feed conversion ratio is more favorable for broilers than for red meat animals and other factors, e.g., labor requirements, can be more easily filled in areas other than feed grain-producing regions.

lowing geographic changes in feed grain production that more recently have resulted from the development of hybrid feed grains, irrigation, and the impact of government crop restriction programs. The Corn Belt became the Corn-Hog Belt. That region also became the major cattle finishing area. This was logical from an economic standpoint. The demand from population concentrations in the east required an eastward flow of livestock and meat. The Corn Belt was strategically located from a geographical (freight cost) standpoint. Subsequently, population concentration increased on the west coast. Coincident with this was increased irrigated feed grain production in California, Arizona, and the Milo Belt of the central and southern Plains States. Farm program acreage restrictions on wheat also encouraged increased production of grain sorghum, especially in the winter wheat region.

Clearly, it made economic sense to feed western-produced grain to cattle for west coast beef consumption. However, in time cattle feeding in California and Arizona exceeded local feeder cattle supplies and feed grain supplies, and importation of both feeder cattle and feed grains became necessary.[3] In this situation the freight rate structure became a significant factor, although climatic conditions as they affect cost of production also is an important factor. The freight rate structure has been such that the combined cost of shipping feeder cattle and grain to the west coast is greater than the cost of shipping meat on an equivalent converted basis. Livestock interests on the west coast periodically petition for a revision in freight rates which would cheapen the cost of imported feed grain but, the Interstate Commerce Commission has been reluctant to grant such a change. Cattle feeders in the Plains area are somewhat vulnerable in this situation and in protection of their interest oppose a change in freight rates which would benefit a competing region. This is an example of institutional factors modifying competition. Population increases on the west coast, southwestern states and Florida, for some years have been greater than in other areas of the United States necessitating greater and greater shipments of meat to those areas.

There is a tendency to visualize the production of various crops in a relatively set geographic pattern. Type-of-farming areas have been delineated on the basis of consistency in geographic production patterns. On a year-to-year basis, or in the relatively short period, this probably is a valid assumption. For certain uses, such analyses serve useful purposes. Over a period of time, however, shifts do occur in feed production patterns which contribute to geographical shifts in livestock production. Other factors also are instrumental in shifting livestock production, but feed production probably is the major one.

---

[3]Urban development in California also has been an important factor in curbing cattle feeding operations. Not only do residential developers outbid crop producers and feedlot operators for land, but feedlot odors are not compatible with residential use.

The development and adoption of hybrid corn in the Corn Belt increased that region's competitive advantage relative to some of its fringe areas. As a case in point, corn and hog production dwindled in Kansas. Kansas still has not regained its pre-hydrid-corn position in hog production. However, the development of hybrid grain sorghum and its early adoption in Kansas and other Plains States has again changed the relative competitive situation (Brandner 1969). The resultant expansion of cattle feeding in the Milo Belt is well known. Not so well known is a reversal of the downtrend in Kansas' hog production. These are examples where break-throughs in plant breeding and development of irrigation triggered changes in livestock production—not only in the extent of feeding operations but also in its location. The potential development of hybrid wheat, possibly a hybrid feed wheat, could result in additional shifts. Somewhere on the horizon, and perhaps in the not too distant future, the availability and cost of irrigation water and energy will become deciding factors. In most of the irrigated area of the Great Plains the water level is declining. In limited areas, underground water already is exhausted. At some point in time feed production and associated livestock production will be affected. On the other hand, parts of Central America, South America, and Africa have the potential for expanding cattle production.

Three types of production and price movements are of concern to livestock market analysis: (1) secular, (2) cyclical, and (3) seasonal. Such movements are observable in inventories, market supplies, and prices. Seasonal variations are those that tend to follow a more or less uniform pattern within the year (i.e., a 12-month period, but need not start in January) and, furthermore, conform to this pattern over a period of years. This means peaks and troughs tend to come at about the same time each year. A cyclical movement, as the name implies, is one that tends to follow a pattern that repeats itself.[4] As defined here, a cycle requires more than 1 year (it tends to repeat itself over a period of years). Secular trends are long-time trends that persist over a period of several cycles. A fourth type of production and price movement is random, or erratic variation. This has no uniformity, hence little or no predictability. Over a period of time it is expected that random variations will cancel out and, hence, usually are ignored.

Price and production data as generated and reported in the markets at any given time represent the combined effect of secular trend, cyclical movement, seasonal movement, and random variation. As mentioned above, random variations usually are ignored. A first task in a study of the others is the problem of isolating the effect of each separately. Statistical techniques are available to do this, although no attempt will be made

---

[4]The term "cycle" implies a degree of regularity that is not actually found in livestock. Cyclical tendencies exist but successive movements are not identical.

here to present the statistical treatment. In the discussion which follows, the notion of secular trend is fairly obvious. Figure 4.1 is presented as an aid in visualizing the concept of cyclical and seasonal movements. Line A represents original data, adjusted only for secular trend. Line A has both cyclic and seasonal influences combined in it. In line B the cyclical effect is isolated, i.e., seasonal influences have been removed. Line C represents the isolated effect of only seasonal variation.

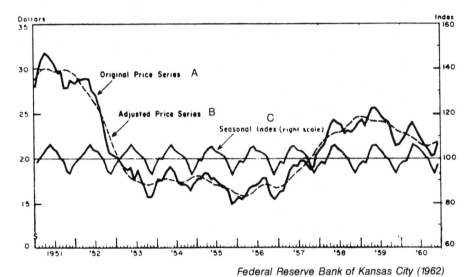

Federal Reserve Bank of Kansas City (1962)

FIG. 4.1.   ORIGINAL, ADJUSTED, AND SEASONAL CHANGES IN AVERAGE PRICES OF ALL CATTLE SLAUGHTERED UNDER FEDERAL INSPECTION

Cattle—Secular Trends

The total of all cattle and calves on hand Jan. 1 of each year since 1867 (the year annual records were begun) is shown in Fig. 4.2. This includes both beef and dairy cattle, and for purposes of this discussion, both should be considered. Over the years the slaughter of dairy stock has contributed substantial quantities of beef, although it has been of declining importance since about the mid-1940's. It is obvious that the secular, or long-time, trend in cattle numbers has been upward. Superimposed on the secular trend are cyclical variations. While the overall trend is upward, the rate of growth has not been uniform. The total period can be logically broken into two segments, with the breaking point in the late 1920's, DeGraff (1960). This is shown in Fig. 4.3.

Production per head also has increased so the secular trend in production has increased at a faster rate than have numbers of cattle. A number

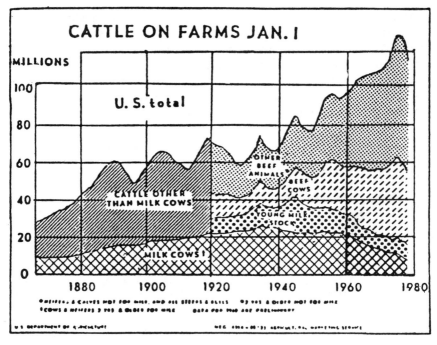

FIG. 4.2.   CATTLE ON FARMS JAN. 1, 1880 TO 1978
Data after 1961 added by author from USDA annual livestock inventory reports.

of factors have contributed to the increase in productivity. Among the more important are:

(1) An increasing proportion of the cattle population is composed of beef cattle, and meat output per head is greater for beef cattle than for dairy cattle. The ratio of beef cows to dairy cows has changed rapidly during the past 25 yr. A generally overlooked fact is that dairy cows outnumbered beef cows by more than 2 to 1 during the 1930's. By the early 1950's, beef cow numbers equaled dairy cows, and in 1978 beef cows outnumbered dairy cows by more than 3½ to 1. By almost any standard, this has been a very rapid change. The number of dairy cows reached a peak about the mid-1940's and has been declining steadily since, while beef cow numbers have been on an increasing trend since the 1930's (see Fig. 4.4).

(2) An increasing proportion of slaughtered cattle is composed of grain-fed cattle. The slaughter of vealers and calves declined steadily and slaughter of "grass-fat" cattle disappeared until a short period of relatively high grain prices in the mid 1970's temporarily halted that trend. Included in the increased numbers and proportion being grain fed are substantially more heifers than previously. Output of meat per head, of

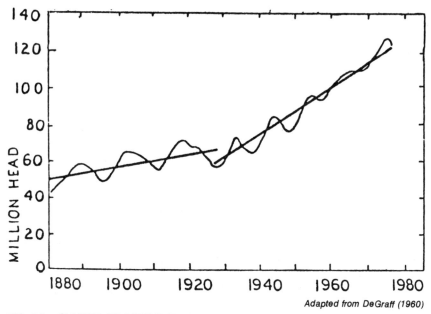

Adapted from DeGraff (1960)

FIG. 4.3.   NUMBER OF CATTLE ON FARMS, JAN. 1, 1880–1978

course, is increased substantially by holding cattle through the growing period, and finishing out on grain.

(3) Improved breeding, feeding, and management are factors in increased productivity. The effects of these improvements show up in many

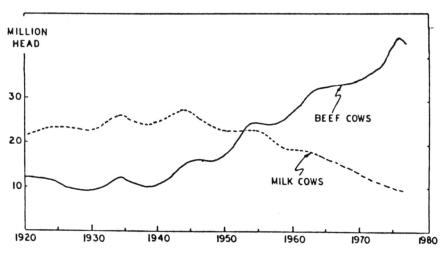

FIG. 4.4.   NUMBER OF BEEF COWS AND MILK COWS ON FARMS AND RANCHES, JAN. 1, 1920-1978

ways, such as size of animal, conformation, carcass yield, yield of lean meat, feed efficiency, rate of gain, death loss, and percentage calf crop.

The net effect of these factors is not only an increasing trend in average dressed weight (Fig. 4.5), but also an improvement in beef quality. In 1970, approximately 80% of the total cattle slaughter was composed of grain-fed beef. It was half that much 30 yr ago.

In an analysis of slaughter cattle prices, Franzmann (1967) calculated an average secular increase of $0.0063 per cwt per month for the period Jan. 1921–Aug. 1967. This was based on monthly prices which had been deflated—a process whereby the effects of inflation and deflation are removed. This long-run increase in price occurred during the time that the quantity of beef supplied (consumed) per person also was increasing. When prices increase simultaneously with quantity over a period of years, it is apparent that demand is increasing, and at a faster rate than supply. These trends have important implications for the beef industry. There is an element of risk in extrapolating any trend into the future, but all statistical projections are based on historical data. There are no statistics available for the future. Used in conjunction with other evidence, it is valid to project a continuation of these trends; and while they do not guarantee a profit to beef producers, an increasing demand must be considered a favorable situation.

Cattle—Cyclical Variations

Cattle cycles have been observed since late in the 19th Century. Over the entire period, cycles have averaged about 12 yr in length, but earlier cycles were longer than later ones. Each cycle may be divided into two

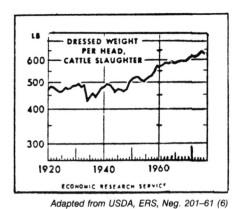

Adapted from USDA, ERS, Neg. 201–61 (6)

FIG. 4.5. MEASURES OF PRODUCTIVITY OF CATTLE INVENTORY—DRESSED WEIGHT PER HEAD, CATTLE SLAUGHTER

phases: (1) the upward or accumulation phase, and (2) the downward or liquidation phase.

The cyclical nature of the cattle industry may be observed in inventory numbers, in slaughter, and in prices. Even a casual observation of Fig. 4.2 reveals the upswings and downswings in cattle numbers. It also can be seen that the peak of each succeeding cycle was higher than the preceding peak, and each trough is higher than the preceding trough. Fig 4.6 is an alternative means of illustrating the cycles. Here the individual cycles are shown separately.

Theory of Self-Generating Cycles.—No two cycles have been identical, but a remarkable similarity exists. This degree of consistency and uniformity suggests some underlying causal forces. There are underlying

CATTLE ON FARMS BY CYCLES

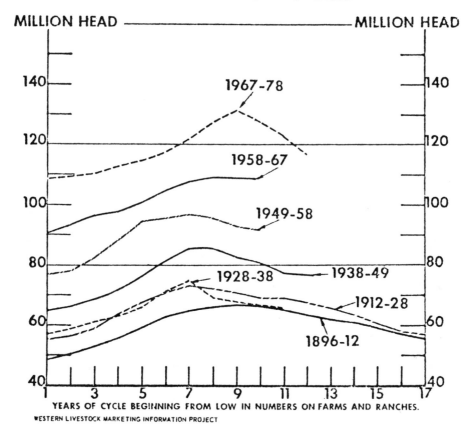

FIG. 4.6.   CATTLE ON FARMS BY CYCLES

causes, but analysts have never reached unanimous agreement on them. The major difference in opinion centers on whether the forces are self-generating within the cattle industry (i.e., endogenous) or arise from forces outside the industry (i.e., exogenous). Analysts in the former camp hold that farmers' and ranchers' price expectations are the key. "The industry (Walters 1965) seems to be seized periodically with 'spontaneous optimism.' During these periods, usually following favorable prices, the growth rate of cattle numbers exceeds an equilibrium growth rate. Cattle numbers in the optimistic periods are increased by adding cows to the basic breeding herds and by keeping feeder cattle longer than usual, consequently to heavier weights. This increasing production finally exceeds the increasing demand, causing first slaughter prices and then feeder prices to decrease.

"With the advent of lower prices, the industry becomes subject to a kind of 'simultaneous pessimism.' The pessimistic cattle producers then reduce the size of their basic breeding herds (because they are less profitable) by selling cows from these herds and not replacing them with heifers. This adds even more to total production and further lowers slaughter prices. Feeders retain the cattle that they are feeding in hopes of better prices. However, after it becomes apparent that prices have stabilized at the lower levels, the fed-cattle producer must then sell his cattle at heavier weights than usual. These cattle are heavier because they have been retained for a longer time on a concentrate ration. Again this adds to the already towering production."

The notion of self-generating cycles has prompted various attempts to explain them in terms of economic behavior. One such effort is the "cobweb" theorem. Original work was done by Ezekiel (1938). Later applications were developed by Harlow (1962) and related work has been done by Ehrich (1966). The cobweb theorem provides an explanation of how, under certain conditions, prices and quantities supplied and demanded move around the hypothesized equilibrium point specified by the intersection of demand and supply curves as producers and consumers react to price changes. One of the conditions inherent in the theorem is a time lag in the response of production (quantity) change to price change. The livestock industry seems to provide such an example. In the case of hogs it takes a year or more before a production change gets on the market as a result of a price change. With cattle the lag is considerably greater.

An abbreviated description of the theorem is as follows. It may be assumed that a system at one point in time is in disequilibrium. In the real world this would not be unusual. We assume a given supply curve SS and a demand curve DD as in Fig. 4.7 with price out of equilibrium at point A. This represents a price well above the level at which producers are willing to produce a considerably larger quantity as would be indicated by a

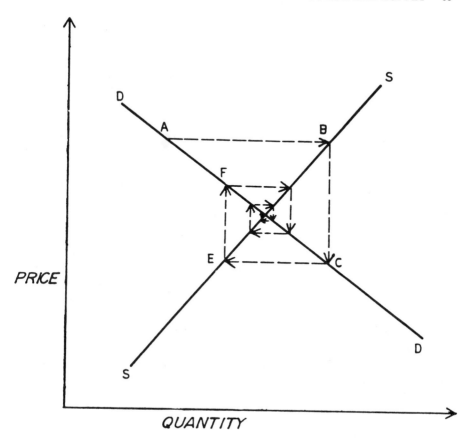

FIG. 4.7.   ILLUSTRATION OF "COB-WEB" PRICE AND OUTPUT MOVEMENTS

horizontal movement to the right until intersection with their supply curve (point B). With a sufficient time lag it may be assumed they would respond in that manner. But that output will be taken by consumers only at a substantially lower price, indicated by a vertical movement downward until intersection with the demand curve (point C). At that price, however, producers are off their supply curve and, in an effort to get back on it, they reduce production (to point E). Consumers, however, will pay a higher price (point F) for that quantity and at the higher price producers will increase production; and so the system goes off on another round. The cobweb appearance is apparent—hence the name. As drawn in Fig. 4.7 the movements are converging toward an equilibrium. It can be shown very easily that, by changing the shape of the curves (a rough indication of relative degree of elasticity), the movements can be changed to an explosive or divergent system.

In a somewhat sophisticated model, Ehrich (1966) concludes that the cobweb theorem is an inadequate model of dynamic economic behavior in either the hog or cattle economy. He states, "The characteristic behavior of production and prices suggests that harmonic motion is an appropriate model of cycle-generating forces in the U.S. beef economy. Essentially, true harmonic motion involves stimulus, response, and 'feed-back' which serves to alter the stimulus after a fixed delay. It appears that the physical characteristics of beef-cattle growth, uncertainty regarding future prices, and producers' behavior in the face of uncertainty, combine to produce harmonic behavior in annual fluctuations in cattle prices and numbers. In particular, producers respond to prices (stimulus) by changing the rate of planned production (increasing or decreasing the breeding herd incrementally), the change in production is realized after a delay (physical growth limitations), and the price stimulus is altered by realized production (prices are unilaterally affected by predetermined supplies).

"Statistical estimates of the relationships among prices, inventory adjustments, and annual production of beef were consistent with the basic model. Of primary importance, the evidence supports the view that producers respond incrementally to deviations of price from equilibrium and it serves to deny the existence of a conventional supply function for beef cattle. For, of course, the conventional concept of a supply function presupposes that producers adjust to a new level of planned output which is independent of the present level of output, in response to a change in price levels, rather than seeking to change the rate of planned output from current levels.

"Parenthetically, other evidence supports harmonic motion as a model of behavior for the hog economy as well as for the cattle economy. The cobweb theorem, which depends on the existence of a conventional supply curve, is then an inadequate model of dynamic economic behavior in either the hog or cattle economy" (Ehrich 1966).

Exogenous Factors in Cattle Cycles.—Others have contended that cycles are generated by stimuli outside the cattle industry. Among the factors cited are wars, inflationary and deflationary price trends, and variation in feed supplies (Hopkins 1926; Burmeister 1949). Most emphasis has been placed on the availability of feed including, of course, range and pasture forage. There is some historical evidence that, on occasion, drouts have coincided with downturns in cattle numbers. However, this is not the case in all cycles.

Breimyer (1954) sums up the situation as follows, "The feed supply may be an important factor at all stages of the cattle cycle. Some think it is. Approximate overall indices of range and pasture condition and crop feed supply have been included in correlation analyses of cattle production, occasionally with positive results. As a rule, though, the supply of feed is not a restraining influence at early stages of expansion because any increase in feed prices is far overshadowed by soaring prices of cattle. Only after the typical cyclical decline in cattle prices has begun are feed supplies and prices watched more closely. From then on the supply of feed can be a controlling factor."

DeGraff (1960) recognizes the effects of cattle prices and feed prices in generating cycles but relates them to the biological timetable of cattle reproduction. He says, "While such influences as a change in demand or in feed supplies may initiate a cycle, they do not explain the sequence of events which follows. The reason why a cycle follows its standardized pattern is found, not in economics, but in biology. Changes in cattle production, whatever caused their beginnings, are converted into a cyclical pattern by the natural biology of the cattle species.

"The life-span of cattle is long. They reproduce and grow slowly. If a bred heifer is kept for breeding instead of being sent to slaughter, her first calf does not reach the market until nearly 3 yr later. This is indeed a long delay in economic response. To say that cycles in cattle originate largely within the industry itself is not to say that producers are either ignorant or indifferent to the consequences of their decisions. The slow-moving biology of the species is the factor that extends the period between decision and consequence and leads to the patterned nature of the cattle cycle."

Price Cycles.—Coincident with cycles in the cattle inventory have been the cyclical trends in prices. Price cycles, typically, are the inverse of inventory cycles (as numbers advance, prices tend to decline and vice versa). However, the turning points do not occur at identical times. Studies have shown that the turning point in beef production (slaughter) lags the turning point in cattle numbers by about 2 yr (McCoy 1959; Ehrich 1966). Some lag is to be expected. For example, if cattle numbers are declining and prices rising, the decline in numbers will be halted by a withholding of heifers from market—the heifers to be retained for breeding purposes. This withholding will start an increase in inventory, but another immediate effect will be a further reduction in already-declining meat supplies and continued strengthening in prices. When it becomes apparent that beef supplies are (or will be) increasing from the buildup in cattle numbers then prices turn downward—but this does not occur immediately with initiation of the buildup in numbers. The turning point in cyclical changes in cow numbers also lags the turning point in feeder cattle prices by zero to 2 yr (Ehrich 1966). Thus, beef production (slaughter) turning points lag prices by zero to 4 yr. The reverse situation has occurred on increasing phases of the cattle cycle.

The inverse relation between cattle numbers and prices may be seen in Fig. 4.8. During war periods (World War I, World War II, the Korean War, and the Vietnamese War) and the period 1972–1974 associated with USSR grain purchases, cattle prices moved upward with increasing cattle numbers due to exceptionally strong inflationary pressures; but generally at other times prices and numbers moved in opposite directions. The cyclical nature of cattle prices is more readily observable when prices are adjusted for effects of price level changes (when prices are deflated, shown in Fig. 4.9). It should be noted, however, that the price line in Fig. 4.9 still shows the effect of the long-time gentle upward trend and the effect of seasonal ups and downs. The latter give the price line the jagged, saw-toothed appearance. Both secular and seasonal trend effects are removed statistically in calculating the true cycle.

Franzman (1967) reported that, "On the average, the adjusted price paid decreased $2.35 per cwt as the industry progressed from a peak to a trough in the cycle. Conversely, when the industry went from trough to a peak, price increased by $2.35 per cwt."

The alternating economic booms and busts that have accompanied cattle cycles have been a matter of great concern in the cattle industry. Generally, members of the industry subscribe to the proposition that in a market economy one of the functions of price is to guide production. However, the biological lag, together with lack of precise market knowledge and other market imperfections, have resulted in drastic variations in income—from prosperity to bankruptcy in many individual cases. For a number of years the industry has championed self-discipline as a means of smoothing drastic price variations.

There was some evidence that cattle inventory cycles were dampening until the liquidation of 1975–78. In previous cycles, the liquidation phase had shortened and the magnitude of decline had lessened. The consequences of cyclical variation have been serious enough to warrant a brief review of the characteristics of past cycles.

Seven cycles have occurred since 1896. These are shown graphically in Fig. 4.6. Annual data for the years 1896–1978 are shown in Table 4.3 with a period divided into alternating accumulation and liquidation phases. Two characteristics of the accumulation phases are apparent: (1) The length of accumulation phases has been remarkably uniform. Accumulation lasted 6 yr during 3 cycles, 7 yr during 2 cycles, and 8 yr in 2 cycles—a range of 6–8 yr. (2) The number of head of cattle added to the inventory also was remarkably uniform. All were within the range 17.2–21.2 million head. The percentage increase in cattle has declined with successive cycles, ranging down from 35% in the first cycle to 19.5% in the second from last. This follows from the fact that a fairly constant increase in absolute number has been added to an increasing base.

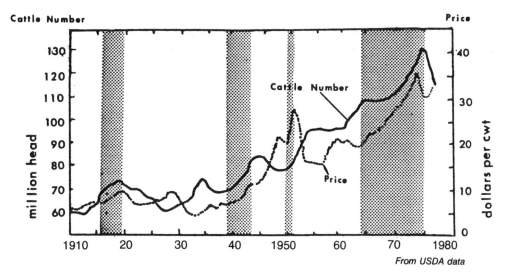

FIG. 4.8.   NUMBER OF CATTLE ON U.S. FARMS JAN. 1, 1910—1977, AND PRICE
RECEIVED BY FARMERS FOR CATTLE

Until the unusually severe reduction of 1975–1978 liquidation phases
were characterized by (1) a declining trend in length of liquidation phase;
(2) a declining trend in number of head liquidated; and (3) a declining
trend in percentage of the inventory liquidated. During early cycles the
liquidation phases lasted 8–10 yr and in later cycles, 2–3 yr. During early
cycles the liquidations amounted to 11–16 million head, while in 1965–66 it
was less than ½ million head. Liquidation during that cycle was so small it
can hardly be called a true downturn. Considering limitations on the
accuracy of reported inventories (reported numbers are estimates, not
actual counts), it appears that cattle numbers actually about leveled off in
1965 and held approximately constant until 1969, then started another
accumulation phase which ended in 1974.

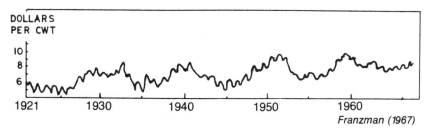

FIG. 4.9.   DEFLATED PRICE OF FEDERALLY INSPECTED SLAUGHTER CATTLE,
JAN. 1921–AUG. 1967

TABLE 4.3.   PERIODS OF EXPANSION AND CONTRACTION IN CATTLE NUMBERS
FOR ALL CATTLE AND CALVES ON U.S. FARMS, 1896–1978

Year	No. on Hand Jan 1. of Indicated Year (1,000 head)	Increase or Decrease During Year (1,000 head)	Increase or Decrease During Year (%)	Period Increase or Decrease (%)
		Accumulation		
1896	49,205	1,242	2.5	
1897	50,447	2,421	4.0	
1898	52,868	3,059	5.8	
1899	55,927	3,812	6.8	
1900	59,739	2,837	4.7	
1901	62,576	1,842	2.9	
1902	64,418	1,586	2.5	
1903	66,004	438	0.7	
Period total		17,347		35.0
		Liquidation		
1904	66,442	− 331	−0.5	
1905	66,111	−1,102	−1.7	
1906	65,009	−1,255	−1.9	
1907	63,754	−1,765	−2.8	
1908	61,989	−1,215	−2.0	
1909	60,774	−1,781	−2.9	
1910	58,993	−1,768	−3.0	
1911	57,225	−1,550	−2.7	
Period total		−10,767		−16.2
		Accumulation		
1912	55,675	917	1.6	
1913	56,592	2,869	5.1	
1914	59,461	4,388	7.4	
1915	63,849	3,589	5.6	
1916	67,438	3,541	5.3	
1917	70,979	2,061	2.9	
Period total		17,365		31.2
		Liquidation		
1918	73,040	− 946	−1.3	
1919	72,094	−1,694	−2.3	
1920	70,400	−1,686	−2.4	
1921	68,714	81	0.1	
1922	68,795	−1,249	−1.8	
1923	67,546	−1,550	−2.3	
1924	65,996	−2,623	−4.0	
1925	63,373	−2,797	−4.4	
1926	60,576	−2,398	−4.0	
1927	58,178	− 856	−1.5	
Period total		−15,718		−21.5
		Accumulation		
1928	57,322	1,555	2.7	
1929	58,877	2,126	3.6	
1930	61,003	2,027	3.3	
1931	63,003	2,771	4.4	
1932	65,801	4,479	6.8	
1933	70,280	4,089	5.8	
Period total		17,047		29.7
		Liquidation		
1934	74,369	−5,523	−7.4	
1935	68,846	− 999	−1.5	
1936	67,847	−1,749	−2.6	
1937	66,098	− 849	−1.3	
Period total		−9,120		−12.3

TABLE 4.3. *(Continued)*

Year	No. on Hand Jan 1. of Indicated Year (1,000 head)	Increase or Decrease During Year (1,000 head)	Increase or Decrease During Year (%)	Period Increase or Decrease (%)
		Accumulation		
1938	65,249	780	1.2	
1939	66,029	2,280	3.5	
1940	68,309	3,446	5.0	
1941	71,755	4,270	6.0	
1942	76,025	5,179	6.8	
1943	81,204	4,130	5.1	
1944	85,334	239	0.3	
Period total		20,324		31.1
		Liquidation		
1945	85,573	−3,338	−3.9	
1946	82,235	−1,681	−2.0	
1947	80,554	−3,383	−4.2	
1948	77,171	− 341	−0.4	
Period total		−8,743		−10.2
		Accumulation		
1949	76,830	1,133	1.5	
1950	77,963	4,120	5.3	
1951	82,083	5,989	7.3	
1952	88,072	6,169	7.0	
1953	94,241	1,438	1.5	
1954	95,679	913	1.0	
Period total		19,762		25.7
		Liquidation		
1955	96,592	− 692	−0.7	
1956	95,900	−3,040	−3.2	
1957	92,860	−1,684	−1.8	
Period total		−5,416		−5.6
		Accumulation		
1958	91,176	2,146	2.4	
1959	93,322	2,914	3.1	
1960	96,236	764	0.8	
1961	97,000	3,369	3.5	
1962	100,369	4,119	4.1	
1963	104,488	3,415	3.3	
1964	107,903	1,097	1.0	
Period total		17,824		19.5
		Liquidation		
1965	109,000	− 138	−0.1	
1966	108,862	− 217	−0.2	
Period total		− 355		−0.3
		Accumulation		
1967	108,783	588	0.5	
1968	109,371	644	0.6	
1969	110,015	2,354	2.1	
1970	112,369	2,209	2.0	
1971	114,578	3,284	2.9	
1972	117,862	3,672	3.1	
1973	121,534	6,136	5.1	
1974	127,670	4,156	3.3	
Period total		23,043		21.2

TABLE 4.3.   *(Continued)*

Year	No. on Hand Jan 1. of Indicated Year (1,000 head)	Increase or Decrease During Year (1,000 head)	Increase or Decrease During Year (%)	Period Increase or Decrease (%)
		Liquidation		
1975	131,826	−3,846	−2.9	
1976	127,980	−5,170	−4.0	
1977	122,810	−6,545	−5.3	
1978	116,265			
Period total[1]		−15,561		−11.8

		Summary		
Years (Inclusive)	No. Years	Increase or Decrease (1,000 head)	%	
		Accumulation Phases		
1896–1903	8	17,237	35.0	
1912–1917	6	17,365	31.2	
1928–1933	6	17,047	29.7	
1938–1944	7	20,324	31.1	
1949–1954	6	19,762	25.7	
1958–1964	7	17,824	19.5	
1967–1974	8	23,043	21.2	
		Liquidation Phases		
1904–1911	8	−10,767	−16.2	
1918–1927	10	−15,718	−21.5	
1934–1937	4	− 9,120	−12.3	
1945–1948	4	− 8,743	−10.2	
1955–1957	3	− 5,416	− 5.6	
1965–1966	2	− 355	− 0.3	
1975–1977	3	−15,561	−11.8	

[1]Preliminary.

Cattle—Seasonal Variations

As defined earlier, seasonal movements, or variations, are those that follow a more or less uniform pattern within the period of a year. Seasonal price movements are a direct reflection of seasonality in marketings and, to a lesser degree, seasonality in demand. Seasonality in marketings is related to biological factors and management practices. The latter are partially a matter of habit but more importantly are related to production costs. The marketing of calves reaches a peak during the fall months at the close of the grazing season. The marketing of cull cows occurs at about the same time as many farmers and ranchers cull their herds during the roundup for calf weaning. These events are related to biological processes associated with the cow-calf production programs as well as those of range and pasture forage growth. Calf prices and cull cow prices normally are seasonally low during the period of heavy marketings.

Owners, of course, could hold their cull cows and calves for later sale, or they could sell them ahead of the normal close of the range and pasture season. Some owners do vary time of sale, but most do not. Many market about the same time each year from habit. For those who do not, the decision rests on a market analysis of expected price changes versus associated costs and their effect on net profit. Herein lies the basis for our concern with seasonal analysis. Some classes of livestock have a relatively pronounced and consistent seasonal price pattern; some do not. It does not follow that the producer who adjusts his program to hit the seasonal high price will automatically maximize his profits. Some may be able to make profitable adjustments. For others, cost of production increases may more than offset higher prices. The producer must know the seasonality of his costs as well as seasonality of prices.

Some producers may view prices with greater uncertainty in particular years. But even if future prices were certain, farmers and ranchers would still respond differently to given price situations because of differences in amount and kind of resources available. Seasonal patterns can be broad guides in planning an individual business or marketing program. A producer can alter his marketing program to take advantage of certain favorable price periods without affecting the general seasonal pattern. However, if a large number of producers make the same adjustment to a seasonal trend, they may find the seasonal pattern changing. It will be shown later that such changes have occurred.

Use of Index Numbers.—It will be noted in the following discussion of seasonal movements that the data are presented as index numbers. This is a particularly useful device. It can be used on both prices and production. It is possible to calculate average seasonal price patterns in dollars, and this sometimes is done, but the price level can change rather rapidly even though the pattern of movement may not change. In this case, calculations based on dollars would be of limited use. Conversion of data to index numbers is not a difficult process. Reconversion of index numbers to dollars is a simple calculation, so that once the indices have been developed they may be applied regardless of price level changes. If the seasonal pattern changes, then, of course, a new set of indices must be calculated.

Indices are convenient ways of showing the combined pattern of a large number of seasons. An average month is given an index of 100. High months then will have indices that are greater than 100, such as 105, 110, or 125. Months in which the seasonal lows occur will have indices of less than 100. One can get an idea of how to use a particular seasonal index by looking at a graph of the index. See Fig. 4.10 for an example. On the graph there is a horizontal straight line which represents an average for the entire year. In other words, the indices for the months of the year are all added together and divided by 12. This value is called 100 merely as a

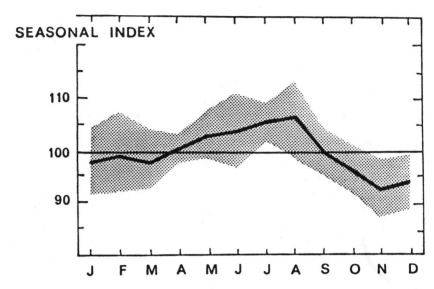

SEASONAL INDEX

FIG. 4.10.   SEASONAL VARIATION IN PRICES OF CHOICE SLAUGHTER STEERS,
900–1100 LB, OMAHA, 1971–1976

base to compare the indices for other months. The actual value for the index for each month is shown by the heavy black line going up or down from month to month. This is called the seasonal trend line.

A shaded band appears around the seasonal trend line. Technically the shaded band is called an "index of irregularity." This band represents the extremes that the seasonal trend line may be expected to stay within in the majority of years. The width of the band gives some idea of the reliability that can be placed in a given seasonal trend. Other things being equal, the narrower the band, the more reliable the seasonal trend. That is, if the band (shaded area) is quite narrow, one can expect prices in the future to vary only slightly from the trend line.

While a seasonal index gives an indication of the changes that can be expected throughout the year, a more direct application is made in estimating prices in some future month (for example) with current prices. The estimate is obtained by dividing the current price by the index for the current month and multiplying the results by the price index for the future month:

$$\frac{\text{Present Monthly Price}}{\text{Present Monthly Index}} \times \text{Future Monthly Index} = \text{Estimated Future Monthly Price}$$

For example, if the present price for feeder steers is $29.00 per cwt and the current index is 92, then the estimated price 3 months from now (when the seasonal index is 105) would be:

$$\frac{\$29.00}{92} \times 105 = \$33.10$$

One can use the index of irregularity to compute an interval in which prices in the future months might be expected to fall in about 60% of the time.[5]

The method of calculating this interval can be shown by using the feeder steer prices previously computed. The procedure is as follows:

(a) Multiply the estimated future price by the index of irregularity for that price series:

$$\$33.10 \times .028 = \$0.93$$

(b) Add the value obtained to the estimated future price to get the upper end of the interval. Subtract the same value to get the lower end of the interval:

$$\$33.10 + \text{and} - \$0.93 = \$34.03 \text{ to } \$32.17$$

This method of forecasting, of course, depends on the absence of disturbances that seriously alter the general price level, the supply, or the demand situation. It might lead to erroneous results in making estimates for a time when conditions are changing rapidly.

Choice Slaughter Steers.—The seasonal price pattern for choice slaughter steers, 900–1100 lb, at Omaha is shown in Fig. 4.10. This shows a tendency for price strength during the summer months with weakness during fall months. However, it must be noted that this pattern is relatively weak. The extent of variation (the amplitude) above the average line (i.e., the index = 100 line) at its highest point during the summer high amounts to only 7 percentage points and the variation below the average line at the lowest point during the period of weakness is just 7.5 percentage points. It also is noted that the monthly index of irregularity is relatively large, indicating a high degree of irregularity. Choice slaughter steer prices in recent years have exhibited a weak, irregular seasonal tendency. This means that little confidence can be placed in this seasonal trend. It is of limited use as a marketing tool. This pattern was influenced considerably by highly unstable prices during the 1973–75 period. The continued expansion of commercial feedlots, with a strong incentive for year-round feeding, tends to level out marketings. There is little doubt that this will continue; and with the leveling, there will be less irregularity in price movements and at the same time a reduction in the amplitude of seasonal trends.

The current pattern differs substantially from earlier periods. Wilson

---

[5]Computed indices of irregularity are not presented here, but the shaded band shows approximate magnitudes.

and Riley (1950), using the 1924–1941 period, reported a seasonal weakening of prices through spring and early summer with prices declining to an index of about 94 in June. Their analysis showed a seasonal recovery from the June low to a high of 105.6 in mid-November. This pattern was comparatively reliable at that time, and earlier. Prior to the era of commercial feedlots the customary procedure was to graze feeder cattle until the end of the grazing season. Some were slaughtered at that time as grass-fat cattle; the remainder went into feedlots—usually in the Corn Belt. The length of feeding period was geared to the quality of the cattle. So-called "plain" cattle were fed the shortest period and their marketings were bunched just prior to marketing of the next higher grade, and so on to the highest. Thus, the heaviest seasonal marketings and corresponding lowest seasonal prices for successively higher grades came in successive waves through the spring and summer. With the passing of marketing peaks, for the particular grades, price recovery would begin. The peak for plain cattle would come in the spring or early summer and following next would come the price peak for successively higher grades. This is a matter only of historical interest now, but it did illustrate the relationship between seasonality in marketings and prices. With present conditions tending more and more to level out marketings, there is reason to believe the seasonal price pattern for slaughter steers will tend to flatten out in the future.

Choice Feeder Steers (Yearlings).—The marketing of feeder steers historically has exhibited a relatively consistent seasonal pattern. As mentioned in the last section, sales were geared to the ending of the grazing season. Purchases by graziers likewise were geared to the beginning of the grazing season. Consequently, a spring high price and fall low could be depended upon with about as much reliability as the movements of a clock. This basic pattern still persists, but at a somewhat reduced level. The increasing demand by feedlots for feeder cattle on a year-round basis has been instrumental in the change. Figure 4.11 shows the current seasonal pattern for choice feeder steers, 600–700 lb, at Kansas City. The spring through summer high is followed by a fall low. The range from high to low is about 8 percentage points. With prices in the neighborhood of, say, $45.00 at the fall low, one could expect a price of about $48.75 for animals of that same weight and grade during spring. An animal purchased at a given weight in the fall usually will be in a different (heavier) weight classification by the following spring. Seasonal price patterns are based on a given weight and grade. To know the probable price differential applicable to animals as they progress through successive increases in weight, one needs an analysis of price margins. This will be presented later.

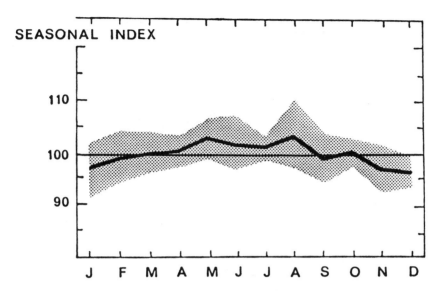

FIG. 4.11.   SEASONAL VARIATION IN PRICES OF CHOICE FEEDER STEERS, 600–700
LB, KANSAS CITY, 1971–1976

There are no absolute standards for evaluating indices of irregularity. On the basis of reasonableness the index of irregularity for yearling feeder steers is relatively small, indicating that a considerable amount of confidence can be placed in expectations that prices in any given year will follow this pattern.

Choice Feeder Steer Calves.—Seasonal prices for steer calves follow essentially the same pattern as for yearling feeder steers (Fig. 4.12). However, the range from low to high (amplitude) is greater and the index of irregularity is slightly less for most months. The evidence indicates that one can depend upon this pattern with a substantial amount of confidence.

Utility Slaughter Cows.—The seasonal movement of utility slaughter cow prices is shown in Fig. 4.13. As mentioned earlier, this pattern is the inverse of seasonality in marketing of cows. Cow prices range from a low index of about 89 in November to a high of around 107, which normally occurs about May. This can be classed as a pronounced seasonal variation. The index of irregularity is relatively small. Again, substantial confidence can be placed in the occurrence of this price pattern.

Other Cattle and Calf Seasonals.—Calculations have been made for many other classes, grades, and weights of cattle and calves, but the four

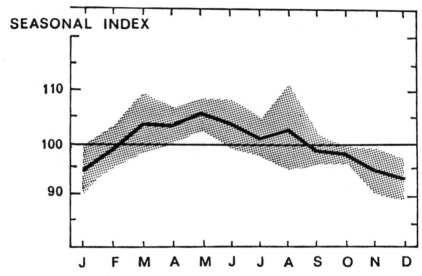

FIG. 4.12.   SEASONAL VARIATIONS IN PRICES OF CHOICE FEEDER STEER CALVES, 400–500 LB, KANSAS CITY, 1971–1976

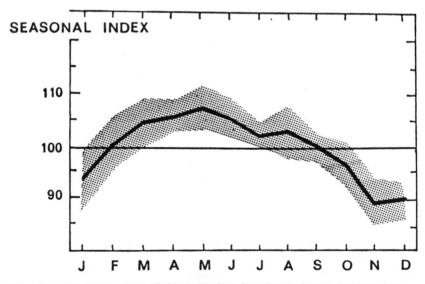

FIG. 4.13.   SEASONAL VARIATION IN PRICES OF UTILITY COWS, OMAHA, 1971–1976

illustrations demonstrate the basic patterns to be found. It could be shown that prices of "good" grade slaughter steers have a slightly greater amplitude than do prices of choice steers, and that prices of "utility" grade slaughter steers have a higher amplitude than do prices of "good" grade steers, but the highs and lows of each occur at approximately the same time of year. This was not the case in earlier times, as was pointed out previously. The seasonal price pattern for heifers, for both slaughter classes and feeder classes and feeder classes, follows the same general pattern as steer prices of comparable grade.

Hogs—Secular Trends

The long-run trend in hog numbers makes an interesting comparison with that of cattle. As may be observed in Fig. 4.14, growth in the hog inventory was steadily upward (ignoring the effect of cyclical variations) from 1867 to the early 1920's. A comparison with the trend in cattle numbers (Fig. 4.2) shows that the rate of growth was about the same for both cattle and hogs up to that point and, furthermore, the number of each species was not greatly different. A significant change occurred during the 1920's. The growth pattern for hogs became irregular, with a severe drop during the drouth-depression period of the 1930's, a sharp increase during World War II, and a sharp decline following the Korean War of early 1950's. A secular trend line through the 1920–1978 period is essentially horizontal. It will be recalled that the rate of growth in cattle numbers increased during that period.

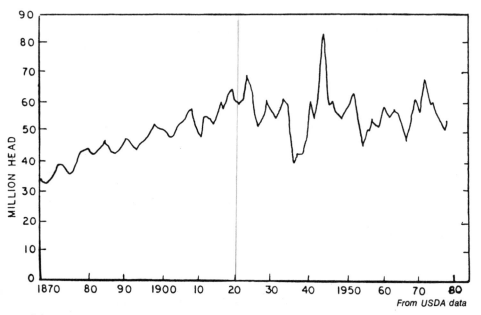

FIG. 4.14.   NUMBER OF HOGS ON U.S. FARMS JAN. 1, PRIOR TO 1968; BEGINNING 1968 THE NUMBER ON FARMS DEC.1

The long-time trend in pork production (this is tonnage of pork produced) shows a different picture (see Fig. 4.15). The drouth-depression effect and World War II effect logically can be ignored on the grounds that they were abnormal occurrences. The secular trend in pork production from 1900 to 1978 comes close to a linear trend with a growth rate of slightly less than 100,000,000 lb of pork per year. Even if one broke the trend around 1930 to obtain a better fit to the two separate periods, the trend during the latter definitely would be upward.

The apparent difference between the trend in pork production and hog numbers during the latter period is due to several factors. Improved breeding and production practices have increased productivity per head. In addition, improved feeding and management practices have speeded up the production and marketing process by finishing hogs at a lower age than formerly, i.e., increased rate of offturn. In spite of these increases in production, supplies of pork per person have trended slightly downward (due to human population increases) on a pork excluding lard basis (Fig. 4.16). This is the so-called "old consumption series" published by USDA until 1977. A new series of pork consumption "including lard" was initiated in 1977 and calculated for back years. That series shows a decided downward trend, due largely to development of hogs with a progressively lower fat to lean ratio.

Hog prices over the long period have followed the general trend of cattle and lamb prices, and prices of all three species have followed the pattern of the "all commodity" wholesale price index (see Fig. 4.17 and 4.18). The effects of inflation and deflation, however, camouflage the trend in "real" prices. Figure 4.19 shows the trend in "real" hog prices (adjusted for price level changes). Visual appraisal of the secular trend may be made by ignoring the short-period ups and downs (which are

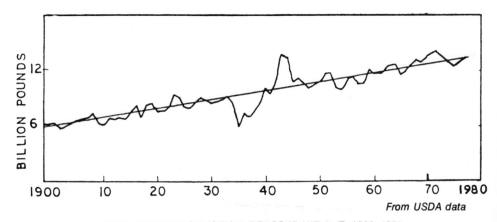

From USDA data

FIG. 4.15.   TOTAL PORK PRODUCTION, DRESSED WEIGHT, 1900–1978

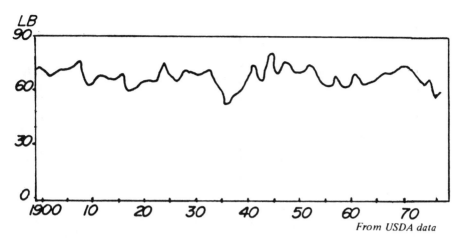

From USDA data

FIG. 4.16.   PORK SUPPLIES (CONSUMPTION) PER CAPITA, CARCASS WEIGHT.

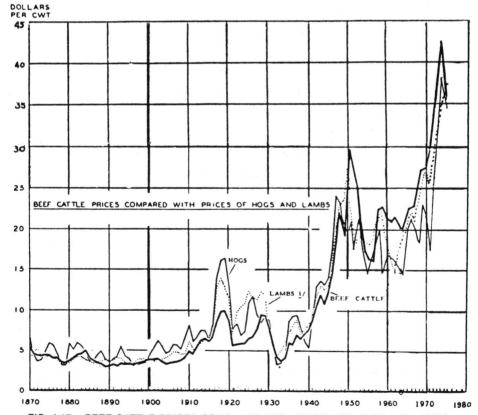

FIG. 4.17.   BEEF CATTLE PRICES COMPARED WITH PRICES OF HOGS AND LAMBS
Years 1870–1949 reproduced from materials of the American Meat Institute; later years
added by author.
[1]Prior to 1910: Sheep and Lambs.

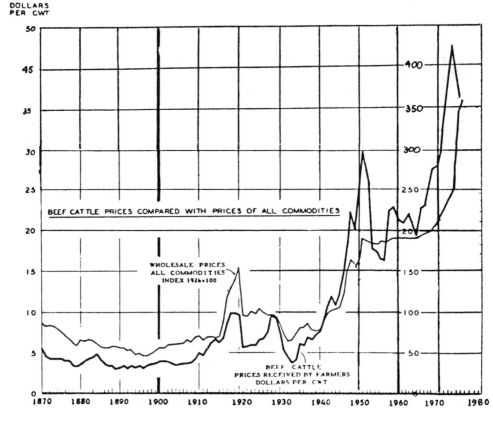

FIG. 4.18.   BEEF CATTLE PRICES COMPARED WITH PRICES OF ALL COMMODITIES
Years 1870–1949 reproduced from materials of the American Meat Institute; later years
added by author.

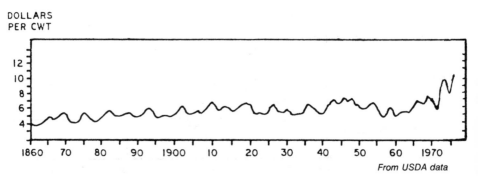

FIG. 4.19.   HOG PRICES, PEORIA, ILL. (CHICAGO PRICES PRIOR TO 1970) ADJUSTED
FOR PRICE LEVEL CHANGES, 1910–1914 BASE

cyclical variations). It is apparent that the trend has been persistently upward although the rate of increase was slight. Real prices increased an average of approximately $0.0025 per cwt per month until about 1960 when the rate increased noticeably. The recent increased rate is attributed to an increase in demand for pork. Producers have made significant improvements in the leanness and quality of pork as will be pointed out in the following chapter. Much promotional work has been done to increase the demand for pork. Their efforts have paid off.

Hogs—Cyclical Variations

Over the years hog inventories, slaughter, and prices have followed cyclical patterns (see Fig. 4.14 and 4.19). A comparison of Fig. 4.2 with Fig. 4.14 indicates that cycles in hog production have been less regular than in cattle. Prior to World War I the cycles occurred with greater regularity than they have since that time. Hog cycles average about 4 yr, and this holds whether based on earlier or later years. This averages 2 yr of expansion and 2 yr of liquidation. The period 1941–1975 encompassed 8 complete cycles (Fig. 4.20) which varied from 3 to 7 yr. It is obvious that for market planning or price forecasting the cycle has limited usefulness. However, when used with supplemental information, a knowledge of hog cycles can be a useful tool.

Hog cycles have been characterized to some extent by alternating major peaks and minor peaks. That is, there is a tendency for a high peak to be followed by a low peak. This may be seen in the price line of Fig. 4.19 especially during the period from about 1890 to 1940. It shows up again in the first two inventory cycles shown in Fig. 4.20.

Figure 4.21 illustrates that prices and production move in opposite directions. Due to the relatively inelastic nature of the demand for pork, the variation in prices is proportionately greater than variation in pork production. In other words, a given change in quantity produced (and marketed) will precipitate a relatively greater change in price.

The theory of the cyclical nature of hog production and prices is based on the supposition that producers respond to prices, and prices reflect quantity produced—an interaction process. When hog prices are relatively high (profitable) farmers respond by increasing production. As production increases, prices tend to decline. Inherent lags in the biology of the production process, lags in reaction by producers, and imperfect knowledge of the situation result in overreaction—an overshooting of equilibrium on both upturn and downturn.

The "hog-corn price ratio," usually shortened to hog-corn ratio, has been used for many years as an indicator of profitability in hog production and as a tool in forecasting turning points in production and price cycles.

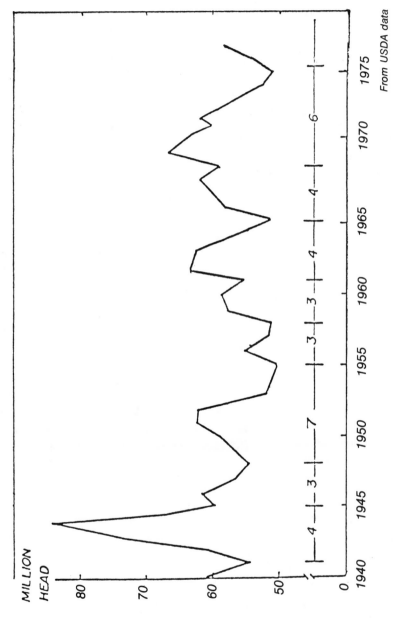

FIG. 4.20.   NUMBER OF HOGS ON FARMS JAN. 1, 1940–1977—WITH LENGTH OF CYCLE INDICATED

From USDA data

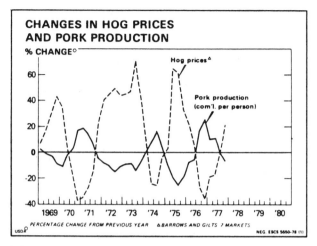

FIG. 4.21.   CHANGES IN HOG PRICES AND PORK PRODUCTION

The hog-corn ratio is the ratio between the price of hogs and the price of corn, i.e., the price per hundredweight of live hogs divided by the price per bushel of corn. The quotient (ratio) actually gives the number of bushels of corn equal in value to 100 lb of hog:

$$\frac{\text{Price per cwt live hogs}}{\text{Price per bu corn}} = \text{Hog-corn price ratio}$$

It is necessary to specify (1) the grade and weight of hogs, (2) the grade of corn, and (3) the market for which the prices apply. The needed information usually is easy to obtain and calculations are simple. For many years the USDA has calculated and reported hog-corn ratios. One that was widely used on a weekly basis as representative of the Corn Belt was the ratio for No. 1–3 hogs, 220 lb, and No. 2 yellow corn at Chicago markets. Since closing of the Chicago public terminal market (hogs in May 1970, cattle and sheep in August 1971), USDA has used Omaha prices—a weighted average price of barrows and gilts, and a simple average of No. 2 yellow corn. Both hog-corn and steer-corn price ratios are published weekly in USDA's Livestock, Meat, Wool Market News. In other publications average hog-corn ratios are quoted for the US farm level and for designated areas. Any producer can calculate a hog-corn ratio by using applicable market prices for his location.

The hog-corn ratio is only a rough approximation of the profitability of raising hogs. It is based on the notion that feed is the most important production cost and that corn is the most important feed. This undoubtedly is the case. But it does not reflect the situation as well now as it did in former years. Corn (or for that matter, any grain used in the ration) is a smaller proportion of total production costs than formerly.

The hog-corn ratio in recent years has averaged around 18. Prior to World War II the ratio averaged about 12. The records show that in this earlier period a ratio higher than 12 tended to induce an expansion in hog production. A ratio less than 12 induced a cutback in production—with a lag in both cases. Analyses typically compare, for example, number of sows farrowed in the spring, with the hog-corn ratio during the previous fall. At that time the ratio was a reliable indicator of the direction of change in farrowing (Harlow 1962). This is observable in Fig. 4.22. In the upper part of the chart a horizontal line is drawn at a ratio of about 12. This was the average for the period shown, 1924 to 1960. The actual fall hog-corn ratio is plotted as it varied above and below the average. When above average the area is cross-hatched, when below it is left open. The lower part of the chart shows change in sow farrowing during the next spring. It is readily apparent that above average fall hog-corn ratios were followed by increases in number of sows farrowed the next spring, and vice versa for periods of below average hog-corn ratios. The logic of this type of analysis is that it is not the absolute price of hogs or corn that is associated with changes in number produced, but the relation between prices of hogs and corn.

In recent years, however, the relationship between the hog-corn ratio and sow farrowings has not been as consistent as formerly. This is due in part to the declining relative importance of grain in total production, as mentioned above. In addition, the government price support program has tended to reduce (but not eliminate) variations in corn prices that formerly were associated with variations in production. Another factor is changing structural characteristics of the hog industry. Many small producers still raise hogs, but the trend is toward larger and fewer operators. The investment required in larger and more specialized units makes those operators more inclined to maintain production near optimum levels for the design of the unit. Producers still make some adjustments in hog production according to the degree of profitability, but output is not as sensitive to relative changes in corn prices as in former years. There is evidence that producers now respond more than formerly to absolute hog price and corn price changes. It also is likely that progressive producers attempt to gear production to long-run expectations of supply and demand conditions.

Hogs—Seasonal Variations

Hog prices follow a distinctive seasonal pattern which is directly related to marketings of hogs. This, in turn, is related to time of farrowing

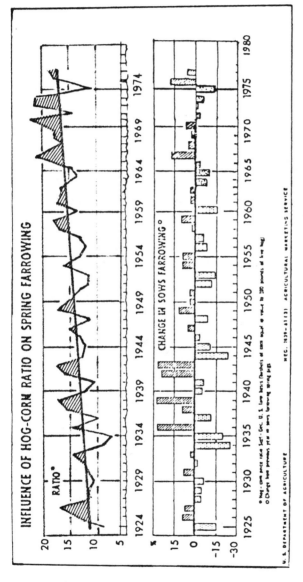

FIG. 4.22. INFLUENCE OF HOG-CORN RATIO ON SPRING FARROWINGS.
Years 1960–1977 and trend line in top part added by author.

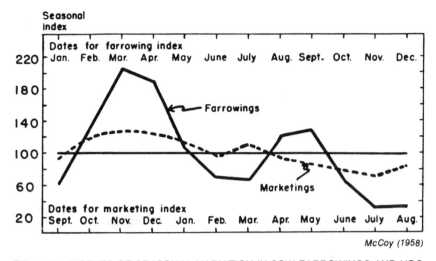

Seasonal
index

Dates for farrowing index
Jan. Feb. Mar. Apr. May June July Aug. Sept. Oct. Nov. Dec.

Farrowings

Marketings

Dates for marketing index
Sept. Oct. Nov. Dec. Jan. Feb. Mar. Apr. May June July Aug.

McCoy (1958)

FIG. 4.23. INDEXES OF SEASONAL VARIATION IN SOW FARROWINGS AND HOG
MARKETINGS FOR THE UNITED STATES, 1953–1957
Note that the same scale is used for both indices.

and to a lesser degree of feeding and breeding programs. While farrowings are distributed throughout the year, they tend to be concentrated in certain months. This leads to the concentration of marketings. The period of most pronounced concentration in farrowings is during March and April for the spring pig crop and during September for the fall pig crop, with the spring crop being the larger of the two, but this difference also is lessening. The spring crop amounted to 61% of the annual production during the early 1940's. Now they are essentially equal.

Figure 4.23 shows the seasonal pattern of farrowings and marketings, using the same scale for both, but the index line for marketing lags (seven months) so that the peak for marketing which occurs in November is matched with the peak in farrowings which occurs in March. Monthly farrowings are not available for updating seasonal farrowings. Undoubtedly it has flattened out to a considerable degree with increased year-round multiple farrowings, but spring and fall peaks still exist. It is clear that marketings are more evenly distributed than farrowings, but this is a relative matter. Marketings are far from uniform throughout the year.

Slaughter Barrows and Gilts, No. 2–3, 200–220 Lb.—The seasonal price for barrows and gilts is shown in Fig. 4.24. This is a movement of considerable proportions. The range is 20 percentage points. Prices reach a seasonal peak during the summer (August) when relatively few hogs are marketed. From that peak, prices decline to a fall low (November) as marketings of hogs from the spring pig crop reach a peak. As this peak in

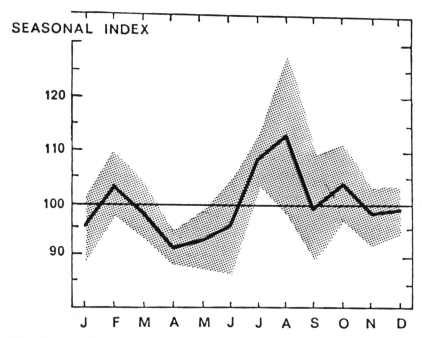

FIG. 4.24.   SEASONAL VARIATION IN PRICES OF U.S. NO. 1–2 SLAUGHTER BAR-
ROWS AND GILTS, 200–220 LB, INTERIOR IOWA AND SOUTHERN MIN-
NESOTA, 1971–1976

marketings passes, prices recover to a minor peak in February, then
decline as hogs from the previous fall pig crop are marketed. This produces a second seasonal low in April. Seasonal highs and lows do not
always hit exactly the months indicated, as particular characteristics of
individual years cause some minor shifting forward or backward. However, the seasonal pattern for hogs is more consistent than that for most
farm products. In earlier years the index of irregularity was relatively
narrow, but large, erratic price movements of recent years have widened
the index. From the standpoint of seasonality, however, it is a rare year
when hog prices do not hit a high in the summer and a low in the fall. The
winter and spring periods are less consistent.

The price pattern for barrows and gilts has changed over the years, as
shown in Fig. 4.25. In the pre-World War II period the summer peak
occurred in September. During the immediate post-World War II period
it moved to August. Currently it tends to occur in August, although in
some years it has been as early as June. This forward movement of the
peak probably is a reflection of production changes made by hog producers
in their attempts to cash in on higher summer prices. However, the peak
of the latest period shown (i.e., 1971–75) moved back to August which may

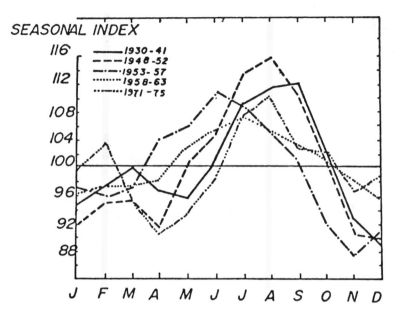

FIG. 4.25. SEASONAL PRICE VARIATION IN U.S. NO. 2 SLAUGHTER BAR-
ROWS AND GILTS, 200–220 LB, KANSAS CITY, FIVE PERIODS

indicate enough shifting of marketings to earlier months to depress those prices and simultaneously relieve pressure on the August period.

Another noticeable change in the seasonal price pattern is a decided reduction in amplitude—a tendency to flatten out. This undoubtedly has resulted from more uniform farrowings (increased multiple farrowings) and more uniform marketings. There is reason to believe this trend will continue in the future.

Slaughter Sows, No. 2–3, 330–400 Lb.—The price pattern for sows is essentially the same as for barrows and gilts (Fig. 4.26). That is, it exhibits a summer high and a fall low. The summer high typically occurs in August. The fall low (which might more appropriately be called a winter low) is in January. A secondary peak occurs in February and this is followed by a low in June. The amplitude is less than for barrows and gilts.

Feeder Pigs, 40 Lb.—Feeder pig prices, on the average, have a decided peak in early spring, then decline to a summer low (Fig. 4.27). This is followed by partial recovery in fall and another dip during the winter. This general pattern reflects the scarcity of feeder pigs on the market during early spring months which, of course, is the main farrowing period. As the marketing of these pigs increases, prices decline. A second

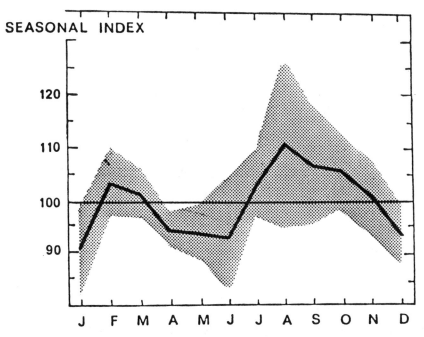

FIG. 4.26. SEASONAL VARIATION IN PRICES OF U.S. NO. 1–3 SOWS, 300–400 LB,
INTERIOR IOWA AND SOUTHERN MINNESOTA, 1971–1976

period of relative scarcity occurs in the fall about the time fall farrowings are highest.

While this seasonal has a high degree of amplitude, it also has a high degree of irregularity during the second half of the year.

Sheep—Secular Trends

The long-time pattern of growth in sheep inventory is shown in Fig. 4.28. The pattern illustrated here is a contrast to that of either cattle or hogs. Sheep numbers were substantially greater than cattle and slightly greater than hog numbers when annual records first became available (1867). The overall trend since 1867 may be described as irregular with no decidedly general upward or downward tendency until World War II. At that time inventories began a precipitous drop that continued until the 1950's. After a minor recovery, numbers again turned downward. The number of sheep in the United States now is at the lowest point in more than 100 yr.

The trend in total numbers conceals some definite changes within the sheep industry. As may be seen in Fig. 4.28, sheep numbers in the "native" sheep states declined generally throughout the period, while num-

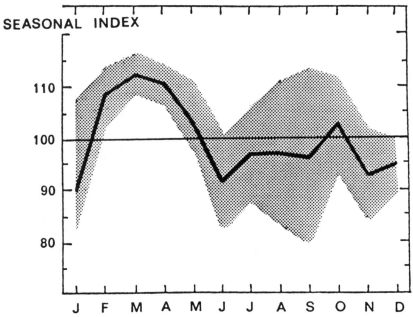

FIG. 4.27.   SEASONAL VARIATION IN PRICES OF U.S. NO. 1–2 FEEDER PIGS, 40 LB, SOUTHERN MISSOURI, 1971–1976

bers increased in the western states until the 1930's. The drop during the 1940's hit all areas, but was relatively greater in western states. Reasons for this decline are varied, but the ones most usually cited are (1) difficulty in obtaining labor, particularly sheepherders; (2) more attractive alternatives (in some cases a more profitable alternative enterprise; in other cases, a less profitable enterprise but with offsetting noneconomic amenities); (3) problems in predator control (partially related to the labor problem); and (4) reductions in grazing permits on federal lands.

Increases have been made in sheep productivity which have partially offset declining numbers. Death losses have been reduced, the percentage lamb crop has been improved and average slaughter weight has increased. The latter has been accompanied by packer and merchandiser criticism and price discounts. Research by animal and meat scientists at Kansas State University, however, has shown that weight alone is not a valid criterion of lamb values. Improved meat type lambs can be fed to heavier weights without undesirable marketing characteristics.

Increasing importation of lamb and mutton also have partially offset declining U.S. production. During 1977 the importation of lamb amounted to approximately 10% of U.S. lamb production. The importation of foreign wool and foreign fabrics together with increased domestic production of synthetic fibers have kept competitive pressures on the domestic wool market.

The U.S. sheep industry appears to be at a critical point but it is not necessarily a life-or-death matter. It would take a massive effort to turn production around to the extent that lamb would be made generally available on a regular basis throughout the United States. The sheep industry is cognizant of the situation (Anon. 1976). In 1976 the American Sheep Producers Council was instrumental in developing a "Blueprint for Expansion." The rate of decrease in sheep numbers has tapered off, indicating a degree of success with the plan.

From a practical standpoint lamb is a specialty product. This will not mark the end of the sheep industry. Lamb is preferred among certain people and in certain areas, thus assuring a market, even though limited in volume.

Sheep—Cyclical Variations

A semblance of cyclical movements can be seen in Fig. 4.28, although they are not as obvious as in the case of cattle. Prior to World War II there were 7 discernable cycles with an average length of about 10 yr. The sheep cycle is longer than that of hogs, but shorter than cattle, and this is logical in view of the biological timetable of sheep relative to the other species.

The rather drastic changes in sheep numbers since World War II have virtually obliterated cyclical trends. Nevertheless, the changes in pro-

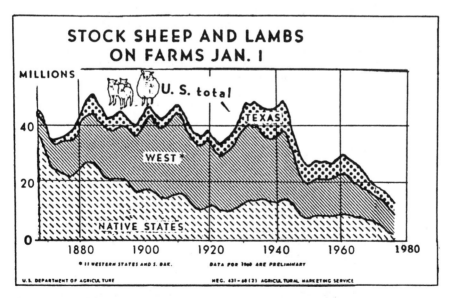

FIG. 4.28.   STOCK SHEEP AND LAMBS ON FARMS JAN. 1, 1880–1978
Data 1961 and later added by author from USDA annual livestock inventory reports.

duction are reflected in corresponding price movements (see Fig. 4.29). If domestic production continues to decline and the importation of lamb increases, variation in domestic production will have less and less influence on prices—except when superior quality of domestically-produced lamb may uphold a specialty market. For the foreseeable future, pronounced or definite cyclical movements are not likely, but variation in domestic production will continue to be the major factor offsetting prices.

Sheep—Seasonal Variations

There are two classifications of slaughter lambs (1) spring lambs and (2) fed lambs (also known in the trade as old crop lambs). In market quotation reports USDA does not use the term fed lambs but quotes their prices under headings of wooled and shorn lambs. None of these terms is entirely descriptive of the characteristics of the two classes of slaughter lambs. A major distinction between the classes is in the feeding programs (Cox 1965).

"Spring lambs" are marketed without having been weaned. They are milk-fat lambs. In addition many receive grain, usually by creep feeding arrangements. The production program is such that these lambs begin to reach markets in late spring and marketings increase from that time into the summer.

"Old crop" or "fed" lambs are lambs that have been weaned and grain-fed in feedlots. Weaning normally takes place in late summer or fall at the end of the grazing season. These lambs may be "rough fed" for a period or

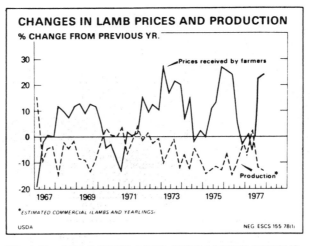

FIG. 4.29.   PER CAPITAL LAMB CONSUMPTION (SUPPLY) AND LAMB PRICES, 1940–1977

placed in feedlots at weaning time. They normally are marketed in greatest volume during the winter months. The pattern of marketings for each class is instrumental in setting the pattern of seasonality in price movements of the respective classes.

There is some overlapping of marketings of spring lambs and fed lambs which, if excessive in the spring months, can be detrimental to prices of both classes. The trade considers the two classes to be essentially two separate products. Other things being equal, consumers have a distinct preference for spring lamb. Merchandizers claim that it is difficult to move fed lambs once spring lambs arrive on the market in volume.

Choice Slaughter Lambs.—The seasonal price patterns for both spring lambs and fed lambs are shown in Fig. 4.30. The period April–September is dominated by spring lambs and the remainder of the year by fed lambs. The peak in spring lamb prices usually occurs in May but occasionally will vary one month forward or backward. A high at this time of year is very dependable and is due to relative scarcity of marketings. Demand for lamb also is strong at this season. Almost invariably spring lamb prices decline from the early high throughout the summer months.

At the beginning of the season for fed lambs, prices usually start at a level which is about average for the season then decline to a fall low. Following this, a moderate advance sets in as marketings taper off. The seasonal variation of fed lamb prices is not as pronounced as that of spring

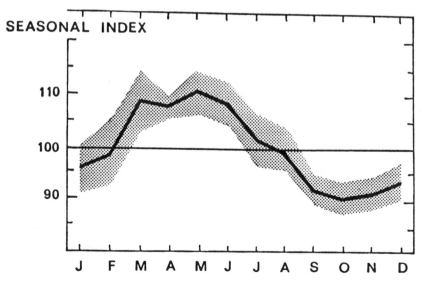

FIG. 4.30.   SEASONAL VARIATION IN PRICES OF CHOICE SLAUGHTER LAMBS, FORT WORTH, 1971–1976

lambs. A relatively high degree of seasonal regularity is indicated by the narrow index of irregularity.

Choice Feeder Lambs.—Prices for feeder lambs follow a distinctive pattern (Fig. 4.31). Typically a high occurs in the late winter-early spring period. This is followed by a rather drastic decline to early summer (June)—with a continuing gentler decline to fall, then a definite rising trend to the late winter high. Prices are considerably more irregular during late winter than the remainder of the year.

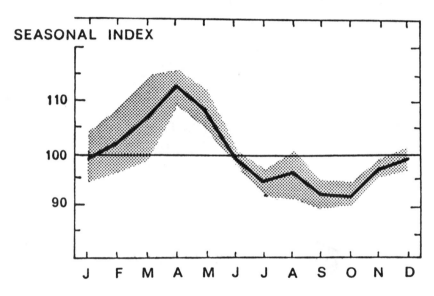

FIG. 4.31. SEASONAL VARIATION IN PRICES OF CHOICE FEEDER LAMBS, SAN ANGELO, TEXAS, 1971–1976

Seasonal price movements have not been calculated for slaughter ewes. Normally, culling takes place at the end of the grazing season. Substantial numbers go directly to slaughter but many go to feedlots in the "beet belt" (western Nebraska, northern Colorado, Wyoming, and Montana) for fattening. It is a common practice to buy them by the head and sell them by the pound (Fowler 1961).

EMPIRICAL SUPPLY STUDIES

The concept of supply, i.e., quantities which producers stand willing to put on the market at various prices during a given time period, leaves considerable leeway in specifying the time period. For some purposes analysts assume the quantity available for market during, say, the current year to be independent of current prices. The reasoning is that the number of animals available for current marketing was predetermined by prices (and possibly other factors) in a previous period. There is little doubt that greater flexibility—greater response to price—exists in changing quantities for market as the time period is lengthened. The time required for reproduction constrains the domestic production response to price and price changes. Imports, however, may be substituted within limits for domestic production.

Within a period of anything less than a reproduction cycle, some supply changes, even though limited, can be made. Producers can (1) liquidate or add to their inventories and (2) decrease or increase average slaughter weights by adjustment in feeding programs. Generally, it is assumed that the coefficient of price elasticity of supply is positive. That is, producers presumably will sell increased quantities as prices increase and reduced quantities as prices decrease. Given sufficient time this probably is almost universally the case. In the short-period there are exceptions. Livestock feeders with animals approaching market grade and weight often react to a price decline by holding the animals longer than they otherwise would, in hope that prices will rebound. Temporarily, quantities marketed move in the expected direction, but the animals withheld gain weight and the end result is an increase in quantity with decreased prices. But after a period of unprofitable low prices, production will decline. Withholding to build up breeding herds, in response to a price rise, may also appear to be a contradiction to the law of supply. Several researchers have developed a theoretical explanation on the reasoning that, at the discretion of the owner, livestock are simultaneously capital goods and consumption goods (Reutlinger 1966, Myers et al. 1970, Marsh 1977, and Nelson and Spreen 1978). As a capital good the livestock may be kept for additional growth or herd expansion. If marketed immediately they enter consumption channels. Marsh (1977) found that, "A priori, respective increases in livestock market prices in the current time period would reduce slaughter; the higher market price would induce additional build-up of beef and hog breeding herds. . . . The elasticities of supply for both are negative and very inelastic. If cattle and hog prices rise 10 percent, then cattle slaughter and hog slaughter are reduced 2 percent and .93 percent, respectively." However, this short period phenomenon is followed in time by increased tonnage from growth of given animals or increased supplies from larger herds.

Derivation of a true supply function is complicated by the fact that farmers and ranchers respond to price increases over time, not only by increasing the quantity of inputs, but they also usually adopt new and improved technology. The observed production response then may be the result of a compounding not only of price but of other factors. Some researchers make no attempt to isolate the effect of price change, but instead calculate a "response function." In some respects this is more descriptive of the real world, especially as the time span increases. It has been observed that after a prolonged price increase, producers do not reduce production correspondingly to an equivalent decline in prices. That is a logical reaction under conditions where more or less fixed investments have been made in technological developments associated with the price rise. Thus, the response curve is not presumed to be reversible—where the supply curve is considered to be reversible.

Among the empirical livestock supply analyses is an Iowa study which reported an elasticity coefficient of 0.60—0.65 for spring farrowed pigs (Dean and Heady 1959). This means that a 10% change in hog prices is associated with (in this case it could be said "would result in") a 6–6½% change, in the same direction, in pigs farrowed—all other factors unchanged. This same study also indicated a higher degree of elasticity in post-World War II years than in prior years. A USDA study reported the supply elasticity for spring farrowings to be slightly higher (0.82) than the Iowa study and confirmed an increase in supply elasticity in recent years. In addition, the USDA study noted lower elasticities for fall farrowings than for spring farrowings (Harlow 1962). Tweeten (1970) estimated short-run elasticities of supply of 0.5 for beef and 0.6 for hogs.

Several researchers have investigated the relationship between feed supplies and livestock production. These are supply-related studies. Our previous discussion of the hog-corn ratio, with references cited, is an example. Swanson (1961) reported that a 1% change in the available quantity of feed concentrates (not counting government storage stocks) was associated with a 0.9% change (in the same direction) in feeds fed to livestock. Beyond that he found that a 1% change in quantity of feed fed was associated with about 0.33% change (in the same direction) in livestock production.

Hassler (1962) found an inverse relationship between the quantity of feed concentrates fed and the price of feed concentrates. He reported that a 1% change in concentrate (grain) prices at the farm level was associated with an inverse change of about 0.2% in quantity of concentrates fed to beef cattle and about 0.1% in quantity fed to hogs—given sufficient time for producer response. These findings are consistent with the discussion presented in connection with producer response to the hog-corn ratio. In other words, as feed prices rise relative to livestock prices, fewer livestock will be produced, hence less feed will be fed. It also is likely that producers will attempt to conserve concentrates as concentrate prices rise. The same argument applies to beef as well as hogs. The comparatively low coefficient found by Hassler, however, indicates that the feeding sector is not particularly sensitive to feed concentrate prices.

BIBLIOGRAPHY

ANON. 1976. Blueprint clearing hurdles to profit with sheep. American Sheep Producers Council. Denver, Colo. July.

ASKARI, H. and CUMMINGS, J. T. 1976. Agricultural Supply Response, a Survey of the Econometric Evidence. Praeger Publishers, New York.

BICKEL, B. W. 1975. Seasonality of agricultural prices. Federal Reserve Bank of Kansas City. Monthly Review, June, 10–16.

BARKSDALE, H. C. et al. 1975. A cross-spectral analysis of beef prices. Amer. J. Agr. Econ. Vol. 57. No. 2. 309–315.

CHOI, W. 1977. The cattle cycle. European Review of Agr. Econ. Vol. 4 (2), 119–136.

COX, R. 1965. Private communication. Professor Emeritus, Dept. Animal Science and Industry, Kansas State Univ., Manhattan.

DAHL, D. C. and HAMMOND, J. W. 1977. Market and Price Analysis, the Agricultural Industries, McGraw-Hill Book Company, New York.

DEAN, G. W. and HEADY, E. O. 1959. Changes in supply functions and supply elasticities in hog production. Iowa Agr. Expt. Sta. Res. Bull. 471.

DeGRAFF, H. 1960. Beef Production and Distribution. University of Oklahoma Press, Norman.

EHRICH, R. 1966. Economic analysis of the United States beef cattle cycle. Wyoming Agr. Expt. Sta. Sci. Monograph 1.

EZEKIEL, M. 1938. The cobweb theorem. Quart. J. Econ. 52, 255–280.

FEDERAL RESERVE BANK OF KANSAS CITY. 1961. Is the cattle cycle changing? Monthly Rev. Apr., 3–9.

FOWLER, S. H. 1961. Meat production and consumption. In The Marketing of Livestock and Meat, 2nd Edition, Interstate Printers & Publishers, Danville, Ill.

FRANZMAN, J. R. 1967. The trend in slaughter cattle prices. Oklahoma Farm Econ., Dept. Agr. Econ., Oklahoma State Univ. Dec.

GARDNER, B. L. 1976. Futures prices in supply analysis. Amer. Jour. Agr. Econ. Feb. 81–84.

GRUBER, J. and HEADY, E. O. 1968. Econometric analysis of the cattle cycle in the United States. Iowa Agr. Expt. Sta. Res. Bull. 564.

HARLOW, A. A. 1962. Factors affecting the price and supply of hogs. USDA Econ. Res. Serv. Tech. Bull. 1274.

HASSLER, J. B. 1962. The U.S. feed concentrate-livestock economy's demand structure, 1949–59 (with projections for 1960–70). Nebraska Res. Bull. 203. Also North Central Regional Publ. 138.

HAYENGA, M. L. and HACKLANDER, D. 1970. Monthly supply-demand relationships for fed cattle and hogs. Amer. J. Agr. Econ. 52, 535–544.

JONES, G. T. 1965. The influence of prices on livestock population over the last decade. Jour. of Agr. Econ. 16, 420–432.

JORDAN, W. J. 1975. The short-run supply of veal. Amer. Jour. Agr. Econ. Nov. 719–720.

KEITH, K. and PURCELL, W. D. 1976. The beef cattle cycle of the 1970's Okla. Agr. Expt. Sta. Bull. B-271.

KERR, T. C. 1968. Determinants of regional livestock supply in Canada. Agr. Econ. Res. Council of Canada. Bull. 15.

LARSON, A. B. 1964. The hog cycle as harmonic motion. J. Farm Econ. May, 375–386.

MAKI, W. R. 1962. Decomposition of beef and pork cycles. J. Farm Econ. Aug., 731–743.

MARSH, J. M. 1977. Effects of marketing costs on livestock and meat prices for beef and pork. Montana Agr. Expt. Sta. Bull. 697.

MARTIN, L. and ZWART, A. C. 1975. A spatial and temporal model of the North American pork sector for the evaluation of policy alternatives. Amer. Jour. Agr. Econ. 57, Feb. 55–66.

McCOY, J. H. 1958. Trends in hog prices. Kansas Agr. Expt. Sta. Circ. 368.

McCOY, J. H. 1959. Characteristics of cattle cycles. Unpublished data, Dept. Econ. Kansas State Univ.

MEILKE, K. D. et al. 1974. North American hog supply: a comparison of geometric and polynomial distributed lag models. Canadian Jour. Agr. Econ. 22, 324–336.

MYERS, L. H. et al. 1970. Short-term price structure of the hog-pork sector of the United States. Indiana Agr. Expt. Sta. Res. Bull. 855.

NELSON, G. and SPREEN, T. 1978. Monthly steer and heifer supply. Amer. J. Agr. Econ. 60, 117–125.

NERLOVE, M. 1958. The Dynamics of Supply; Estimates of Farmers' Response to Price. Johns Hopkins Univ. Press, Baltimore, Md.

POWELL, A. A. and GRUEN, F. H. 1967. The estimation of production frontiers: The Australian livestock cereal complex. Australian Jour. of Agr. Econ. 11, 63–81.

PURCELL, J. C. 1965. Sources of beef and veal supplies and prices of cattle and calves. Georgia Agr. Expt. Sta. Mimeo Ser. N.S. 233.

REUTLINGER, S. 1966. Short run beef supply response. Amer. J. Agr. Econ. 48, 909–919.

SWANSON, E. R. 1961. Supply response and the feed-livestock economy. In Agricultural Supply Functions. Iowa State Univ. Press, Ames.

TOMEK, W. G. and ROBINSON, K. L. 1972. Agricultural Product Prices, Cornell Univ. Press, Ithaca, N.Y.

TRYFOS, P. 1974. Canadian supply functions for livestock and meat. Amer. J. Agr. Econ. 56, 107–113.

TWEETEN, L. G. 1970. Foundations of Farm Policy. Univ. of Nebr. Press, Lincoln, Nebr.

USDA. 1977A. Foreign agriculture circular. USDA Foreign Agr. Serv. FLM 10-77.

USDA. 1977B. Foreign agriculture circular, USDA Foreign Agr. Serv. FLM 12-77.

WALTERS, F. 1965. Predicting the beef cattle inventory. USDA Agr. Econ. Res. XVII, No. 1, 10–18.

WILSON, C. P. and RILEY, H. M. 1950. Seasonal variation in prices and supplies of livestock. Kansas Agr. Expt. Sta. Unpublished mimeo data, Nov.

Meat Consumption and
Related Demand

5

In a technical sense consumption is not synonymous with demand. As explained in an earlier chapter, demand connotes a relationship between quantity and price. More specifically, demand is the schedule of quantities that purchasers are willing to buy at alternative prices during some given time period. Consumption is defined as the quantity of a product used (consumed) during a specified period without necessarily relating to price.

During a period of 1 yr, for example, the quantity of meat consumed is essentially equal to the quantity available. The quantity available is primarily determined by current production. Storage stocks and net import-export balances are modifying factors. The quantity carried over in storage from 1 yr to the next is only a negligible fraction of total consumption. Some meat is imported and some is exported. While substantial for certain meats and, at particular times, net import-export balances on the average are a relatively small fraction of total meat consumption. International trade in livestock products will be covered in a later chapter.

The price consumers are willing to pay for a given quantity at any particular time is a function of their incomes, tastes and preferences, expectations about future prices, and prices of competing products. Over the long period, the quantity made available by producers at a given time is a function of prices and profit expectations at a previous time.

Consumption trends will be examined, as will factors which influence demand and consumption.

WORLD MEAT CONSUMPTION

Meat consumption in other countries is a matter of interest and economic importance to U.S. producers. The United States imports and exports substantial quantities of meat and animal products. Our exports and imports are affected by consumption elsewhere. Within certain limits the world is one big market for meat. In the countries for which data are available beef and veal account for almost 57% of the total red meat consumption, pork accounts for 37% and lamb, mutton, and goat meat account for 6%. The distribution among countries, however, varies to a high degree. The demand for meat is highly related to income level and affluence which are associated with economic development.

The United States ranks high among countries of the world in per capita meat consumption, but is far from the top. In 1977 Australia was the leader with 105 kilos (231 lb) per person (Table 5.1). This is a position historically held by Uruguay. Following in order were New Zealand 104 kilos (228 lb), and Argentina 101 kilos (222 lb). The United States was in fourth place with 88 kilos (194 lb). Of the twelve leading meat consuming countries nine increased consumption between the early 1960's and mid-

TABLE 5.1.   LEADING RED MEAT CONSUMING COUNTRIES, AVG 1968–1972, 1973, 1974, 1975, 1976, AND 1977[1]

	Consumption per Person					
	Avg 1968–1972 (Kilos)	1973 (Kilos)	1974 (Kilos)	1975 (Kilos)	1976 (Kilos)	1977[2] (Kilos)
Australia	96.8	91.3	99.5	108.3	108.2	104.6
New Zealand	108.8	104.1	97.7	103.0	99.9	103.6
Argentina	90.6	79.1	86.6	99.1	101.3	100.9
United States	88.1	81.0	86.6	93.1	88.6	87.8
Canada	71.4	70.9	72.6	73.2	77.5	79.4
Belgium-Luxembourg	61.9	70.3	76.1	74.9	75.0	74.5
Ireland	60.1	59.7	64.3	66.5	63.8	60.6
Germany, F.R.	63.8	65.3	67.2	67.9	69.6	70.3
Uruguay	108.1	84.6	100.0	111.9	100.9	80.1
Austria	60.6	63.4	65.2	67.8	69.8	71.5
France	62.9	63.2	64.9	66.2	68.5	69.5
Switzerland	62.7	65.8	64.3	64.4	66.3	69.1

Source: USDA (1978).
[1]Carcass weight basis, including horsemeat.
[2]Preliminary.

1970's. The 112 kilos consumed by Uruguayans in 1975 was far from a record. In 1953 the people of that same country consumed 137 kilos (302 lb) per capita. Argentina and Australia have exceeded 100 kilos in previous years.

Total meat consumption by major consuming countries, including those already mentioned, is shown in Table 5.2. Beef and veal consumption by country is given in Table 5.3. Uruguay and Argentina have been obvious leaders over the years. Cyclical factors tend to distort the trend effect in the short time period shown; however, Australia, New Zealand, and Japan are experiencing a definite increasing trend. Japan has the greatest percentage increase. The long-time trend in the United States was upward to the mid-1970's. Severe financial losses by the cattle industry at that time associated with cyclical liquidation prompted serious questions about further per capita gains, at least in the immediate future.

Consumption of pork in top consuming countries—i.e., West Germany

TABLE 5.2.   TOTAL RED MEAT[1] PER CAPITA CONSUMPTION IN SPECIFIED COUN-
TRIES, AVG 1961–1965, ANNUAL 1973, 1974, 1975, 1976, AND 1977

Continent and Country	Avg 1961–1965 (Kilos)	1973 (Kilos)	1974 (Kilos)	1975 (Kilos)	1976 (Kilos)	1977[2] (Kilos)
North America						
Canada	64	71	73	73	78	79
United States	76	81	87	83	89	88
Mexico	18	21	22	22	23	23
South America						
Argentina	90	79	87	91	101	101
Brazil	24	30	27	27	27	28
Chile	29	24	28	28	29	24
Colombia	25	21	22	23	24	25
Paraguay	61	27	31	30	—	—
Peru	16	12	12	12	11	10
Uruguay	108	85	100	112	101	80
Venezuela	22	25	25	26	28	31
Europe						
Western						
EEC						
Belgium and Luxembourg	53	70	76	75	75	75
Denmark	51	53	51	54	59	59
France	65	63	65	66	69	70
Germany, F.R.	54	65	67	68	70	70
Ireland	50	60	64	67	64	61
Italy	26	45	43	42	43	45
Netherlands	44	49	53	56	57	56
United Kingdom	65	59	58	58	57	58
Austria	54	63	65	68	70	72
Finland	35	49	49	53	51	48
Greece	25	40	42	44	47	48
Norway	36	40	42	45	46	48
Portugal	19	29	29	30	31	31
Spain	21	35	37	36	37	38
Sweden	46	46	52	52	55	53
Switzerland	51	66	64	64	66	69
Eastern						
Bulgaria	31	39	38	40	41	41
Czechoslovakia	45	59	59	58	55	55
Hungary	38	48	48	51	47	53
Poland	37	49	74	72	69	69
Yugoslavia	20	29	35	37	18	38
USSR	30	40	44	45	39	42
Africa						
South Africa, Republic of	36	29	33	30	30	31
Asia						
Japan	6	18	16	17	18	19
Philippines	11	12	14	14	12	12
Oceania	59	100	108	108	108	105
New Zealand	112	104	98	103	100	104

Source: USDA (1969) and USDA (1978).
[1]Carcass weight basis including horsemeat.
[2]Preliminary.

and Austria—at 46 kilos (101 lb) per person is just slightly more than
one-half the consumption of beef in the top beef consuming countries
(Tables 5.3 and 5.4). Eastern and Northern European countries generally
have the highest levels of pork consumption.

While lamb, mutton, and goat meat are very important in some countries,

TABLE 5.3.   BEEF AND VEAL[1] PER CAPITA CONSUMPTION IN SPECIFIED COUN-
TRIES, AVG 1961–1965, ANNUAL 1973, 1974, 1975, 1976, AND 1977

Continent and Country	Avg 1961–1965 (Kilos)	1973 (Kilos)	1974 (Kilos)	1975 (Kilos)	1976 (Kilos)	1977[2] (Kilos)
North America						
Canada	38	43	45	49	52	53
United States	45	51	54	57	61	59
Mexico	11	13	14	15	16	16
South America						
Argentina	77	65	74	86	89	88
Brazil	18	23	20	20	19	20
Chile	20	14	20	23	21	18
Colombia	22	16	17	19	20	21
Paraguay	49	27	31	30	—	—
Peru	9	6	6	6	5	5
Uruguay	81	65	73	84	76	66
Venezuela	18	19	19	20	23	26
Europe						
Western						
EEC						
Belgium-Luxembourg	24	28	31	31	29	28
Denmark	15	16	15	16	16	16
France	33	28	30	30	31	31
Germany, F.R.	22	24	23	23	24	23
Ireland	14	18	23	29	25	24
Italy	16	28	25	23	23	24
Netherlands	20	19	20	21	22	21
United Kingdom	26	23	24	26	25	25
Austria	20	23	25	26	26	26
Finland	20	23	23	25	24	23
Greece	9	14	17	18	20	21
Norway	15	16	17	19	19	20
Portugal	7	12	13	13	13	14
Spain	7	13	12	14	13	13
Sweden	20	16	19	20	20	19
Switzerland	24	26	25	25	26	26
Eastern						
Bulgaria	9	12	11	10	11	11
Czechoslovakia	18	28	27	24	23	23
Hungary	10	9	8	9	9	12
Poland	12	15	19	20	24	24
Yugoslavia	6	9	11	12	14	13
USSR	13	22	25	25	24	25
Africa						
South Africa, Republic of	25	21	23	20	20	22
Asia						
Japan	2	4	4	4	4	4
Philippines	3	3	5	5	4	4
Oceania						
Australia	44	47	59	71	72	71
New Zealand	50	49	68	54	56	57

Source: USDA (1969) and USDA (1978).
[1]Carcass weight basis.
[2]Preliminary.

world production and consumption has declined in recent years. From the
standpoint of per capita consumption Australia and New Zealand are in a
class by themselves with 20 kilos (44 lb) and 34 kilos (75 lb), respectively
(Table 5.5). This is a substantial drop from previous levels. People of the

TABLE 5.4. PORK[1] PER CAPITA ANNUAL CONSUMPTION IN SPECIFIED COUNTRIES, AVG 1961–1965, ANNUAL 1973, 1974, 1975, 1976, AND 1977

Continent and Country	Avg 1961–1965 (Kilos)	1973 (Kilos)	1974 (Kilos)	1975 (Kilos)	1976 (Kilos)	1977[2] (Kilos)
North America						
Canada	24	26	27	23	24	26
United States	29	29	32	26	27	28
Mexico	5	7	6	6	6	7
South America						
Argentina	8	10	9	10	9	9
Brazil	6	7	7	7	7	7
Chile	5	6	5	3	3	3
Colombia	3	4	4	4	4	4
Paraguay	12	—	—	—	—	—
Peru	4	3	4	4	3	3
Uruguay	9	8	10	9	5	6
Venezuela	4	5	5	5	5	5
Europe						
Western						
EEC						
Belgium-Luxembourg	24	38	41	40	40	41
Denmark	35	37	36	37	42	42
France	28	30	31	31	32	33
Germany, F.R.	32	41	44	44	45	46
Ireland	25	31	31	27	29	27
Italy	8	15	16	17	18	19
Netherlands	22	27	30	32	32	32
United Kingdom	28	28	27	24	24	26
Austria	34	40	40	41	44	46
Finland	15	25	25	27	27	25
Greece	5	10	12	12	13	13
Norway	15	19	21	21	22	22
Portugal	9	14	14	15	15	15
Spain	10	18	20	18	20	20
Sweden	25	29	32	31	33	34
Switzerland	25	38	38	37	39	41
Eastern						
Bulgaria	15	18	18	22	22	22
Czechoslovakia	27	31	32	33	32	32
Hungary	27	38	39	41	37	41
Poland	24	33	53	31	43	44
Yugoslavia	11	17	22	23	21	22
USSR	13	14	15	16	12	14
Africa						
South Africa, Republic of	3	4	4	3	3	3
Asia						
Japan	3	11	9	11	11	11
Philippines	8	9	9	9	9	8
Oceania						
Australia	10	17	14	12	13	14
New Zealand	16	12	11	12	11	13

Source: USDA (1969) and USDA (1978).
[1]Carcass weight basis.
[2]Preliminary.

Middle East have a traditional preference for these meats. Residents of Greece consume about as much lamb, mutton, and goat meat as either beef or pork. Average consumption in the United States has declined for several decades and nationwide stands at 1 kilo. However, it is estimated that consumption in New York is about 5.5 kilos (12 lb) and is almost at that level in large cities of California.

Horse meat is a minor component of world consumption. However, in recent years people of the Belgium-Luxembourg area are reported to consume about 3.2 kilos (7 lb) per person. That is as much total meat as is consumed by people of some less-developed countries.

In total pounds of meat consumed (that is per capita consumption times population) the United States leads all other countries. U.S. consumption accounted for approximately ⅓ of all beef and veal, slightly more than ¼ of all pork and about $1/_{10}$ of all lamb, mutton, and goat meat. This has not changed substantially in recent years.

During the period from the early 1960's to the mid 1970's all but seven of the countries shown in Table 5.2 recorded gains in meat consumption. Four of the seven showing decreased consumption are in South America. Growth in meat consumption is closely associated with income level, and income level in turn is closely associated with economic development.

The less-developed countries of Africa, Asia, and Central America hold vast numbers of people who consume only a few pounds of meat per person per year. Aside from those whose religion forbids the consumption of certain meats (e.g., Moslems and Hindus) these people want more meat. They would purchase more meat if their incomes were higher. The potential of this immense latent market challenges the imagination, just as the formidable array of barriers to economic growth challenges those who are so vigorously pursuing it. Progress will be made and eventually these countries will be brought into the world market economy. Some Central American countries have made significant strides in exportation of beef in recent years. Additional efforts can be expected in the future. In the immediate future, however, foreseeable developments will not significantly affect meat consumption in the less-developed countries. This is a long-range situation.

U.S. MEAT CONSUMPTION

Per capita consumption of red meat was high in the United States in the early days, which is typical of frontier countries. Around 1900, however, total meat consumption began a slight downturn as population increased faster than meat supplies. This trend was reversed following the drouth-depression period of the 1930's and the upward trend established then still in continuing. A decline in beef consumption from the early 1900's to mid-1930's largely accounted for the decline in total meat consumption. In terms of absolute quantity, beef consumption now is far ahead, although this has not always been the case. It is not apparent from

TABLE 5.5. MUTTON, LAMB AND GOAT MEAT[1] PER CAPITA CONSUMPTION IN SPECIFIED COUNTRIES, AVG 1961–1965, ANNUAL 1973, 1974, 1975, 1976, AND 1977

Continent and Country	Avg 1961–1965 (Kilos)	1973 (Kilos)	1974 (Kilos)	1975 (Kilos)	1976 (Kilos)	1977[2] (Kilos)
North America						
Canada	2	2	1	1	1	1
United States	2	1	1	1	1	1
Mexico	1	1	1	1	1	1
South America						
Argentina	5	4	4	4	3	4
Brazil	1	1	1	1	1	[3]
Chile	4	3	3	3	2	2
Colombia	—	1	1	[3]	[3]	[3]
Paraguay	[3]	—	—	—	—	—
Peru	4	3	2	2	2	2
Uruguay	18	11	18	20	20	8
Venezuela	[3]	[3]	[3]	[3]	[3]	[3]
Europe						
Western						
EEC						
Belgium-Luxembourg	[3]	1	1	1	1	2
Denmark	[3]	[3]	[3]	[3]	[3]	1
France	[3]	3	3	4	4	4
Germany, F.R.	[3]	[3]	[3]	1	1	1
Ireland	11	10	11	11	10	10
Italy	1	1	1	1	1	1
Netherlands	[3]	[3]	[3]	[3]	[3]	[3]
United Kingdom	11	9	8	9	8	7
Austria	[3]	[3]	[3]	[3]	[3]	[3]
Finland	[3]	[3]	[3]	[3]	[3]	[3]
Greece	12	16	13	14	14	13
Norway	4	5	5	5	5	5
Portugal	3	3	3	2	2	3
Spain	4	4	5	4	4	4
Sweden	[3]	1	1	1	1	1
Switzerland	1	1	1	1	1	1
Eastern						
Bulgaria	8	10	9	7	8	8
Czechoslovakia	[3]	1	1	1	1	1
Hungary	1	[3]	[3]	[3]	[3]	[3]
Poland	1	1	1	1	1	1
Yugoslavia	2	2	2	2	3	2
USSR	4	4	4	4	4	4
Africa						
South Africa, Republic of	8	5	6	7	6	6
Asia						
Japan	[3]	3	2	2	2	3
Philippines	[3]	[3]	[3]	[3]	[3]	[3]
Oceania						
Australia	41	28	27	25	24	20
New Zealand	46	43	39	38	33	34

Source: USDA (1969) and USDA (1978).
[1]Carcass weight basis.
[2]Preliminary.
[3]Less than 500 grams.

Fig. 5.1, but pork consumption generally was higher than beef consumption until the early 1950's. Prior to that, beef consumption had temporarily exceeded pork consumption on several occasions, but since the early

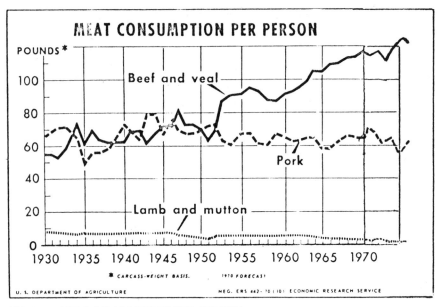

FIG. 5.1.   MEAT CONSUMPTION PER PERSON

1950's beef has dominated. Since the early 1950's beef consumption has increased about 50%. U.S. consumers ate 129 lb of beef per capita in 1976—the peak year since records have been kept. Pork consumption has varied around the 65-lb level most of the time since the early 1950's and only in the most recent years has it shown a tendency to increase.

Veal, lamb, and mutton have always comprised a relatively minor fraction of total meat consumption. Consumption of each was on a slightly increasing trend until about the end of World War II. Since that time the trend of each has been decidedly downward. Consumption of veal amounted to only 3.9 lb per person in 1977, and lamb and mutton consumption was about 1.8 lb. Thus, the gain in total meat consumption since the early 1950's has been virtually all in beef.

While by far the bulk of meat consumed is in the fresh form, Ives (1966) has pointed out that the share of total meat represented by sausage and canned meats has risen more than the overall trend.

Meat Consumption Patterns—Regional and Type of Household

By Specie.—Consumption data are reported on a regular annual basis for the United States as a whole, but regional or state information is available only from special studies which are not made at regularly scheduled intervals. During the spring of 1965, the USDA carried out a nationwide survey[1] during one week which provides some information on regional meat consumption characteristics by type of household and income level (Rizek and Rockwell 1970). A comparable study was made 10 yr earlier, but unfortunately later data are not available on a comparable basis. The 1965 study showed that meat was the most important item in the food budget—accounting for 30.2¢ of each dollar spent for food at home. This was up from 29.5¢ in 1955. Approximately 88% of the meat expenditures were for red meat and the remaining 12% for poultry. Fruits and vegetables ranked second to meat, taking 19¢ of the food dollar.

For purposes of analysis, the United States was divided into four regions—North Central, Northeast, South and West. During the week of the study, "Consumers (Rizek and Rockwell 1970) in the North Central region used the most red meat—3.6 lb per person . . . Those in the West consumed 3.4 lb of red meat; the Northeast, 3.3 lb; while those in the South averaged 3.1 lb" (Table 5.6). These differences may appear to be minor, but on this basis consumption in the North Central would be 17% greater than in the South. Farm families were the biggest consumers of red meats, however, farm consumption also varied considerably among regions. Western farm households ate 47% more red meat than did Southern farm households.

Beef consumption was highest in the Western region, with the North Central a close second. The South was lowest in beef consumption, but highest in pork consumption. The Northeast and West were relatively low in pork consumption, but exceeded the South and North Central in lamb and mutton consumption. It is worth noting that luncheon meats accounted for 13% of total consumption for the United States as a whole.

Per capita consumption of red meat increased ⅓ lb per week between the 1955 and 1965 surveys, with the largest increase occurring in the South.

Beef consumption in the survey week increased from 1.25 pounds per capita in 1955 to 1.65 pounds in 1965 (Rizek and Rockwell 1970). Pork consumption by farm families increased slightly from 1955 to 1965, but this was more than offset by declines in other households, resulting in a net decline for total pork consumption. In this connection it is noted that pork production was 7% higher in 1965 than in 1955, but population increased more than 7% so per capita production was lower in 1965 than in 1955. The per capita consumption of veal and also of lamb and mutton declined from 1955 to 1965.

For the United States as a whole, beef accounted for about ½ of all

---

[1]This section draws heavily on a USDA national survey of household consumption of meat and poultry (Rizek and Rockwell 1970).

TABLE 5.6. USE (POUNDS) OF MEAT PER PERSON, FARM AND NONFARM HOUSEHOLDS, FOR ONE WEEK IN THE SPRINGS OF 1955 AND 1965, BY REGIONS

Region and Household Group	All Meat[1]		Beef		Pork		Veal		Lamb and Mutton		Variety Meats[1]		Luncheon Meats	
	1955	1965	1955	1965	1955	1965	1955	1965	1955	1965	1955	1965	1955	1965
United States														
All	3.02	3.35	1.25	1.65	1.14	1.10	0.08	.05	0.09	0.06	0.10	0.07	0.36	0.43
Urban	3.17	3.43	1.34	1.69	1.13	1.09	0.10	0.06	0.12	0.07	0.11	0.08	0.36	0.43
Rural nonfarm	2.80	3.10	1.10	1.51	1.15	1.06	0.05	0.02	0.03	0.02	0.08	0.05	0.39	0.44
Farm	2.82	3.44	1.18	1.73	1.21	1.25	0.02	0.01	0.02	0.01	0.07	0.05	0.32	0.39
Northeast														
All	3.07	3.34	1.29	1.63	0.98	0.98	0.12	0.10	0.19	0.11	0.13	0.08	0.37	0.44
Urban	3.10	3.42	1.29	1.64	0.95	1.02	0.15	0.11	0.23	0.13	0.14	0.09	0.35	0.43
Rural nonfarm	2.92	3.09	1.23	1.60	1.01	0.88	0.06	0.05	0.09	0.05	0.10	0.04	0.43	0.48
Farm	3.30	3.35	1.54	1.87	1.15	0.95	0.05	0.04	0.07	0.00	0.09	0.06	0.39	0.43
North Central														
All	3.37	3.60	1.51	1.81	1.23	1.17	0.07	0.04	0.05	0.02	0.09	0.06	0.42	0.50
Urban	3.42	3.59	1.52	1.77	1.22	1.18	0.10	0.05	0.08	0.03	0.09	0.06	0.42	0.50
Rural nonfarm	3.17	3.46	1.43	1.76	1.17	1.11	0.05	0.02	0.01	0.01	0.08	0.06	0.43	0.51
Farm	3.45	3.91	1.61	2.16	1.34	1.24	0.02	0.01	0.01	0.01	0.06	0.06	0.40	0.44
South														
All	2.57	3.10	0.85	1.43	1.26	1.18	0.04	0.03	0.02	0.02	0.09	0.06	0.30	0.38
Urban	2.93	3.26	1.09	1.56	1.33	1.16	0.06	0.04	0.03	0.03	0.12	0.08	0.30	0.39
Rural nonfarm	2.32	2.88	0.64	1.27	1.22	1.14	0.03	0.02	0.02	0.02	0.06	0.05	0.34	0.38
Farm	2.18	2.89	0.68	1.17	1.18	1.34	0.01	0.01	0.01	0.01	0.06	0.05	0.23	0.32
West														
All	3.31	3.41	1.62	1.86	1.00	0.95	0.07	0.02	0.13	0.10	0.11	0.08	0.37	0.39
Urban	3.25	3.43	1.52	1.85	1.00	0.95	0.07	0.03	0.17	0.11	0.12	0.09	0.38	0.39
Rural nonfarm	3.58	2.99	1.89	1.66	1.05	0.90	0.12	0.05	0.04	0.02	0.09	0.02	0.38	0.38
Farm	3.15	4.26	1.73	2.60	0.89	1.12	0.03	0.02	0.10	0.02	0.08	0.07	0.31	0.42

Source: Rizek and Rockwell (1970).
[1]Excludes game.

meat, but was considerably higher in the West (54%) than in the South (46%). However, the sharpest increase in beef consumption between 1955 and 1965 was in the South (see Table 5.7). Pork comprised ⅓ of the red meat consumed nationwide. In 1955, pork accounted for 49% of the red meat consumed in the South versus only 33% for beef. By 1965, the positions of pork and beef were just about reversed (Rizek and Rockwell 1970).

Raunikar, *et al.* (1970) found that, ". . . . per capita demand (for pork) is relatively high in the Southern, Lower Mid-Atlantic, Lakes, and Central Plains regions, relatively low in the Middle Mid-Atlantic, New England, Rocky Mountain, and Pacific regions, and near the national average in the North Plains and Upper Mid-Atlantic regions." Raunikar, *et al.* (1969) in a companion study found that, ". . . per capita demand (for beef) is relatively high in the Pacific and Mountain regions, relatively low in the South and near the national average in the Upper Mid Atlantic and New England regions." For both pork and beef these studies projected a reduction in regional differences by 1980, and for both indicated that the most rapidly expanding markets are located along the southern boundaries of the United States from Florida to California.

By Cut of Meat.—When meat was shipped to retail outlets in carcass form, sales and consumption of the various cuts were in about the same ratio throughout the various regions. This generally was the case with beef in 1955.

By 1965 an increase had occurred in shipment of primal-wholesale cuts of beef. As a result, consumption of different cuts varied considerably by region and by type of household. For the United States as a whole, steaks comprised a larger percentage of total beef consumption in 1965 than in 1955—41% compared to 31%. Ground beef was a smaller percentage of total beef consumption—25% compared to 30% (Rizek and Rockwell 1970).[2]

Consumption of steak increased markedly (as a percentage of total beef consumption) in all regions and among all types of households (Table 5.8). The increase was greater in the Northeast and West than the North Central and Southern regions. While the consumption of ground beef increased in absolute terms during this interval (from 0.38 lb in 1955 to 0.41 lb in 1965) it decreased in relative terms, i.e., as a percentage of the total. Exceptions were in the farm households of the Northeast, North Central and South. Relative declines also occurred in all regions for roasts and "other" beef cuts.

Changes in consumption of various pork cuts by regions and type of household are shown in Table 5.9. Here, wide differences are noted

---

[2]Part of this change was attributed to possible changes in retailing practices and survey reporting procedures.

TABLE 5.7. BEEF, PORK, AND OTHER MEATS AS A PERCENTAGE OF TOTAL RED
MEAT CONSUMPTION, ALL HOUSEHOLDS BY REGIONS, FOR ONE WEEK IN THE
SPRINGS OF 1955 AND 1965

Regions	Beef 1955 (%)	Beef 1965 (%)	Pork 1955 (%)	Pork 1965 (%)	Other Meats[1] 1955 (%)	Other Meats[1] 1965 (%)
United States	41	49	38	33	21	18
Northeast	42	49	32	29	26	22
North Central	45	50	36	32	19	18
South	33	46	49	38	18	16
West	49	54	30	28	21	18

Source: Rizek and Rockwell (1970).
[1] Veal, lamb, variety meats, and luncheon meats.

TABLE 5.8. PERCENTAGE OF BEEF CONSUMED AS VARIOUS CUTS, BY REGION
AND TYPE OF HOUSEHOLD, FOR ONE WEEK IN THE SPRINGS OF 1955 AND 1965[1]

Region and Household Group	Steaks 1955 (%)	Steaks 1965 (%)	Roasts 1955 (%)	Roasts 1965 (%)	Ground 1955 (%)	Ground 1965 (%)	Other[2] 1955 (%)	Other[2] 1965 (%)
United States								
All	31	41	28	26	30	25	11	8
Urban	32	44	28	27	29	22	11	7
Rural nonfarm	28	36	26	25	34	31	12	8
Farm	32	36	29	25	27	31	12	8
Northeast								
All	31	48	28	27	30	17	11	8
Urban	33	51	28	27	28	14	11	8
Rural nonfarm	24	43	27	25	36	26	13	6
Farm	27	29	29	28	31	34	13	9
North Central								
All	31	36	29	27	30	30	10	7
Urban	32	37	30	28	29	28	9	7
Rural nonfarm	29	33	27	25	35	35	9	7
Farm	32	34	29	26	29	33	10	7
South								
All	31	39	25	25	30	27	14	9
Urban	31	41	26	26	30	25	13	8
Rural nonfarm	30	36	21	24	34	31	15	9
Farm	33	37	27	22	23	30	17	11
West								
All	32	44	29	26	29	21	10	9
Urban	31	46	30	25	29	20	10	9
Rural nonfarm	34	27	26	31	29	33	11	9
Farm	33	44	28	24	27	26	12	6

Source: Rizek and Rockwell (1970).
[1] Excludes quantities of those meats consumed as luncheon meats.
[2] Corned, chipped dried, stewing and canned beef.

TABLE 5.9. PERCENTAGE OF PORK CONSUMED AS VARIOUS CUTS, BY HOUSEHOLD GROUP AND REGION, FOR ONE WEEK IN THE SPRINGS OF 1955 AND 1965[1]

Region and Household Group	Loin and Chops		Ham[2]		Sausage		Bacon		Salt Pork		Other	
	1955 (%)	1965 (%)	1955 (%)	1965 (%)	1955 (%)	1965 (%)	1955 (%)	1965 (%)	1955 (%)	1965 (%)	1955 (%)	1965 (%)
United States												
All	23	21	26	25	8	9	22	22	5	2	16	21
Urban	25	23	27	25	7	9	21	21	3	1	17	21
Rural nonfarm	22	18	25	25	9	11	23	25	6	3	15	18
Farm	19	15	24	26	10	10	24	22	8	5	15	22
Northeast												
All	27	26	32	28	5	9	17	17	1	1	18	19
Urban	27	27	32	27	5	9	18	16	1	1	17	20
Rural nonfarm	27	24	36	32	6	8	17	21	1	1	13	14
Farm	22	16	30	36	9	10	18	20	2	1	19	17
North Central												
All	28	22	26	26	8	9	21	21	1	[3]	16	21
Urban	29	24	28	24	7	8	20	21	1		15	23
Rural nonfarm	30	19	24	31	8	10	23	23		1	15	16
Farm	27	17	25	29	10	8	20	21	1	2	17	23
South												
All	16	17	22	22	10	11	25	26	11	4	16	20
Urban	18	19	24	24	9	10	24	26	8	3	17	18
Rural nonfarm	15	14	21	19	11	13	24	28	14	6	15	20
Farm	11	14	21	22	11	11	27	22	17	9	13	22
West												
All	24	21	23	23	7	8	27	25	1	1	18	22
Urban	23	21	24	23	6	8	27	25	1	1	19	22
Rural nonfarm	27	21	20	20	10	10	27	28	2	1	14	20
Farm	18	16	23	22	10	14	30	27	2	[3]	17	21

Source: Rizek and Rockwell (1970).
[1] Excludes quantities consumed as luncheon meats.
[2] Fresh and cured. About 9/10 is cured.
[3] Less than 0.5%.

among regions in both 1955 and 1965. Changes between the two time periods appear rather minor. Consumption of pork chops was particularly low in the South, as was bacon in the Northeast. Consumption of ham was relatively low in the South and West. Consumption of salt pork was far higher in the South than in any other region, but in this region a marked decline occurred between 1955 and 1965.

Factors Related to Meat Consumption

Many factors are related to the quantity of meat consumed and to the amount of money spent for meat. In previous sections attention was called to differences in consumption among nations and among regions within the United States. These differences largely are reflections of more basic factors, of which the more important are price of the product, level of family income, and prices of competing products. Other factors are ethnic background, type of occupation, religious beliefs, personal tastes and preferences, diets, and food fads. In terms of total consumption, the number of people obviously is an important factor. The relationship involved between number of people and meat consumption needs little analysis. Other things being equal, a 1% increase in the number of people will result in about 1% increase in quantity of meat consumed. Our concern primarily is with factors which affect the quantity consumed per person, and emphasis is placed on product prices, incomes, and prices of competing products.

Price of the Product.—The relationship between consumption and price of the product is explained by theories of demand. It is quite obvious to the housewife with a limited budget that she can, and will, buy more meat at lower prices than at higher prices, and vice versa. Prices, of course, are established at various levels throughout the marketing system. An examination of the price-quantity relationship at the retail and farm levels follows.

Retail Demand.—The relationship between prices and quantity consumed involves the notion of price elasticity of demand which was discussed in Chap. 3. A considerable amount of research has been done on price elasticity for various products. The results are not identical, but reasonably consistent. Differences in research results on this problem can arise from differences in the time period upon which the study is based, and differences in research methodology.

Two general conclusions are: (1) demand is less elastic at the farm level than at retail, and (2) demand is less elastic for aggregates of food than for individual items. A number of excellent demand studies have been made.

Two of the most comprehensive involving meats are Brandow (1961) and George and King (1971). In an exhaustive study Brandow found retail price elasticities as follows: beef −0.95, veal −1.60, pork −0.75, and lamb and mutton −2.35. A more recent analysis of similar scope by George and King indicates retail price elasticities of −0.64 for beef, −1.72 for veal, −0.41 for pork, and −2.63 for lamb and mutton. These coefficients are shown in the diagonal position in Table 5.10 (the other data in Table 5.10 are cross-elasticities and will be discussed later). The latter study derived somewhat lower (i.e., less elastic) estimates for beef and pork, and slightly higher estimates for veal and lamb and mutton than found by Brandow. *Know this table.*

TABLE 5.10. DEMAND INTERRELATIONSHIPS AT THE RETAIL LEVEL

Commodities	Elasticity coefficients					
	Beef	Veal	Pork	Lamb and mutton	Chicken	Turkey
Beef	−0.6438	0.0280	0.0826	0.0454	0.0676	0.0075
Veal	0.3593	−1.7177	0.1977	0.0660	0.1736	0.0135
Pork	0.0763	0.0141	−0.4130	0.0602	0.0353	0.0050
Lamb and mutton	0.5895	0.0661	0.8914	−2.6255	0.2336	0.0151
Chicken	0.1971	0.0436	0.1208	0.0546	−0.7773	0.0837
Turkey	0.0976	0.0147	0.0653	0.0178	0.4000	−1.5553

Source: George and King (1971).

Interpretation of the elasticity coefficients, taking beef as an example from Table 5.10, is that a 1% increase in retail beef prices will result in a decrease in quantity purchased 0.64 of 1% (or vice versa for price decrease). This can be classed as moderately inelastic, where Brandow found it to be only slightly inelastic. In an earlier pioneering study, Working (1954) reported that the demand for beef was more price elastic in the long-run (5 or 10 yr) than in the short-run (1 yr). His calculation of short-run price elasticity was −0.90 which is reasonably close to Brandow's figure. Working determined the long-run price elasticity for beef to be −1.40 to −1.50. At this level demand definitely would be elastic. No one disagrees with the proposition that the long-run demand could be more elastic than in the short-run, but not all researchers agree that Working's analysis proved the point. From the standpoint of market control, the degree of elasticity is an important consideration. Supply reduction programs are based on the supposition that demand is price inelastic (that a reduction in quantity would result in a more than proportionate increase in price). This would mean an increase in total revenue and (unless costs increased) presumably an increase in net revenue. But if the demand is elastic, supply reduction would decrease rather than increase revenue. Whatever the degree of price elasticity, the beef cattle industry, through the spokesmanship of its national organization, has never condoned a mandatory supply control program. The opposition to such an approach has never been based on possible negative revenue effects as a result of an elastic demand. It is not known whether industry actions are the reflection of an intuitive feeling that revenue would decline, simply a reflection of rugged individualism, or some other aspect. While the question of degree of price elasticity in the long-run is not settled, there is general agreement that the short-run elasticity of demand for beef is less than unity. As mentioned, Brandow reported −0.95, Working, −0.90. Others (Breimyer 1961; Hassler 1962) have reported somewhat lower elasticities and there is some evidence that the degree of elasticity is decreasing with the passing of time.

The elasticity of demand for veal is −1.72, and for lamb and mutton −2.62 (Table 5.10). In the latter it can be assumed that the coefficient represents primarily the demand for lamb, rather than mutton. Consumers are more sensitive (i.e., respond by adjusting purchases) to price changes of veal and lamb than to price changes of beef or pork.

Pork has the lowest degree of price elasticity of any meat, according to most studies. Table 5.10 shows the coefficient of elasticity to be −0.41. Results of other studies range both higher and lower than this, but most studies put the elasticity of pork less than unity and below that for beef.

Price elasticity coefficients for chicken (−0.78) and turkey (−1.56) also are shown in Table 5.10. The demand for turkey is relatively elastic, meaning quantity purchased is relatively responsive to price change. Thus, it is seen that at retail the two major meats (beef and pork) have a relatively inelastic demand—the others (except chicken) have a relatively elastic demand.

Farm Level Demand.—Producers generally are more concerned with price-quantity relationships at the farm level. Most realize that demand for their products is derived from retail demand, but lags and market imperfections often screen the direct relationships. So far, we have discussed elasticity of demand in terms of the effect on quantity purchased of variations in price. This is the viewpoint of a merchandiser. The same situation viewed in the inverse may be put in the form of a question. What is the effect upon price of variations in quantity? This is more in line with the viewpoint of producers. There is evidence that more and more producers are becoming concerned. Various farm organizations have undertaken programs designed to gain some control over variations in quantity marketed, both in terms of total quantities and cyclical and seasonal variations. If price elasticity is known, this question can be answered by calculating its reciprocal—which gives an approximation of the "price flexibility" coefficient. It follows that price flexibility increases as elasticity decreases. In other words, the more inelastic the demand, the greater effect a given change in quantity will have upon price received. The following discussion of farm level demand for livestock will be in terms of price elasticity. The implications to price flexibility are obvious.

Farm level price elasticity coefficients from Brandow's (1961) study are as follows: cattle -0.68, calves -1.08, hogs -0.46, and sheep and lambs -1.78. George and King (1971) found farm level price elasticities of -0.42 for beef, -0.24 for pork, and -1.67 for lamb and mutton (Table 5.11). In all cases elasticity is less at the farm level than at retail. This follows from

TABLE 5.11.   DEMAND INTERRELATIONSHIPS AT THE FARM LEVEL

| Commodities | Elasticity Coefficients | | | |
	Beef	Pork	Lamb and mutton	Chicken
Beef	-0.4165 —2.4	0.0482	0.0289	0.0524
Pork	0.0493	-0.2409 —4.17	0.0383	0.0274
Lamb and mutton	0.3813	0.5199	-1.6705 —.5986	0.1810
Chicken	0.1275	0.0704	0.0344	-0.6023 —1.660

Source: George and King (1971).

the fact that marketing charges remain relatively fixed when prices vary. A price change at retail will be passed back through marketing channels to farmers in almost its entirety. Since farm prices are lower than retail (by the amount of marketing charges), a given price change has a greater percentage effect on farm price (which is a lower absolute amount than retail price) than on retail price. An example will make this clear. Assume retail price is $1.00 per lb and marketing charges are 40¢ per lb. Farm price equivalent would be 60¢ per lb (i.e., $1.00 − 0.40 = $0.60). Suppose retail prices decrease 10¢ per lb. At the retail level the change amounts to 10%. If the entire 10¢ decrease is taken from the farm price, the change amounts to almost 17% of the farm price. If it is assumed that the percentage change in quantity marketed was the same at both levels, this translates to a lower degree of elasticity at the farm than at retail. This situation appears to be verified by the evidence presented.

The relatively low price elasticity at the farm level for most agricultural products is at the root of the comparatively violent price fluctuations that accompany production variations, e.g., seasonal and cyclical variations in livestock and weather-induced variations in crop production.

A knowledge of price elasticities, coupled with good estimates of impending changes in quantities available for market, provides one basis for forecasting price changes. Take the case of hogs as an example. The coefficient of price elasticity at the farm level is -0.24 (Table 5.11) and its reciprocal is $\frac{1}{-0.24} = -4.17$. This is interpreted to mean that a 1% change in the quantity of hogs marketed will result in a change in price at the farm (in the opposite direction) of 4.17%. Thus, if one had a good estimate that hog marketing will increase 10% next year, he can predict that, other things being equal, prices will decline about 42%. It may be noted that hog price flexibility derived from Brandow's analysis would be substantially less than that. A tentative estimate of hog price flexibility used by many market analysts in past years was −2.5 to −3.0.

This type of information also can be used as a criterion for determining market strategy or market policy. An organization (or public agency) charged with developing a market program would be interested in knowing the effect upon price of reducing supplies some specified percentage. A knowledge of price elasticities is essential in such a situation. It might be noted in this connection, that an organization or agency will be interested not only in the immediate price and income effect but also in the longer period situation. This involves the possibility and probability of competitors coming in with alternate products which may be domestically-produced substitutes or the same product imported from other countries.

The price elasticity of "all meats" taken as an aggregate is less than that of most individual meats. This arises from the reaction of most consumers in general that "meat," in some form, virtually is a necessity. Price elasticity is lower on necessities than on luxuries. It is recognized, of course, that such foods as eggs, fish, cheese, etc., can replace meat to a degree, but to a smaller degree than one meat item can replace another. In addition, the quantity of total meat which consumers will purchase is somewhat limited by the capacity of the human stomach. Once a person's stomach is reasonably full it takes substantial price reductions to induce that person to purchase additional quantities. He will, however, substitute one meat for another as relative prices change.

Price of Competing Products.—The quantity of a product consumed during a given time is affected by prices of competing products. Competing products are defined as goods that can be used as substitutes for the product in question. Most people prefer some change in the meat dish from day to day simply for the sake of variety and enjoyment. Here, however, we are concerned with the degree of response that purchasers will make in substituting one meat for another as a result of price changes. If the price of pork rises, consumers will purchase larger quantities of beef, lamb, chicken, etc.—assuming prices of the competing products remain unchanged. How much effect will a 1% increase in pork prices have upon the consumption of beef? This information is obtained from a determination of cross elasticities.

A summary of coefficients of cross-elasticities is shown in Table 5.10. It will be recalled that the negative numbers in the diagonal of Table 5.10 are price elasticities. The remaining numbers are cross elasticities and all are positive. Under the column headed "beef" one finds 0.36 on the row titled "veal." This is interpreted as follows: a 1% increase in beef prices will result in an increase of 0.36 of 1% in the quantity of veal purchased by consumers—or, as it often is stated, a 10% increase in beef prices will result in an increase of 3.6% in the quantity of veal purchased.

The reaction of consumers in purchasing pork in response to a change in beef prices is relatively low. The coefficient is 0.08. It is noted that the reaction in purchasing beef in response to a change in pork prices is the same (0.08). A change in pork prices has a greater effect on purchases of veal, lamb, and chicken than it does on purchases of beef. Among the red meats, it appears that lamb is substituted more readily than other meats. This may be questioned. However, when one considers the difference in absolute level of lamb consumption compared to beef, pork, or chicken, the lamb cross elasticity coefficients appear more plausible. The consumption of lamb is so low that, for example, a 0.59% increase amounts to a considerably smaller absolute quantity than a 0.08% increase in pork consumption.

In general, the cross elasticities between poultry and red meats are low. This appears to belie a rather common assumption that a lowering of poultry prices will result in a drastic switch from red meat. Over the years, a long-term declining trend in poultry prices has been associated with a long-term increasing trend in poultry consumption. It has been presumed this was at the expense of red meat. The low degree of cross elasticity shown in Table 5.10 might cast some doubt on this presumption. However, over a long period of years even a relatively low cross elasticity could have a considerable cumulative effect. It might also be possible that the degree of cross elasticity declined as consumption of poultry reached higher and higher levels. If this were the case, competition from poultry could taper off in the future.

Cross elasticities also have implications when viewed from the standpoint of the effect that a change in quantity of one meat will have on the price of another—or the effect that a change in quantity marketed of one species of live animals will have upon the price of another. The latter ordinarily is more generally the concern of producers. During the fall of 1977 it was predicted that hog production would expand in 1978 by 10%. What effect would this have upon beef cattle prices? Two things are obvious: (1) increased hog production will result in lower hog prices and lower beef cattle prices (assuming no change in other factors), and (2) the lower the cross elasticity (i.e., the lower the degree of substitutibility) the less will be the effect on beef cattle prices. As may be observed in Table 5.11, the cross elasticities indicate that a 1% increase in hogs (pork) marketed will result in less than 0.1 of 1% decline in cattle (beef) prices. It should be noted here, however, that the cattle price used in this study is an aggregate of all cattle. The effect on slaughter steers and heifers likely would be somewhat greater than the effect on a composite average price for all cattle.

Income.—The relationship of income changes to meat purchases can be expressed as (1) changes in quantity purchased, (2) changes in expenditures (or value), (3) changes in prices paid. A number of studies have developed this type of information. We will draw primarily from just three: (1) a USDA analysis which relied on a national household survey (Rizek and Rockwell 1970), (2) Brandow's (1961) analysis mentioned earlier and (3) George and King (1971). The USDA analysis was based on a limited time span and there is reason to believe that changes have occurred in some of the findings. However, since more recent studies of a similar nature are not available, results are presented here as benchmarks and for comparison with selected aspects of later studies.

The USDA study shows graphic income-consumption relationships for all meat, for beef, and for pork by regions (Fig. 5.2) and veal, lamb and mutton, luncheon meats, and variety meats for the United States as a whole (Fig. 5.3). Data for both of these figures are plotted on double-logarithmic scale. Therefore, the slope of the curves allows a direct indication of relative (or percentage) change regardless of the level of consumption.

It may be observed in Fig. 5.2 that the consumption of all meat increases as income increases, although this tendency is less pronounced in the North Central region than in the others. It also is apparent that the negative relationship between income and pork consumption in the North Central region is the factor that holds down "all meat" consumption in upper income brackets. Beef consumption has a pronounced positive relationship with income in that region. Beef consumption and income are positively related in all four regions. There appears to be a tendency for pork consumption to increase with income at lower income levels, but then decline at upper income levels in both the Northeast and South. Pork consumption overall in the West appears to remain fairly constant with increasing income.

Figure 5.3 shows a positive relationship between income and veal consumption, and also with lamb and mutton consumption, but a negative relationship exists between income and variety meat consumption. Luncheon meat consumption increases with increasing incomes at lower income levels, then decreases with increasing incomes at upper income levels.

Table 5.12 shows that a positive relationship exists between income

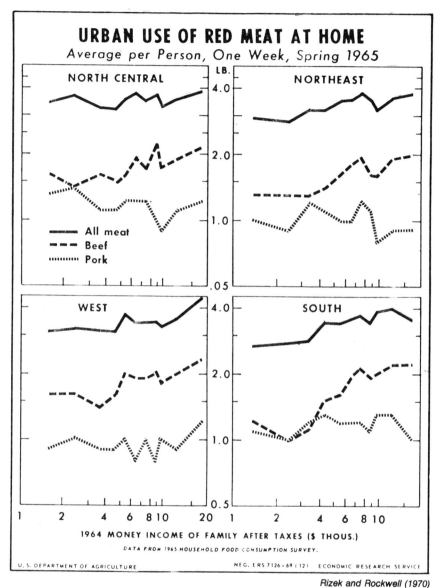

URBAN USE OF RED MEAT AT HOME
Average per Person, One Week, Spring 1965

NORTH CENTRAL

NORTHEAST

——— All meat
- - - Beef
·········· Pork

WEST

SOUTH

1964 MONEY INCOME OF FAMILY AFTER TAXES ($ THOUS.)

DATA FROM 1965 HOUSEHOLD FOOD CONSUMPTION SURVEY.

U. S. DEPARTMENT OF AGRICULTURE   NEG. ERS 7126-69 (12)   ECONOMIC RESEARCH SERVICE

Rizek and Rockwell (1970)

FIG. 5.2.   RELATION OF ALL MEAT, BEEF, AND PORK CONSUMPTION TO INCOME

changes and (1) quantity of beef consumed, (2) value per person (expenditures for beef), and (3) value per pound (price paid for beef). This is confirmed by Brandow (1961), George and King (1971) and a number of other works. Other studies (Breimyer 1961) have shown a positive corre-

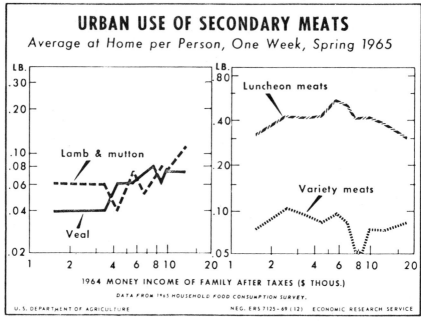

Rizek and Rockwell (1970)

FIG. 5.3.   RELATION OF THE CONSUMPTION OF LAMB AND MUTTON, VEAL, AND LUNCHEON MEATS TO INCOME

lation between time and beef consumption. In a statistical sense, time is a rough indicator of change in personal tastes and preferences. These factors—income, tastes and preferences—go a long way in explaining the increased demand for beef in recent decades.

Consumer purchases of veal and lamb are considerably more sensitive to income changes than in the case of pork or beef. This is indicated by the higher coefficients shown in Table 5.13 and the higher income elasticities in Table 5.14.

The persistent decline in lamb consumption appears to be a direct reflection of reduction in the available supply. Lamb ranks relatively high in both price and income elasticity. Consumers who eat any lamb, generally have a distinct preference for it. There is a presumption that consumers who previously ate lamb when it was available, lose their preference when it is unavailable and will not again purchase lamb. Evidence is lacking on this point, but it is a reasonable hypothesis. It is know that lamb consumption is highest on the East and West Coasts, and extremely light in the heartland of the country.

Veal consumption has declined as the demand for feeder cattle has

TABLE 5.12.   PERCENTAGE DIFFERENCE IN QUANTITY, PRICE, AND VALUE OF MEAT CONSUMED PER PERSON FOR EACH 10% DIFFERENCE IN FAMILY INCOME, FOR ONE WEEK IN SPRING OF 1965[1]

Region and Household Group	All Meat (%)	Beef (%)	Pork (%)	Veal (%)	Lamb and Mutton (%)
Quantity per person					
All households	1.3	2.3	−0.1	4.5	5.6
Urban-all regions	1.0	2.0	−0.3	3.7	3.6
Northeast	1.1	2.0	−0.7	5.1	3.2
North Central	0.1	1.0	−0.8	6.6	−0.4
West	0.6	1.7	−0.2	−5.1	1.1
South	1.5	2.9	0.3	1.2	0.7
Rural nonfarm	1.7	2.8	0.4	3.0	1.2
Farm	1.5	2.7	−0.2	5.3	6.2
Value per person					
All households	2.5	3.4	1.1	5.3	6.0
Urban-all regions	2.1	3.1	0.9	4.4	3.8
Northeast	2.1	2.8	0.6	5.7	3.7
North Central	1.3	2.2	0.2	7.0	0.8
West	1.6	2.9	0.7	−3.8	1.0
South	2.6	3.8	1.3	1.3	0.7
Rural nonfarm	2.8	3.6	1.6	4.3	1.2
Farm	2.0	3.1	0.3	6.0	5.2
Value per pound[2]					
All households	1.2	1.1	1.2	0.9	0.3
Urban-all regions	1.2	1.1	1.2	0.7	0.1
Northeast	1.0	0.8	1.2	0.6	0.4
North Central	1.2	1.2	1.0	0.5	1.3
West	1.0	1.2	0.9	1.4	−0.1
South	1.1	0.9	1.0	0.1	[3]
Rural nonfarm	1.1	0.8	1.2	0.2	0.1
Farm	0.5	0.4	0.5	0.6	−1.1

Source: Rizek and Rockwell (1970).
[1] Not all results will meet tests of significance—primarily veal and lamb and mutton.
[2] Differences in price per pound reflect differences in price paid for each cut or product, quality, plus differences in kinds of meat eaten.
[3] Less than 0.05%.

TABLE 5.13.   INCOME ELASTICITIES OF DEMAND AT RETAIL; PERCENTAGE CHANGES IN QUANTITIES DEMANDED RESULTING FROM 1% CHANGES IN INCOMES

Quantities Demanded of:	Income Elasticities	
	Brandow (1961)	George and King (1971)
Beef	0.47	0.29
Veal	0.58	0.59
Pork	0.32	0.13
Chicken	0.37	0.18
Turkey	0.49	0.77

Source: Brandow (1961) and George and King (1971).

TABLE 5.14. PERCENTAGE DIFFERENCE IN QUANTITY, VALUE, AND PRICE OF POULTRY CONSUMED PER PERSON FOR EACH 10% DIFFERENCE IN FAMILY INCOME, BY HOUSEHOLDS, FOR ONE WEEK IN SPRING OF 1965[1]

Item	All Poultry (%)	Chicken (%)	Turkey (%)
Quantity per person			
All households	0.1	-0.2	6.5
Northeast	0.1	-0.4	9.6
North Central	0.4	0.4	-0.9
South	0.5	0.2	4.4
West	-0.8	-1.0	-1.1
Urban	-0.2	-0.5	6.0
Rural nonfarm	[2]	-0.3	6.6
Farm	-0.3	-0.5	2.4
Value per person			
All households	1.0	0.6	8.0
Northeast	0.9	0.4	7.3
North Central	1.1	0.9	0.6
South	1.2	0.7	6.6
West	0.2	-0.2	1.4
Urban	0.6	0.2	7.0
Rural nonfarm	0.9	0.5	7.0
Farm	0.2	[2]	3.4
Value per pound[3]			
All households	1.1	0.8	1.5
Northeast	0.8	0.8	-2.3
North Central	0.6	0.6	1.5
South	0.7	0.5	2.2
West	0.9	0.8	2.5
Urban	0.9	0.7	1.0
Rural nonfarm	0.9	0.8	0.4
Farm	0.5	0.4	1.0

Source: Rizek and Rockwell (1970).
[1] Not all results will meet tests of significance.
[2] Less than 0.05%.
[3] Differences in price per pound reflects differences in price paid for product, plus differences in kind and quality purchased.

strengthened relative to the veal slaughter market. The demand for beef is strong enough to hold this position. Currently most vealers are grown out for feedlot finishing. Consumption will continue to decline, but most of the possible shift already has been accomplished.

Data in Table 5.12 show that pork consumption (i.e., quantity per person) does not bear the same relationship to income as does beef. Incomes and quantity of pork consumed are negatively related in all except in urban Southern households and rural nonfarm households for the United States as a whole. It should be pointed out that this analysis applies at average incomes and average pork consumption. The coefficient was not calculated at lower levels of income. There were strong indications at that time that the income elasticity for pork was positive at lower income levels. In any event, the negative relationship at average incomes with respect to quantity purchased had serious implications for the pork industry. This would mean that as average incomes increase, consumers would cut back on the quantity of pork they purchase. Increasing average income is taken to be a symbol of economic progress. Average incomes have been increasing over the years and national economic policy is designed to continue the trend.

The USDA (1970) study and many others, including Brandow (1961) and George and King (1971) derived a positive relationship between expenditures for pork and average income—although the relationship is considerably weaker for pork than for other red meats. The income elasticity for pork apparently has changed over time. In a study covering the years 1950–1973, Ray and Young (1977) found, "The elasticity value was negative in the earlier periods of fit, then approached zero and finally became positive and statistically significant around .4 and .5 for the most recent periods of fit."

Income-chicken consumption relationships fall in about the same patterns as that of pork, i.e., negative with relation to quantity, but positive with relation to value of purchases and price paid (see Table 5.14). In the case of turkey, the relationships are positive and at a higher level than for chicken (Table 5.14).

Other Factors.—In addition to income, meat prices, and prices of competing products, meat consumption is influenced by type of occupation, ethnic background, religious beliefs, diets and food fads, and perhaps other factors.

Mechanization and automation not only have influenced type of occupation but also have reduced the amount of heavy, physical labor involved in given occupations. These factors were, and still are, instrumental in shifting large numbers of people from farm residencies and farm occupations to cities and, in many cases, less strenuous occupations. Farm households typically consume more pork than do urban or rural nonfarm households. The only exception to this in the 1965 household consumption survey was in the Northeastern region (see Table 5.6). This factor undoubtedly contributed to the decline in pork consumption experienced in recent decades.

Farm households also tend to eat more beef than do urban and rural nonfarm households. This was the case in all regions except the South (Table 5.6).

Ethnic background shows up particularly in food habits of immigrants. For example, people from the Mediterranean and the Near East areas have a much higher preference for lamb and mutton than many other peoples. These preferences are especially strong during the lifetime of the immigrants. They carry over to a considerable extent to their children, but gradually dissipate with succeeding generations. This phenomenon is considered to be a factor, though not a controlling one, in the decline in lamb consumption in the United States. People from certain areas in Europe, and especially eastern Europe, have a decided preference for rather highly seasoned sausages. This shows up not only as an ethnic characteristic, but also as a geographic characteristic due to the tendency of immigrants to settle near relatives and friends, thus developing geographic concentrations of certain nationalities in the United States. An area south of the Great Lakes at one time was known as the "dry sausage belt" largely for this reason.

Religious beliefs can influence meat consumption. Jewish and Moslem restrictions on pork consumption obviously reduce the average consumption of pork and enhance the consumption of beef, veal and mutton, and undoubtedly other foods. It is estimated that Orthodox Jews consume 40–50% more beef per capita than the national average (Anon. 1953).[3] Jewish kosher[4] as applied to meat necessitates the purchase of more pounds per person to get the same net pounds of beef as nonkosher meats. Only the forequarters are used for the kosher trade, and forequarters contain more waste in the form of bone and fat than do hindquarters. Prior to 1966, Roman Catholics were restricted in the consumption of meat on Fridays and during Lent. The lifting of this ban probably resulted in a small increase in meat consumption, but tradition still induces a considerable restraint on Friday meat consumption.

Widespread publicity and pronouncements from the medical profession on possible linkage of obesity and overweight to diseases of the heart and blood vessels have resulted in some shifting away from fatter to more lean cuts of meat and probably to some extent away from animal fats to vegetable fats. The question of animal fats versus vegetable fats is not resolved. Nevertheless, it has an impact on meat consumption. Pork consumption probably sustained the greatest adverse effect due to the public image of pork as a relatively fat meat. But, that image has improved immeasurably in recent years.

BIBLIOGRAPHY

ANON. 1953. Armour's Analysis. Armour's Livestock Bur., Chicago. Apr.-May.
BRANDOW, G. E. 1961. Interrelations among demands for farm products and implications for control of market supply. Penn. Agr. Expt. Sta. Bull. *680*.

---

[3]An analysis of meat consumption in the Northeast, however, indicated no significant difference in beef consumption among members of Jewish, Catholic, and Protestant faiths.
[4]Kosher is a Hebrew word meaning clean or fit to eat according to Jewish dietary laws.

BREIMYER, H. F. 1961. Demand and prices for meat. USDA Econ. Res. Serv. Tech. Bull. *1253*.

DAHL, D. C. and HAMMOND, J. W. 1977. Market and Price Analysis, the Agricultural Industries. McGraw-Hill Book Company, New York.

FOOTE, R. J. 1956. Price elasticities of demand for nondurable goods, with emphasis on food. USDA Agr. Marketing Serv. *96*.

GEORGE, P. S. and KING, G. A. 1971. Consumer demand for food commodities in the United States with projections for 1980. California Agr. Expt. Sta. Giannini Foundation Monograph Number 26.

GOODWIN, J. W. *et al.* 1968. The irreversible demand function for beef. Oklahoma Agr. Expt. Sta. Tech. Bull. *T-127*.

HASSLER, J. B. 1962. The U.S. feed concentrate-livestock economy's demand structure, 1949-59 (with projections for 1960–70). Nebraska Res. Bull. *203*. Also North Central Regional Publ. *138*.

HEIMSTRA, S. J. 1970. Food consumption, prices, expenditures supplement for 1968. USDA Econ. Res. Serv., Agr. Econ. Rept. *138*.

IVES, J. R. 1966. The Livestock and Meat Economy of the United States. American Meat Institute, Chicago.

LANGEMEIR, L. and THOMPSON, R. G. 1967. Demand, supply and price relationships for the beef sector, post World War II period. Amer. J. Agr. Econ. 49, 169–185.

MANN, J. S. and ST. GEORGE, G. E. 1978. Estimates of elasticities of food demand in the United States, USDA Econ. Stat. and Co-op Serv. Tech. Bull. No. 1580.

MCDONALD, R. F. 1966. Influence of selected socio-economic factors on red meat consumption patterns in the Northeast Region. Maryland Agr. Expt. Sta. Bull. *477*.

RAUNIKAR, R. *et al.* 1969. Spatial and temporal aspects of demand for food in the United States. II. Beef. Georgia Agr. Expt. Sta. Res. Bull. 63.

RAUNIKAR, R. *et al.* 1970. Spatial and temporal aspects of demand for food in the United States. III. Pork. Georgia Agr. Expt. Sta. Res. Bull. 85.

RIZEK, R. L., and ROCKWELL, G. R. 1970. Household consumption patterns for meat and poultry, Spring 1965. USDA Econ. Res. Serv., Agr. Econ. Rept. *173*.

ROY, S. K. and YOUNG, R. D. 1977. Demand for pork: a long-run analysis. Texas Tech Univ., College of Agricultural Sciences Publication No. T-1-153.

TOMEK, W. G. and COCHRANE, W. W. 1962. Long-run demand: a concept and elasticity estimates for meats. J. Farm Econ. 44, 717–730.

TOMEK, W. G. and ROBINSON, K. L. 1972. Agricultural Product Prices. Cornell Univ. Press, Ithaca, N.Y.

USDA. 1969. World meat consumption. USDA Foreign Agr. Serv. Circ. *FLM 12-69*.

USDA. 1977. Per capita red meat consumption. USDA Foreign Agr. Serv. Circ. FLM 2-77.

WALTERS, F. E. *et al.* 1975. Price and demand relationships for retail beef: 1947–1974. Colorado Agr. Expt. Sta. Tech. Bull. 125.

WORKING, E. J. 1954. Demand for Meat. Inst. Meat Packing, Univ. Chicago.

6

Types of Livestock Markets and Marketing

Marketing encompasses a wide range of functions. Among those most directly concerned with livestock are transporting, grading, financing, market news reporting, risk bearing, buying, and selling. This chapter is devoted primarily to institutional arrangements and to agencies concerned with buying and selling. Certain aspects of other marketing functions are treated in subsequent chapters. However, in terms of producer problems, doubts, frustrations, and dissatisfaction, it is the pricing of the product—the outcome of the buying-selling function—which stands out above all others.

It was pointed out in Chapter 2 that growth and development of the marketing system followed an historical pattern of adaptation to evolving geographical settlement, and to technological developments in transportation and meat preservation. Producer dissatisfaction with alleged lack of competition among buyers and lack of market information also were factors which helped shape market development. Many changes have occurred in the system over the years. A number of modifications and innovations have been introduced, but the changes have been of an evolutionary nature. The basic organizational arrangement of earlier days continues to provide the predominant market framework.

Producer concern with deficiencies in the system varies with economic conditions, which in turn are associated with production cycles. During recent years much attention has been given to alternative marketing systems and to possible modifications of the present system (Black and Uvacek 1972, Farris and Dietrick 1975, Forker *et al.* 1976, Wohlgenant and Greer 1974). Extreme financial stress sustained by the industry during the mid-1970's gave considerable impetus to these studies. A persistent upward spiral of production costs coupled with extreme instability of livestock prices leads to a high risk factor, and that provides a powerful incentive for change. In this chapter we examine present markets and marketing arrangements, and discuss selected alternatives which have been suggested to replace or to modify the present system.

TYPES OF LIVESTOCK MARKETS

Producers in most sections of the United States have access to several types of markets. From the standpoint of competition among markets this is a desirable situation. A producer who is dissatisfied with a particular market, if he has alternatives, can patronize another. That tends to generate competition among markets, but if the locality is dominated by one or a few buyers, the degree of competition may still be less than satisfactory (Armstrong 1976). At the same time, some researchers have pointed out the operational inefficiencies of duplicative, small-scale markets in certain parts of the country (Broadbent 1970). These are aspects of market organization that have implications for both pricing efficiency and operational efficiency.

Newberg (1959) distinguished 13 types of livestock markets as follows:

(1) "Terminal Public Markets. These markets are referred to as public stockyards, central public markets, or terminal markets. Livestock is consigned to commission firms for selling at these markets. Two or more commission firms must operate on such a market. A stockyard company owns and maintains the physical facilities, such as yards, alleys, scales, loading and unloading docks, office buildings, and facilities for feeding and watering livestock. Individuals, partnerships, corporations, and cooperative associations operate as commission agencies on terminal public markets.

(2) "Auctions. Auctions also may be called sale barns, community auctions. Livestock auctions receive livestock and sell to buyers on an auction basis. Bidding and selling are open to the public. They may be owned privately by individuals, partnerships, corporations, or cooperative associations.

(3) "Local Markets, Concentration Yards. These may be referred to as local stockyards, union stockyards, etc. At such markets livestock is purchased from farmers on a lot or graded basis, and usually is resorted and sold to slaughterers, order buyers, or to other markets. All have fixed facilities, such as chutes, pens, etc., for handling livestock. Livestock is purchased directly from the farmer at these fixed facilities. Individuals, partnerships, corporations, or cooperative associations may own and operate these markets.

(4) "Country Dealers. These are independent operators who buy and sell livestock. They may resell the livestock to any of the outlets used by farmers. Country dealers may also be referred to as local dealers, truck buyers, traveling buyers, traders, or in some areas as scalpers or pinhookers. Most of their dealing is with farmers. Trading usually is done at the farmer's home. Local markets differ from dealers primarily in the place of purchase. Dealers purchase primarily at the farm, while local markets buy mostly at their own yards.

(5) "Packer Buyers. Packer buyers are employed by slaughterers. They travel in the country and buy livestock from the farmer, usually in his own feedlot. The farmer's check for the stock is drawn on a packing

company. If the buyer issues his own pay check, he is assumed to be acting as a country dealer.

(6) "Packing Plants and Packer Buying Station. Livestock may be sold by a farmer to the slaughtering plant or to yards owned and operated some distance away from the slaughtering plant. The farmer gets the check from the packing company. These outlets are called packing plants or packer buying stations. In some states, packer buying stations are called concentration yards.

(7) "Order Buyers. Order Buyers act as agents of livestock buyers in procurement of livestock. Most commonly they buy through terminal markets or auctions or from dealers and local markets. However, they also occasionally act as agents of the buyer in purchase of livestock directly from farmers. In procuring livestock, order buyers sometimes are authorized to execute a draft on the funds of the purchaser.

(8) "Other Farmers. One farmer may sell breeding or feeding stock to another farmer or to members of calf clubs. This includes auction sales which a farmer may hold when liquidating all his livestock and other farm assets when he is going out of business or for similar reasons.

(9) "Locker Plants and Retailers. Occasionally, farmers sell a few head of livestock to a local locker plant, or to a store which retails the meat itself.

(10) "Pools. In some areas lambs from many farmers are pooled by grades and sold by grades to slaughterers. The farmers then share the receipts from the sales on the basis of each grade they supplied. This type of selling is referred to as a lamb pool.

(11) "Special Type Auctions. Special auctions are held primarily for feeder calves and cattle. Generally, these sales are held at infrequent intervals.

(12) "Cooperative Shipping Associations. These organizations, which are owned and operated by farmers, assemble livestock from farmers, load the livestock and ship cooperatively by rail or truck to a market, usually where the selling function is performed by commission-men. The primary function of cooperative shipping associations is assembling and forwarding livestock.

(13) "Cooperative Selling Associations. These are cooperatives which operate much like the cooperative shipping associations, but generally they perform more services in obtaining bids on livestock, selecting outlets for livestock, and providing information for the farmers. The precise functions they perform vary from one area to another. Where a cooperative actually takes title to the livestock, it is defined as a 'local market,' not a cooperative shipping association. Where the cooperative may only act as a cooperative commission firm, as at a terminal, it is not a cooperative selling association as defined here.

In many cases cooperatives operated auction markets on a certain day or days of the week, purchased livestock directly from farmers outside the ring all week, and also may have handled farmers' livestock essentially as a commission agent selling the livestock to other agencies. Many independently-owned auction markets also regularly purchased livestock directly on non-auction days or outside the ring on auction days."

All of these types are still on the scene, and the descriptions are adequate, although many changes have taken place in use of the various markets. Commercial feedlots, country commission-men, and producer bargaining associations can be added to the list.

There is no completely satisfactory nomenclature of market types that serves all purposes. Some would object to ambiguity and duplication in the above list. For example, order buyers operate at terminal public markets and auctions as well as carrying out free lance country operations. Many terminal public markets have incorporated auctions in their operations. Packer buyers operate at terminal public markets and auctions, and also buy directly from feeders in the country.

TYPES OF MARKETING

In addition to classifying markets, there is merit in a two-way classification of marketing. The differentiation rests on whether the principals to the transaction (sellers and buyers) utilize the services of a professional intermediary, i.e., a market agency, or middleman. From the producer's standpoint, public terminal markets, auctions, country commission-men, bargaining associations on the selling side, and order buyers on the buying side, all are clearly in the class of markets providing or utilizing intermediaries. Other types of markets ordinarily do not use professional intermediaries. In essence, this differentiation of the market process involves the issue of direct versus indirect marketing (although the term indirect is not commonly used in marketing literature). This has been a controversial issue for decades. It involves the question of whether a livestock producer engages the services of a professional to make the sale or purchase transaction, or does the job himself. The issue revolves around the question of whether a professional marketer can obtain a net return for the producer over and above what the producer himself could get, that will at least pay for the services of the marketer—and, hopefully, something in excess of that. In deciding whether to engage such services, the producer must weigh additional costs against additional revenue. The additional costs may include commission, shrinkage, trucking, yardage, feed, insurance, etc.

Trends in the use of various channels over the years reflect a combination of forces, such as changes in technology (communication, transportation, refrigeration) that spurred decentralization of the packing industry, and producer dissatisfaction with existing market channels.

MARKET CHANNELS

Market channels for live animals are defined as the routes, or paths, in the marketing system through which livestock pass as they move from farm or ranch to slaughter. The typical process of finished livestock production consists of three production stages or phases which are followed by slaughter. The phases are: (1) the basic cow-herd, sow-herd and ewe flock operations from which arise increases in livestock numbers and the initial growth that takes place prior to weaning, (2) a growing phase—a period during which the chief objective is growth involving liberal use of roughages and grass, and sometimes limited quantities of grain, (3) a finishing phase—a period of relatively heavy grain feeding during which the chief objective is finishing or fattening, and (4) the slaughtering of the animals. It is recognized that the first three stages are not entirely separate and distinct in all production programs. For example, young animals may be creep fed prior to weaning. This is the normal process of producing spring lambs and baby beef. Hog production often is a more or less continuous full-feeding process from weaning to slaughter. Cattle sometimes are grain fed while on grass. Lambs and cattle may go directly to slaughter off wheat pasture. Lambs often go to slaughter upon weaning at the end of the grazing season. Formerly it was normal to slaughter large numbers of grass-fat cattle directly off native pasture. That practice virtually disappeared, but made a temporary comeback during the mid-1970's when relatively high grain prices made grain fattening unprofitable. Subsequent declines in grain prices resulted in a return to the former trend of declining slaughter of grass-fat cattle.

Cull livestock normally are not put through a finishing stage. However, as Fowler (1961) reports, feeders in the Beet Belt (western Nebraska, through northern Colorado, Wyoming, and Montana) follow a practice of buying broken mouth ewes and fattening them. Also, cull cows sometimes are fattened on wheat pasture or in feedlots.

While the above list of exceptions and modifications may appear to be large, the bulk of meat produced from each of the three species is finished. A relatively large proportion of cattle and lambs change ownership during the production process, while a large proportion of hogs are grown out and finished without a change of ownership.

Change in ownership as livestock moves through successive production phases necessitates the use of markets. Figure 6.1 is a simplified diagram

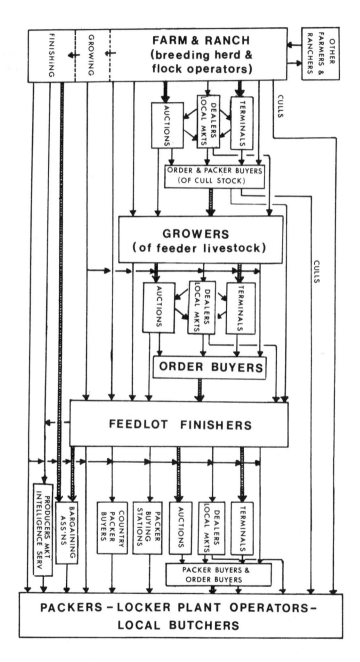

FIG. 6.1.   SCHEMATIC DIAGRAM OF LIVESTOCK MARKET CHANNELS
Wide arrows are channels in which professional marketing assistance is used by livestock producers.

of marketing channels. The least complicated channel arises in the case where the original herd or flock owner grows out and finishes his own output, and then sells the finished animals directly to a packer. In many cases the channels are more complicated than any combination shown. This may arise where livestock pass through ownership of several dealers, for example, moving from one phase to another. There also may be a change of ownership while livestock are within a given phase of production. Nevertheless, the diagram gives a general picture of types of markets and a number of connecting linkages that together portray the bulk of channels used. It also shows the links where professional intermediaries ordinarily are used. In this discussion, use of the latter will be referred to as indirect marketing.

INDIRECT MARKETING

Public Terminal Markets

For brevity, public terminal markets will be referred to simply as terminal markets. Key elements in the definition indicate this is a public market. Anyone has the privilege of buying or selling on a terminal market. However, as will be pointed out later, patrons usually find it advantageous to engage commission firms or order buyers to make the transactions. Physical facilities are owned by a stockyard company.

A brief sketch of the evolution of terminal markets was given in Chapter 2. They were a logical outgrowth of the pattern of railroad transportation and the concentration of packing plants at rail terminal points.

In 1978 there were 30 terminal livestock markets in operation. That number has declined persistently from a high of 80 in the 1930's. Terminals are located primarily in the North Central states, although some are found in other areas, particularly Southern Plains and Mountain States. Decentralization of the packing industry and technological developments in communication and transportation have had an adverse effect on terminals. The impacts still are being felt and undoubtedly will be felt for some time to come. The most notable recent development was the closing of the Chicago Union Stockyards. After more than a century of operation this landmark stopped marketing hogs in May 1970, and in August 1971 it was closed for all livestock. For decades Chicago had been the prime example of terminal marketing. Its organization and operation were cited in textbooks, research publications, and popular articles as the pattern for terminals. The Chicago market is now closed, but this does not necessarily mark the end of terminal markets. Receipts had been declining at Chicago for a number of years. The yards were located on relatively high-priced land—land that had other, and more lucrative, alternative uses. Some other terminals are in this same situation and may in time succumb to declining receipts.

Powers and Bendt (1968) make the following observation: "Technological changes have made obsolete the idea that space is an important element in the definition of a market. Today, widely scattered buyers and sellers can be in instant communication with each other via telephone, radio, and television. Transportation facilities are such that supplies can be quickly and easily distributed to areas of greatest demand. In short, there is less need for large centralized markets to serve as collection points for livestock and there is less need for buyers and sellers to be in close physical proximity to have keen competition in a market system . . .

". . . the terminals must now adjust to the new realities, the new demands of a changed marketing system operated by new participants with new technologies and new demands."

The trend in receipts of livestock at terminal markets is shown in Fig. 6.2. Recent data are shown in Table 6.1. The decline in terminal hog marketings has been greater than for other species in both absolute numbers and in percentage of total marketings. Numbers of calves and sheep

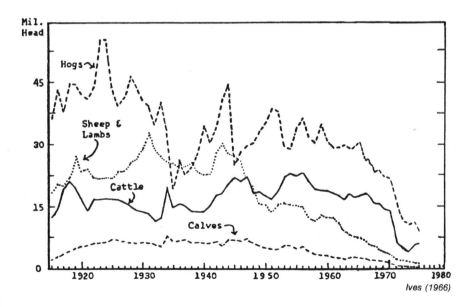

Ives (1966)

FIG. 6.2. TOTAL LIVESTOCK RECEIPTS AT TERMINAL MARKETS, 1915–1977
1965–1977 added from USDA data.

TABLE 6.1.   TERMINAL MARKET PURCHASES BY PACKERS, 1960–1976

Year	Cattle (1000 Hd)	(%)	Calves (1000 Hd)	(%)	Hogs (1,000 Hd)	(%)	Sheep (1,000 Hd)	(%)
1960	9,987	45.8	1,538	25.4	23,356	30.3	5,020	35.4
1965	10,162	34.0	1,127	16.5	17,375	23.4	3,321	25.5
1970	5,919	18.4	449	11.4	13,863	17.1	1,453	15.1
1971	5,126	15.9	291	8.6	15,329	16.9	1,339	13.6
1972	4,216	13.2	214	7.7	13,533	16.3	1,343	13.7
1973	3,667	11.9	181	8.2	13,186	17.3	1,103	12.3
1974	4,509	13.9	181	6.5	13,707	17.6	1,007	11.5
1975	5,155	14.4	369	8.3	11,090	16.3	773	10.0
1976	4,791	12.9	377	7.7	11,710	17.1	642	9.8

Source: USDA (1977) and earlier issues.

have dropped, but total numbers slaughtered also have dropped. The number of head of cattle marketed through terminals held fairly constant until recent years (Table 6.1). However, the total number slaughtered has been on an inclining trend so the percentage of total slaughter cattle moving through terminal markets has steadily declined.

There is no question of the declining relative importance of terminals as a class. However, there is a considerable difference among the remaining markets, as some are holding their own very well and will be in the picture for the foreseeable future.

Organizational Arrangements.—The organizational setup at all terminals is similar but not identical. As indicated in the definition above, physical facilities are owned by a stockyards company—usually a corporation. The stockyards company does not engage in buying and selling of livestock. Its remuneration is derived from charges for use of the yards, sale of feed, office rental, and charges for various services such as branding, castrating, dehorning, vaccinating, and dipping livestock.

The yards physically are divided into divisions by species, cattle, hogs, and sheep. The stockyards company assigns certain pens to specific commission firms for their exclusive use. Scales are owned by the stockyards company and operated by its employees. The scales, however, are inspected and tested from time to time and are under the supervision of government personnel of Packers and Stockyards, an agency of the Agricultural Marketing Service of the USDA. This agency formerly was Packers and Stockyards Administration. It is commonly known simply as P&S.

While buying and selling is open to the public, the selling of patrons' livestock is done almost exclusively by personnel of commission firms. Buying is done by a wide range of types of buyers: packer buyers, order buyers, yard dealers, wholesale provisioners, local butchers, farmers, ranchers, and feedlot operators.

Packer buyers are employed by a particular packing firm (usually located in relatively close proximity to the market) and receive a salary from the employer.

Order buyers are not salaried. They buy livestock which normally is shipped to some more or less distant point. Order buyers usually specialize in a particular class of livestock and have clientele who phone them orders for livestock that meet certain specifications—both slaughter and feeder livestock. Order buyers do not take title (ownership) to the livestock they purchase. They simply act as agents in filling the orders. Their remuneration is a commission received from the parties who place the orders.

Yard dealers (also known as yard traders, speculators, and scalpers) buy livestock with the objective of reselling at a higher price. The term "yard trader" has, for some people at some times, carried a derogatory connotation. To a large extent, this is a misunderstanding. Yard dealers perform a function of always being in the market. At certain times other buyers may be relatively scarce or inactive, particularly on odd lots. In the absence of yard dealers these odds and ends might be in distress situations. Yard dealers often can accumulate enough livestock through a number of such transactions to form larger lots of uniform quality. Such lots then are attractive to buyers and the enhanced price for the larger, more uniform lot provides a profit incentive for yard dealers.

Wholesale provisioners often like to go into the yards to personally select slaughter livestock for a specific trade. They have the livestock slaughtered on a custom basis. Local butchers, who may operate a retail store or locker plant, buy limited numbers of slaughter livestock. Farmers, ranchers, and feedlot operators do some buying at terminals, but most purchasing for this group is done through their order buyers.

From an organizational standpoint, sellers and buyers (including primarily commission firms, order buyers, and packer buyers, and sometimes yard dealers) usually form a "livestock exchange." In some cases yard dealers form a separate "dealers exchange." The livestock exchange serves as a self-regulatory organization. It also makes recommendations on changes in commission rate charges, promotes various public relations activities, and cooperates with conservation organizations in the interest of reducing loss from bruising, crippling, disease, and death.

An original distinction of terminal markets was the transacting of business by "private treaty." Private treaty simply means that price is arrived at by buyer and seller bargaining in privacy. The outcome (price or other terms of the transaction) is known only to the participants. This is in contrast to an auction where bidding is done by public outcry and terms of the transaction are known to everyone in attendance. With the introduction of auction selling at a number of terminals, private treaty no longer is a singular distinction. Most terminals which have introduced auctions also maintain use of the private treaty method.

Operational Features.—Terminals ordinarily transact business Monday through Saturday, although activity is light late in the week (Cramer 1958). The markets are open seven days a week to receive livestock. Hog receipts are relatively uniform Monday through Friday, although some markets deviate from this pattern. Cattle receipts at terminals traditionally have been concentrated early in the week—with more than 40% of the week's receipts on Monday and about 20% on Tuesday. From an operational standpoint this is an inefficient, under-utilization of physical facilities and manpower during the latter part of the week. Among factors associated with early week concentrations on terminals are more active buying by packers and order buyers, influence of truckers, and tendency on some markets for prices to be higher early in the week (Cramer 1958).

Many commission agents are not opposed to early receipts since they like to spend the latter part of the week in the country visiting patrons and soliciting business. By holding an auction late in the week, some terminals have attempted to spread out receipts of feeder cattle. In the operation of its auction, the Wichita market is somewhat of an exception in that it operates the auction Monday through Thursday—handling slaughter cattle as well as feeder cattle. Slaughter cattle receipts still are heaviest early in the week. The Wichita yards are also open for private treaty sales throughout the week.

A producer who wishes to sell livestock at a terminal market consigns his livestock to a commission firm. Most producers who patronize terminals have such acquaintanceship. Anyone who has not established a contact probably should visit the market for this expressed purpose prior to shipping livestock. Most commission firms send their personnel to the country several days a week to solicit business and advise with patrons about the condition of their livestock, the market situation, and the probable value of their livestock.

The number of commission firms varies with the size of market. Most terminals have enough firms so that shippers have some choice. Competition among commission firms for producer patronage is the major force that makes a market go. Likewise, competition among markets makes the market system effective.

Some commission firms specialize in certain species. Others handle all species, and larger firms have specialists who devote their time exclusively to a single class of a given specie.

A shipper ordinarily notifies the commission firm in advance of his intentions to ship—giving approximate time of arrival. He also should give any special instructions he may have regarding feeding or handling of the livestock at the market. Upon loading of the livestock at the feedlot, farm, or ranch, the trucker is given the name of the commission firm. The name of this firm, together with a count and description of the livestock, is entered on a ticket by the trucker. This serves several purposes. Upon arrival at the market, it indicates the commission firm to which the cattle are to be delivered. It also provides a check on the count and description, and it provides the commission firm with the name of the trucker. When the livestock are sold, trucking charges are deducted from the proceeds and remitted directly to the trucker by the commission firm.

In this process the shipper of the livestock "consigns" his livestock to the specified commission firm. The shipper, in effect, authorizes personnel of that commission firm to act as his agent in selling the livestock. For this service the firm charges a fee or "commission." Each market has its specified schedule of charges for selling and buying livestock. These schedules are known as "tariffs." There can be differences in selling and buying charges among terminal markets, but there is no difference in charges among commission firms at a given market. This is enforced by federal government regulation. The logic of this arrangement is that competition among firms at a given market must be on the basis of services performed rather than on the amount of commission charged. It is presumed this enhances the services received by producers.

In 1978 P&S ruled that commission charges could be on either a per head basis or a percentage of the proceeds. From time to time it had been argued that logically a difference should exist in the commission on, say, a choice slaughter steer selling for $500 and a canner cow selling for $200. While these are points pro and con, the basic rationale had been decided on the proposition that the time, effort, and knowledge required to handle the various classes is reasonably equivalent, and a charge by the head keeps competition on the basis of service performed. The courts had upheld this ruling on the principle that charges based on value of the animals are unfair and discriminatory, and bear no relation to costs of providing services.

Upon notification by a customer of an impending shipment, the commission agent issues a feeding order to stockyard company personnel in accordance with instructions received from the consignor (shipper). In the absence of instructions, the commission agent issues feeding instructions in accordance with his own judgment.

Livestock delivered before opening of the market is sorted for uniformity in weight, grade, color, etc., by the commission agent, if, in his judgment, that will enhance their sale. When the market opens the commission agent stays at his pens, shows the livestock to prospective buyers, offers them for sale, and attempts to sell them at the most advantageous terms from the standpoint of the seller. This involves considerable bargaining over a single price for the entire lot versus sale of the majority at an agreed price with the privilege of cutting back specified animals at a lower price.

The skill and expertise of a commission agent comes in knowing when he has the best possible offer for a pen of livestock on that particular day. If he is unable to obtain a reasonable offer, he may hold the livestock over for the following day's market, but this involves some expense and risk and is not common operating procedure.

It is possible for a shipper to specify that the agent notify him of bids before consummating a sale. However, this ordinarily is not done. It tends to severely restrict the actions of the commission agent unless the shipper has accompanied his livestock and is at the pen when the bargaining takes place. A bid by a prospective buyer is good only at the time it is made. Once the bidder leaves the pen, his bid, so to speak, "goes with him." An owner who has specified that the commission agent must call him for an approval on bids, could lose a sale from a bidder who will not wait for the commission agent to place a long-distance call.

Upon completion of sale, the livestock are weighed and delivered to pens of the buyer, or holding pens, pending shipping instructions. Title of ownership passes to the buyer upon the weighing.

Details of the transaction (price, weight, and name of buyer) are transmitted to the commission firm's office. An "account of sale" is prepared for the shipper giving the information and calculation of gross proceeds, itemized charges, and net proceeds. This, along with a check for the net proceeds, is transmitted to the shipper. Itemized charges include the trucking charge (which, as mentioned earlier, is forwarded directly to the trucker), commission charges which are retained by the commission firm for its selling services, a small deduction per head which is transmitted to the National Livestock and Meat Board and used for industry promotional purposes, brand inspection (applicable in some states), a small charge for transit insurance, and charges assessed by the stockyards company. Included in the latter are yardage, feed, and insurance applicable at the yards. The deduction for the National Livestock and Meat Board will be refunded to a consignor upon request, otherwise it is forwarded to the Board's central office in Chicago. All commission firms are required by federal regulation to carry bond which helps assure a shipper that he will receive payment.

A shipper who prefers to have his livestock sold by auction at a terminal would consign the livestock to a commission firm in the manner described above (for private treaty), but would specify he wanted them sold at auction. The commission agent again would sort and shape up lots which in his judgment would enhance the sale. At most terminal auctions a member of each commission firm provides the starting price for livestock consigned to that firm. Some terminals assess a "ring fee" in addition to yardage on livestock sold through the auction.

An individual who wishes to purchase livestock at a terminal may, if he chooses, deal directly with a commission agent by private treaty or by personal bidding at the auction. On the other hand, if he wishes to engage the services of an order buyer he simply conveys to the order buyer specifications of the particular class of livestock he desires, including price constraints. In addition to firms and individuals who limit their activities exclusively to order buying, most commission firms also act as order buyers in filling requests of their patrons.

Types of Receipts.—Three different types of livestock receipts are handled at terminal stockyards: (1) saleable, (2) direct, and (3) through receipts. "Saleable receipts" are livestock delivered to the market for sale at that market. "Direct receipts" are livestock which have been purchased elsewhere by a local packing plant which does not have sufficient holding pen space. They are delivered to the terminal yards for temporary holding prior to slaughter. A yardage fee, somewhat smaller than the regular charge, is assessed for direct receipts. "Through receipts" are livestock which are unloaded for feed, water, and rest, then reloaded for further shipment. Ordinarily, "throughs" are not assessed yardage, but are charged for feed.

Regulation and Supervision.—Terminal markets are subject to regulation at two levels: (1) self-regulation imposed by rules of livestock (and dealers) exchanges, and (2) federal regulations—specified by the Packers and Stockyards Act of 1921 and amendments.

Rules and regulations of livestock exchanges are drawn up voluntarily by members in accordance with the constitution and by-laws of the exchanges. The overall objective is a matter of self-interest in maintaining a viable and progressive market. Rules are designed to maintain a high level of ethical business activity and provide a mechanism for settling controversies.

Prior to 1958, only stockyards involved in interstate trade and having 20,000 or more square feet of pen space came under jurisdiction of P&S. The P&S Act was amended in 1958 to include all persons and firms engaged in marketing and meat packing in interstate or foreign commerce. This amendment substantially expanded the number of markets covered. The Act prescribes rules of fair competition and fair trade practices. Terminal markets (and auctions) are "posted," which means that a notice is posted in three conspicuous places at the market by P&S personnel giving notice that this is a posted market and setting forth a schedule of charges for services performed. Posting gives P&S jurisdiction over the following: (1) Changes in charges and fees of agencies operating at that market. (This applies to commission charges, yardage, etc. To change the established rates, agencies concerned must file recommended changes in writing with P&S at least 10 days before the change is to be effective, and present evidence substantiating reasons for the change.)[1] (2) Trade practices which may be construed as fraudulent or unfair competition. (3) Maintenance of reasonable and adequate facilities. (4) Testing of scales to the capacity used, at least twice a year. (5) The requirement that individuals and agencies operating on the market carry a prescribed minimum bond and report annually all transactions of livestock and payments as well as all other operating accounts.

Auctions

Although the auction method of selling livestock has a long history, it was not until the 1920's and 1930's that auction marketing really caught on. As illustrated in Fig. 6.3, growth during the 1930's was especially pronounced. Expansion continued through the 1940's and early 1950's, though the rate of increase gradually tapered off. A peak in numbers (about 2500) was reached in the early 1950's. Since that time a decline has occurred in the number of auctions. The volume of livestock handled by auctions is not available on a national basis (except for packer purchases),

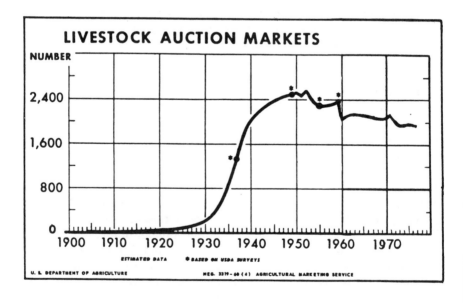

FIG. 6.3.   LIVESTOCK AUCTION MARKETS, 1904–1976
1960–1976 added from USDA data.

---

[1]A series of public hearings in 1978 produced a recommendation from Packers and Stockyards that market operators be allowed to change selling charges after 30-day notice—without the previously required clearance from that agency.

TABLE 6.2. NUMBER OF KANSAS AUCTIONS AND VOLUME OF LIVESTOCK HANDLED AT SELECTED INTERVALS, 1947–1975

Fiscal Year	No. of Licensed Auctions	No. of Cattle	No. of Hogs	No. of Sheep and Goats	No. of Horses and Mules	No. of Animal Units
1947–48	160	1,081,636	551,077	20,547	32,490	1,253,350
1949–50	165[1]					
1951–52			1,001,378[1]			
1952–53	153	1,448,960	583,668	26,228	13,519	1,611,019
1957–58	132	1,887,191	557,344	38,976	3,428	2,033,853
1962–63	128	2,871,930	935,904	68,939[1]	11,380	3,124,180
1965–66		3,098,541[1]				
1967–68	109	2,754,184	995,377	44,140	14,328	3,021,770
1972–73	99	2,860,911	1,222,254	37,189	11,830	4,132,184
1974–75	95	2,832,996	1,370,033[1]	63,673	14,867	4,281,569[1]

Source: Beaton and McCoy (1970) and Gurss (1977).
[1]Peak years.

but the data shown for Kansas in Table 6.2 probably are typical. This shows a decline in number of auctions but an increase in volume of livestock handled.

Auction markets are distributed fairly uniformly throughout the United States relative to the density of livestock production. However, the proportion of livestock handled by auctions varies considerably by region as shown in Table 6.3. While this table shows only marketing through posted markets, it reveals a high level of use of auctions in the Mountain and Pacific regions and in Southern regions—an exception being hogs in the Southern Plains.

Organization.—From a business standpoint, auctions may be organized as corporations, partnerships, or individual proprietorship. A limited number are organized as cooperatives. Most of the larger ones are corporations and many small ones are owned and operated by individuals. In contrast to terminals, where ownership and operation of yards, scales, building, etc., are separate from the selling agencies, the ownership of physical facilities and the selling function of auctions are all under the same management.

Auction markets are used primarily, but not exclusively, for feeder livestock and cull animals. While not a major avenue for slaughter livestock sales, packers nevertheless obtain substantial numbers through auctions. Table 6.4 shows packer purchases through auctions in recent years. These purchases in 1976 amounted to approximately 21% of the cattle, about 62% of the calves, 12% of the hogs, and 15% of the sheep slaughtered by packers.

Operation.—Auctions are distinguished by public bidding in response to an auctioneer's chant—with final sale to the highest bidder. Most auctions sell livestock just 1 day per week, however a few operate 2 or more days per week.

TABLE 6.3. LIVESTOCK MARKETED THROUGH POSTED PUBLIC MARKETS, BY TYPE OF MARKET AND BY REGION, 1976

Kind of Livestock and Region	Number of Head			Percentage		
	Terminals (1,000 Hd)	Auctions (1,000 Hd)	Total (1,000 Hd)	Terminals	Auction	Total
Cattle and Calves						
North Atlantic	12	104	116	10.3	89.7	100.0
East North Central	87	587	674	12.9	87.1	100.0
West North Central	1,004	981	1,985	50.6	49.4	100.0
South Atlantic	4	237	241	1.7	98.3	100.0
South Central	6	88	94	6.4	93.6	100.0
Southern Plains	78	1,288	1,366	5.7	94.3	100.0
Mountain	0	335	335	0.0	100.0	100.0
Pacific	0	313	313	0.0	100.0	100.0
Alaska and Hawaii	0	0	0	0.0	0.0	0.0
United States	1,191	3,933	5,124	23.2	76.8	100.0
Hogs						
North Atlantic	214	577	791	27.1	72.9	100.0
East North Central	3,371	3,498	6,869	49.1	50.9	100.0
West North Central	7,812	8,980	16,792	46.5	53.5	100.0
South Atlantic	65	3,380	3,445	1.9	98.1	100.0
South Central	252	2,278	2,530	10.0	90.0	100.0
Southern Plains	842	477	1,319	63.8	36.2	100.0
Mountain	0	301	301	0.0	100.0	100.0
Pacific	0	263	263	0.0	100.0	100.0
Alaska and Hawaii	0	0	0	0.0	0.0	0.0
United States	12,556	19,754	32,310	38.9	61.1	100.0
Sheep and Lambs						
North Atlantic	12	104	116	10.3	89.7	100.0
East North Central	87	587	674	12.9	87.1	100.0
West North Central	1,004	981	1,985	50.6	49.4	100.0
South Atlantic	4	237	241	1.7	98.3	100.0
Southern Plains	78	1,288	1,366	5.7	94.3	100.0
Mountain	0	335	335	0.0	100.0	100.0
Pacific	0	313	313	0.0	100.0	100.0
Alaska and Hawaii	0	0	0	0.0	0.0	0.0
United States	1,191	3,933	5,124	23.2	76.8	100.0

Source: USDA (1977).

A livestock owner consigns his livestock to an auction in a manner similar to that described for terminals. Though not a requirement, it is good business practice for an owner to contact the auction manager several days prior to sale day and notify him of intentions to deliver. It is a common practice for auction managers or their representative to visit patrons on request, inspect their livestock and advise the patron regarding condition of the livestock, market conditions, and probable value of livestock.

Upon delivery, the livestock are penned or marked in a way to maintain identity for each individual owner. Operators usually follow a consistent order in sale by species. That is, hogs may be sold first, followed by sheep, and then cattle. Within a given specie, most auctions follow the practice of presenting livestock for sale in the order in which they are received. Some auctions, however, have a considerably greater lag in

TABLE 6.4. AUCTION MARKET PURCHASES BY PACKERS, 1960–1976

Year	Cattle (1,000 Hd)	(%)	Calves (1,000 Hd)	(%)	Hogs (1,000 Hd)	(%)	Sheep (1,000 Hd)	(%)
1960	3,393	15.6	1,940	32.1	6,695	8.7	1,493	10.6
1965	6,235	20.9	3,373	49.2	10,151	13.7	1,571	12.1
1970	5,265	16.3	2,139	54.6	11,586	14.3	1,192	12.4
1971	4,992	15.5	1,998	59.0	12,543	13.8	1,213	12.4
1972	4,681	14.6	1,680	60.7	11,085	13.3	1,171	12.0
1973	4,653	15.1	1,341	60.9	9,432	12.4	1,319	14.7
1974	5,309	16.4	1,755	63.1	9,681	12.4	1,184	13.5
1975	7,068	19.7	2,651	59.3	8,260	12.1	1,204	15.6
1976	7,749	20.8	3,037	62.3	7,879	11.5	980	15.0

Source: USDA (1977) and earlier issues.

time from delivery to sale than do other auctions, which can be a factor in weight shrinkage.

Auction operators usually sort an owner's livestock for size, grade, or other characteristics to obtain uniformity as the animals are presented in the sale ring. This is intended to enhance sale of the entire lot. As the animals arrive in the ring it is common practice for the manager, or his representative, to provide a starting bid. This is not done in all parts of the United States, but does speed up movement through the ring. The starter attempts to put the initial price just slightly below what he thinks the livestock will sell for. To the extent he is able to judge this value, the procedure obviates the necessity of needless bidding. Following the auctioneer's chant, bidding progresses to higher levels until no one is willing to advance the bid. The livestock is sold to the highest bidder. Depending upon local custom and/or class of livestock involved, the selling may be done on a per head basis or by weight.

Several different procedures are followed in obtaining sale weight. Many modern auctions are constructed so the sale ring is the scale platform. In addition many have circuitry to an electronic display which shows total weight, number of head, and average weight in full view of the audience. Some markets, however, locate the scales on the "in" side of the sale ring and the livestock are weighed prior to entering the ring. In this case the weight can be announced or flashed on a lighted sign. This eliminates the need for buyers estimating weight. It is not uncommon, however, where the scale is located on the "out" side, for the operator to "catch" the weight of a draft of animals or a part of a draft prior to final sale so bidders will have an estimate of average weight while bidding. However, in this case the official weight is taken after sale. At some markets, weights are taken as livestock is unloaded at delivery to the auction, although this is not common practice in the midwest.

After the livestock are sold they are penned for loading out. Large volume buyers who wish their purchases penned by sex, weight, grade or other characteristics, may specify as the livestock leaves the sale ring, how they wish each draft penned.

Packers and Stockyard Act regulations allow auction operators to purchase livestock that otherwise would sell at a price below what would be considered a competitively established level for that particular class, grade, and weight of animal at that time. This is known as "market support," and is designed to provide a degree of price protection to sellers at times when buying activity is slack. A Kansas study (Beaton and McCoy 1970) based on 1965 data indicated that market support was carried out on only 3.7% of the livestock handled. This study showed that auctions averaged a loss on livestock bought in their market support activities. The losses ranged from approximately 12¢ per animal unit in smaller auctions to 7¢ in the larger auctions when the losses were spread over all livestock handled.

Other means of price protection also are available to consignors. One method is to set up a reservation price with the auction operator prior to sale. This price may be noted on the ticket which accompanies the livestock into the sale ring. A reservation price means that if bids do not reach that price the livestock are not sold. Another method is for the consignor to declare "no sale" at the time the auctioneer has received the last bid. This necessitates the consignor or his representative being present during the sale. A third alternative, one which definitely is frowned upon, is for the consignor to bid on his own livestock, or arrange for a cohort to do the bidding. This is known as "by-bidding." Charges assessed for "no-sale" livestock vary, depending upon the policy of particular auction operators (Abel and Broadbent 1952).

Costs of marketing at an auction fall in essentially the same categories as at a terminal, although there are slight differences among states. The major items are commission and yardage. Lesser deductions may be made for such items as insurance, feed, state inspection, state fee, National Livestock and Meat Board, and brand inspection depending upon the particular state. A considerable variation exists among auctions in the determination of commission charges. Some auctions assess commission on a per head basis, others on a percentage of the proceeds, and some on a combination of the two.[2]

An innovation in auction marketing is the electronic auction. There are three major types of electronic auctions—based upon technology used: (1) telephone, (2) teletype, and (3) cathode ray TV-like display which may also include voice and computerized matching of bids and offers. Telephone and teletype auctions commonly are known as Tel-O-Auctions and Teleauctions. Tel-O-Auctions consist of an interlocking system of telephone conference call set-ups whereby bidders in geographically separated locations can bid on livestock as described over the voice system.

---

[2]As a result of a series of nationwide public hearings in 1978 the administration of Packers and Stockyards recommended that all public market selling commissions be on a per-head basis only.

The livestock may be assembled at selected points or left on the producer's farm. Sorting and grading must be very precise because buyers are bidding on description—sight unseen. They must have confidence in the grading system. Livestock producers in Virginia and neighboring areas have successfully operated a Tel-O-Auction since 1963. Originally this market was used for slaughter hogs, but in recent years it has been used primarily for slaughter lambs, feeder lambs, and feeder cattle.

A hog marketing installation is located in Wisconsin. In the Wisconsin setup producers deliver their hogs to a designated point. Here the hogs are identified by owner, weighed, and graded by recognized experts. They are then comingled and sold in lots of uniform weight and grade. Each lot is described over the communication system by number of hogs, their weight, and grade. Bidding can be done by ringside buyers as well as by telephone participants. Farmers pay a fixed fee per head for the selling service. The Wisconsin Tel-O-Auction is part of a larger program, the Modern Merit Market, in which an attempt is made to establish a market outlet that recognizes top-quality hogs and rewards farmers for producing them. In addition, it is anticipated that packer procurement costs will be reduced.

The Missouri Farmers Association (MFA), a cooperative marketing organization, has operated a successful Tel-O-Auction for feeder pigs since December 1965. In January, 1968, MFA initiated Tel-O-Auction sales for slaughter hogs. The slaughter hog Tel-O-Auction is operated in conjunction with their already existing (since 1958) dealer operations in slaughter hogs (producers are given the option of selling their hogs outright to MFA's dealer operations or consigning them for sale through the Tel-O-Auction). Essential success has been achieved, but additional time will be required to determine whether farmers will patronize the Tel-O-Auction in preference to alternative outlets.

A slaughter lamb telephone auction operates in eight northwestern states and Iowa. In that system the grading, originally done by state extension personnel, now is done by knowledgeable producers (Holder 1974). The American Sheep Producers Council has recommended the organization of four regional telephone auctions which would serve the entire sheep industry of the 48 contiguous states.

From a technological standpoint it would be possible to operate one national teleauction for lambs. However, on that scale it probably would be more feasible to use teletype or computerized cathode-ray equipment. Declining sheep numbers of recent decades has resulted in the closing of many sheep slaughtering plants and severe decline in market outlets. Under these circumstances the teleauction is an alternative worthy of consideration.

Considerable research has been devoted to potential use of electronic auctions for livestock marketing (Armstrong 1976, Holder 1974, Johnson

1972, Ward 1977). Most studies point to possible improvements in both operational and pricing efficiency. To the extent that livestock can be adequately described, bidding can be done by buyers without their assembly or the assembly of livestock at some centralized location. Thus, competitive forces can be centralized from a wide area without the physical presence of either the bidders or the livestock. It is likely that electronic auctions will receive greater attention in the future.

At this time there are no installations of computerized cathode-ray units for livestock marketing in the United States. The cotton industry operates such an installation—known as TELCOT. A seller places his offering price for a specifically described lot of cotton in a given terminal from which it is disseminated through an interconnected network to any number of units located near or far from the originating unit. Prospective buyers enter bids in a like manner. The computer matches offers with the highest bid, and adjustments can be made for transportation cost if the seller is responsible for delivery.

Another innovation in livestock marketing is the auctioning of slaughter cattle on a carcass basis. This has not progressed beyond the experimental stage in the United States (Bowen and Thomas 1970), but has been used on a limited scale in Ontario, Canada. In this approach bidding is done on an agreed upon base grade (in the U.S. that was Choice, Yield 3). When the animals were slaughtered price differentials were applied as quality and yield grade deviated from the base. While this method has the potential of improving pricing efficiency it has not been accepted in the United States at this time.

Practically all auctions in the United States operate on ascending bids. An auctioneer calls for higher and higher bids until no one is willing to raise the last bid. There is another approach known as the Dutch (or Danish) auction. In a Dutch auction the starting price is slightly above the price at which a sale is expected. The price then is lowered in predetermined increments at regular intervals until a buyer accepts. This system is used in some western European countries. It has been used in Canada's Ontario hog market under a teletype auction arrangement since 1961 and in several additional Canadian province hog markets for lesser periods. (Hawkins *et al.* 1972, and Lowe 1968). The Dutch auction requires that someone provide a starting price, but that is about the only way it resembles a regular auction. Once the starting price is provided, the incremental lowering can be done mechanically, or electronically, as well if not better than by human means. A starting price and a series of lower prices can be prepunched on a tape for an electronic display. As the tape is activated, prices can be made to show up on a lighted board or printed out on a receiving tape. Prospective buyers can be provided with a button to push when the price reaches an acceptable point. The first person who presses a button is the buyer. This system can be carried out without an auctioneer and in so doing can avoid human error in taking bids, but does not have the flexibility of auctioneering in such things as cutting certain animals from a lot (Eckert 1965).

Regulation and Supervision.—Auctions which handle livestock that move interstate are considered to be public markets. They are posted and are under the jurisdiction of the Packers and Stockyards Act. The applicable provisions are those discussed previously under terminal markets. Marketing charges, once established, cannot be changed without following procedures set out by P&S. Scales must be tested twice a year. The auction must (1) refrain from trade practices that limit or restrict competition, (2) must carry bond to help assure payment to patrons, and (3) report annually all transactions and payments.

Auctions not regulated and supervised under P&S may be regulated under state provisions. Many states require bonding to help assure payment to consignors. Many states require health inspection by qualified veterinarians; however, few auctions assume any responsibility or grant any warranty regarding health or condition of livestock sold. Many auctions plainly display a sign disclaiming guarantees of any kind. In addition, it is common in case of an animal in doubtful condition, for the auctioneer to declare that the animal is being sold "as is."

Country Commission-man

The term "country commission-man" is not well established in marketing literature, but such agents do exist. A country commission-man simply is a commission agent who operates in the country rather than at an established market. He represents producers in the sale of their livestock. This may be a case of the proposition that "if producers will not take their livestock to the commission-man, then the commission-man will go to the producer." It also probably is a recognition on the part of some producers that selling is a specialized function, and some would rather specialize in other aspects of production than attempt to be a jack-of-all-trades.

These commission-men tend to operate in areas of highly concentrated production as are found in the cattle feeding areas of the Milo Belt, Colorado, Arizona, and California. They travel with packer buyers to the feedlots. Final transactions may be made at the feedlot or elsewhere.

This movement has not been underway long enough to know if it will develop into significant proportions. It would appear to have merit for producers, large and small, who need the service. On the other hand, it is of doubtful feasibility where production is widely scattered.

Order Buyers

Order buyers act as agents or intermediaries for purchasers. They buy livestock of certain specifications to fill orders which they have received from clients. The specifications normally detail weight and grade. Some orders specify maximum price—others leave price to the discretion of the order buyer. But, good order buyers attempt tu fill their orders at the lowest price possible consistent with getting that order filled. This, of course, is what keeps an order buyer in business. He works in the interest of the purchaser.

Order buyers operate at terminal markets and at auctions. They also will buy from dealers, from local markets, and directly from farmers and ranchers in the country. Many buyers are specialists in a particular class and many operate at only one type of market. Specialists in buying slaughter cattle, slaughter hogs, and slaughter lambs are found at most terminals purchasing for packers located at distant points. Feeder cattle order buying specialists also operate at terminals and, as mentioned earlier, commission agents at terminals act as order buyers when requested by patrons. Commercial feedlots utilize the services of order buyers at all types of markets. The demand for feeder cattle to stock the feedlot industry has created problems in filling these orders.

Many slaughter lambs are bought by order buyers for forwarding to packers. The number of packers who slaughter lambs has gradually declined, so buying has become more scattered and lambs must be shipped longer distances. Many hogs are bought on order at midwestern markets for shipment to the west coast and other points.

The true order buyer does not take title to the livestock he handles, but instead merely acts as agent for the purchaser. The order buyer typically is authorized by the purchaser to draw a draft on the purchaser's bank in payment for the livestock. His reimbursement is a fee for the service performed. In most cases the fee is a flat rate per head, which varies by class of animal. Some order buyers mix dealer operations with their order buying (i.e., they sometimes buy livestock outright, take title, and attempt to sell at a higher price for a profit). There are no regulations to prevent an individual or a firm from engaging in both activities simultaneously. There are no regulations to prevent the filling of an order with personally-owned livestock. However, most purchasers prefer to deal with order buyers who stick strictly to order buying. The mixing of dealer and order buying activities tends to diminish the confidence of purchasers. Experience has shown that one of the biggest problems of livestock growers and finishers is the purchase of stocker and feeder animals. Order buyers of feeder livestock, with widely accepted reputation, are much in demand.

Bargaining Associations

Bargaining associations are organizations set up with the express purpose of bargaining for producers in the sale of their products or purchase of inputs. Bargaining often covers terms of trade in addition to price. For a number of years, milk producers in Federal Order Milk Markets have had an organization for bargaining with milk handlers and processors. These are set up under federal market orders and agreements. The federal enabling act specifies application only to certain commodities. Livestock, meat, and wool are not included in the federal act.

Some states have statutes authorizing market orders and agreements for specified commodities. But again, most states do not include livestock, meat or wool. Colorado authorized a marketing order for lambs in 1969 but it was discontinued in 1970.

Bargaining associations are not limited to governmentally sponsored organizations. There also are privately organized bargaining associations. Probably the best known example at this time which includes livestock in its operations is the National Farmers Organization—commonly known as NFO, organized in 1955. The NFO is a farmer organization which bargains with packers in the sale of members' slaughter livestock. This organization operates as a cooperative under provisions of the Capper-Volstead Act which authorizes farmers to organize and bargain collectively. Members pay an annual membership fee and agree to turn over to the organization the job of bargaining for price and other considerations.

The NFO gained wide attention and some notoriety during several "withholding" actions. These were attempts to use withholding tactics as a leverage in negotiating contracts with packers. Early attempts were not successful in forcing the signing of contracts. Subsequently, it has been reported that contracts have been signed, but the organization has not revealed the extent of its contracts nor the terms.

It is generally held among economists that the withholding approach used by NFO (the withholding of a perishable or at least semiperishable product like finished livestock, after the livestock are produced and ready for market) does not give the leverage needed for effective bargaining power. Withholding at this point results in even greater tonnages to be marketed at the end of the withholding since additional weight is gained during withholding.

It is possible, however, that some packers are willing to enter into a contract if, in return, the association guarantees scheduled delivery of uniform weight and grade of livestock. Such a contract would reduce the packer's procurement costs which would make possible a higher paying price. Effective withholding, however, necessitates the power to control production. Private, voluntarily organized, bargaining associations have problems in maintaining the necessary discipline to gain production control. If some members abide by an organization's production control plan, a powerful incentive develops for individuals to drop out of the association and increase production. It is difficult for a private organization to enforce stringent enough penalties to prevent this. Nevertheless, the desire and enthusiasm for bargaining power among producers has gained considerable momentum in recent years. This is an area that will attract increasing attention. It is a complex problem. The trend toward fewer and bigger farm operators enhances the possibility of forming bargaining associations, from an organizational standpoint. The number of livestock producers, however, still is relatively large. Production is widely dispersed and producers have not exhibited a complete unanimity of interests.

Recent antitrust decisions indicate a recognition that farmers legally have the power to form collective bargaining associations (Clodius 1962). At the same time, these decisions make it fairly clear that undue enhancement of prices would be construed as a violation of antitrust laws. Even if a bargaining association legally gained sufficient control to raise prices above otherwise prevailing levels, economic limits still would exist on the extent to which this power could be exercised long term. Unduly high prices would tend to encourage consumers to switch to substitute products.

DIRECT MARKETING

As used in this discussion, direct marketing is the marketing of livestock without the services of an intermediary (i.e., without the services of commission agents, order buyers, auctioneers, or personnel of a bargaining association).

Historically, direct marketing has been defined as the use of any channel, or channels, which by-pass terminal markets. This definition is too restrictive for many purposes. The range of markets that would be included in the nonterminal classification still would require subclassification to differentiate markets within that group. From the producer's standpoint it appears that the chief point of distinction is whether he himself performs the functions of selling and buying or whether he engages the services of someone else.

Live Versus Carcass Grade and Weight Direct Marketing

Two major arrangements are practiced in direct marketing of slaughter livestock: (1) sale on live grade and weight, and (2) sale on carcass grade and weight. The latter is also commonly known as rail grade and weight, "on the rail," or "in the meat." Live grade and weight is the historical arrangement whereby price is negotiated on the basis of live animals. Whatever grading is done is an estimated grade from inspection of the live animals. Weight and carcass yield are estimated on live animals. Weight is estimated at the time price bargaining is taking place. However, where sale is made on weight instead of by the head, the pay weight is a scale weight (sometimes reduced by a pencil shrink which is discussed later).

In sharp contrast, under sale by carcass grade and weight the price is not finally determined until the animals are slaughtered. Pay weight is taken from the carcass weight as it hangs on the rail. The carcass is graded either by a packing plant employee or a government grader. Fill is no factor since weight is based on the carcass. Price is established from a predetermined scale of prices based on weight and grade. This method requires that each carcass be identified as to producer ownership. Hog carcass identity can be maintained rather successfully by tattooing where carcasses are not skinned. Cattle and sheep carcasses can be tagged, but there is some risk of tags becoming lost. The same problem occurs where hog carcasses are skinned, and skinning is done in the more modern plants. Either procedure necessitates a cost not incurred in live selling. Additional record keeping for settlement purposes also adds to cost. Grading by packing plant employees has raised the question of possible bias, as has weighing on packing plant scales by plant employees. Objections also have been raised due to delayed payment since final settlement cannot be made until the livestock are slaughtered, weighed, and graded. Some firms, however, give an advance payment upon delivery of the animals. It is obvious in carcass grade and weight selling that practically no opportunity exists for bargaining after the livestock have been delivered and slaughtered. Bargaining must precede delivery.

Due to lack of uniformity among packers in carcass grade and weight buying procedures, the USDA in 1968 established regulations as follows: (1) packers must divulge to sellers, either orally or in writing, terms of the purchase contract prior to sale; (2) identity of each carcass must be maintained; (3) sufficient records must be maintained to verify settlement with producers; (4) purchase and payment must be made on the basis of carcass prices; (5) weight must be established on the hot carcass (not chilled); (6) all hooks, rollers, gambels, and other equipment must be uniform in weight for each species and only this weight be deducted as tare; (7) payment may be made on USDA grades or other grades, but if the latter, the seller must furnish written specifications, and (8) carcasses must be graded no later than the close of the second business day following slaughter. These regulations, although somewhat obnoxious to many packers, have established some needed uniformity among packers which permits better comparison by producers of carcass price offers than formerly was the case.

Research results, although not completely unanimous, have indicated that this method has the potential for improving pricing efficiency (Engelman et al. 1953). If the price applied to various grades and weights truly reflects market value, each producer would receive true value for his livestock. Each animal would be valued separately (not an average grade and weight for an entire lot). Thus, a producer would receive a differential for the more desirable grade and weight animals. Presumably, such a differential would encourage production of the more desirable animals and discourage production of those less desirable.

It is possible, of course, to sort live animals and apply price differentials according to grade and weight. This is done to some extent in all types of direct and indirect markets. Commission agents and auction operators normally do some sorting. It is standard procedure for buyers to cut, or attempt to cut, less desirable animals from a lot and apply a price discount. Two restrictions, however, limit the extent of sorting on a live basis. First, it generally is agreed that grading of carcasses is more precise than grading on the hoof (Naive et al. 1957). Second, market agents and buyers are reluctant to make extensive sorts. A common argument is that sorting, while possibly resulting in higher prices for some animals, will at the same time necessitate lower prices for others, so the average price may be no better than selling unsorted. This ignores the issue of pricing efficiency. While the arguments pro and con go on, producers have shown a tendency to increase direct marketing. Table 6.5 shows the number and percentage of packer purchases by carcass grade and weight in recent years. The most significant increase occurred in cattle. In 1963 (for all packers) only 8% of the cattle were purchased by this arrangement. By 1976 the proportion had increased to slightly more than 23%. Smaller increases are shown in carcass grade and weight purchases of calves, hogs and sheep.

Controversial Issues in Direct Marketing

It is impossible to state unequivocally whether a producer should use direct marketing. It is possible, however, to state some criteria upon which producers can make a decision. As mentioned earlier, the critical consideration is net return. Net return is gross return less marketing costs. Involved here are livestock prices, out-of-pocket costs, and non-cash costs. Out-of-pocket costs such as commission, yardage, trucking, etc., are relatively easy to determine. In direct marketing, commission, yardage and other marketing costs, as enumerated on published market

TABLE 6.5. LIVESTOCK PURCHASED ON CARCASS GRADE AND WEIGHT BY TYPE OF PACKER, 1963-1976[1]

	CATTLE			CALVES			HOGS			SHEEP & LAMBS		
	Total Purchases (1000 Hd)	Grade and Weight Purchases (1000 Hd)	(% of Tot)	Total Purchases (1000 Hd)	Grade and Weight Purchases (1000 Hd)	(% of Tot)	Total Purchases (1000 Hd)	Grade and Weight Purchases (1000 Hd)	(% of Tot)	Total Purchases (1000 Hd)	Grade and Weight Purchases (1000 Hd)	(% of Tot)
10 Major Packers												
1963	8,122	667	8.2	1,990	94	4.7	43,845	1,529	3.5	9,362	590	6.3
1965	9,668	1,240	12.8	2,359	186	7.9	39,877	1,453	3.6	8,169	379	4.6
1970	11,091	2,323	20.9	742	51	6.9	39,023	2,776	7.1	5,544	567	10.2
1971	11,379	2,693	23.7	589	36	6.1	42,024	3,067	7.3	5,621	596	10.6
1972	12,002	3,211	26.8	426	26	6.1	37,742	3,199	8.5	5,376	707	13.2
1973	11,526	3,513	30.5	261	11	4.2	35,598	3,295	9.3	4,897	486	9.9
1974	11,842	3,538	29.9	270	32	11.9	38,267	4,303	11.2	4,855	552	11.3
1975	11,280	3,654	32.4	597	175	29.3	32,631	4,108	14.7	4,150	468	11.3
1976	12,751	3,847	30.2	617	149	24.1	33,779	5,598	16.6	3,350	381	11.4
Other packers												
1963	16,285	1,254	7.7	3,746	49	1.3	35,770	538	1.5	5,810	199	3.4
1965	20,184	2,143	10.6	4,492	128	2.8	34,262	489	1.4	4,850	277	5.7
1970	21,107	3,686	17.5	3,178	127	4.0	41,823	1,108	2.7	4,087	380	9.3
1971	20,873	3,907	18.7	2,799	161	5.8	48,801	1,384	2.8	4,176	129	3.1
1972	20,045	4,049	20.2	2,343	159	6.8	45,562	1,160	2.5	4,423	112	2.5
1973	19,302	3,693	19.1	1,941	113	5.8	40,482	1,083	2.7	4,055	224	5.5
1974	20,504	3,885	18.9	2,512	141	5.6	39,773	1,024	2.6	3,935	225	5.7
1975	24,605	5,081	20.7	3,873	223	6.0	35,445	1,247	3.5	3,573	345	9.7
1976	24,504	4,840	19.8	4,257	259	6.1	34,842	1,574	4.5	3,198	242	7.6
All packers (no. of firms)												
1963 (295)	24,407	1,921	7.9	5,736	143	2.5	79,615	2,067	2.6	15,172	789	5.2
1965 (339)	29,852	3,383	11.3	6,851	314	4.6	74,139	1,942	2.6	13,019	656	5.0
1970 (358)	32,198	6,009	18.7	3,902	178	4.5	80,846	3,884	4.8	9,631	947	9.8
1971 (333)	32,252	6,600	20.5	3,388	197	5.8	90,825	4,451	4.9	9,797	725	7.4
1972 (317)	32,047	7,260	22.6	2,769	185	6.7	83,304	4,359	5.2	9,799	819	8.4
1973 (329)	30,828	7,206	23.4	2,202	124	5.6	76,080	4,377	5.8	8,952	710	7.9
1974 (304)	32,346	7,423	22.9	2,782	173	6.2	78,040	5,328	6.8	8,790	777	8.8
1975 (311)	35,891	8,735	24.3	4,470	408	9.1	68,076	6,048	8.9	7,723	813	10.5
1976 (314)	37,255	8,687	23.3	4,874	408	8.4	68,621	7,172	10.5	6,548	623	9.5

Source: USDA (1977).

[1]Summarized from annual reports of packers filed with the Packers and Stockyards Administration, USDA.

Note: Data includes livestock where the amount of settlement was determined after slaughter on the basis of carcass weight and on carcass grade and weight.

"tariffs," are avoided. Trucking may be less in direct marketing or it may be more depending upon the specified delivery point and upon whether seller or buyer provides the trucking service.

Shrinkage is a major noncash marketing cost. What is needed is the difference in shrinkage that would be incurred under direct marketing as compared to the indirect method. Some shrinkage is incurred regardless of the method of marketing. As will be seen in a later chapter, shrinkage is affected by a number of variables. Some research is available that provides guidelines, but research application is not precise to a given farm situation. Producers with an adequate set of scales on the premises can, over a period of several shipments, develop applicable shrinkage standards for their particular operation.

The question of price level under direct versus indirect marketing is an unresolved controversy. Terminal market personnel and auction operators who see direct marketing as a threat to their existence have opposed this method, contending that producers will not get full price due to lack of competition when a producer deals directly with professional market personnel, such as packer buyers, order buyers, dealers and local market operators.

Lack of Competition.—The contention that competition is lacking in direct marketing appears to rest upon the assumption that sellers, buyers, and the livestock must be physically present to have effective competition. Economic theory, however, does not substantiate this argument. What is needed are well-informed buyers and sellers with an adequate communication system. It may be assumed that most buyers are well informed. Whether a producer is well informed is largely up to him. Market news reports are available from both public and private sources. The decline in receipts at some terminal markets makes quotations from those sources less reliable than formerly. The USDA Market News Service has extended coverage to some auctions and interior markets. Some of these offices have up-to-date market information taped on an automatic telephone hookup. With direct dialing anyone can easily call, either locally or long distance, and obtain current information almost instantly. Further changes may be necessary in the future to obtain adequate information from both producers and packers on terms of direct sales.

Producers' organizations may play a greater part in the future in developing market intelligence. The National Cattlemen's Association has a market information gathering system under the name CATTLE-FAX. With this system producers furnish information which is summarized and interpreted in a central office, then disseminated to subscriber members. The National Farmers Organization has an organizational arrangement which funnels market information to its central office. Many commercial feedlot operators subscribe to one (or more) of several available market wire services. These services provide for private installations which print out on tape almost continuous up-to-the-minute market reports. Such installations are available to individual producers who are willing and able to pay the price. There also are a number of reputable research-consultant market services available on subscription. Some of these provide continuous toll-free telephone advisory service.

These types of market news services generally are beyond the capability of any one individual, but where such information can be obtained on a subscription basis a producer can be well-informed. Reasonably adequate communications systems are available in radio, television, telephone, teletype, etc. Improvements in electronic processing and in communications are constantly improving the availability of information. Nevertheless, it probably is true that many producers are ill-informed either through neglect or in ignorance of the availability of information. If so, they are not in position to be competitive in direct marketing. However, operators of most commercial type feedyards and many progressive farm feeders probably are as well-informed as the buyers.

Research studies have not verified the contention that direct marketing has lowered prices received by producers (USDA 1935; Stout and Feltner 1962).

Producers often claim a stronger competitive position in direct marketing at their premises, in that they know what the terms of the transaction are before committing themselves to a sale. Producers who have delivered their livestock to a market any substantial distance from home are just about committed to sell. Technically they may have a reservation price and may take the stock back, or to another market if dissatisfied, but from a practical standpoint this usually is not feasible. It may be more feasible at local auctions, but even there it is not a common practice.

Lower Quality at Terminals.—On occasion, opponents of direct marketing have argued that direct buyers take higher quality stock, leaving the lower quality to be marketed at terminals, and since direct purchase prices often are based on terminal prices, lower quality animals are setting the standards. No recent studies are available to test this statement. It was tested in 1935 (USDA 1935). The results of the study showed no evidence of a difference in quality of hogs marketed at public (terminal) markets and interior markets. The latter represented direct markets. The same study found no difference in quality of hogs marketed at the Chicago terminal and hogs sold direct to packers.

While information is not available on the current situation, past studies have shown that producers generally tend to be relatively firmly opinionated in regard to market preferences. A 1956 study (Newberg 1959) covering the entire North Central Region revealed that "four fifths of the the farmers interviewed indicated that in selling livestock they had only one outlet where they sold all their major class."

Noncash Marketing Costs.—Out-of-pocket marketing costs are less in direct marketing than in the indirect methods since no commission and yardage are charged. Opponents of direct marketing claim, however, that producers overlook the importance, particularly, of shrinkage concessions. Most producers would argue that shrinkage is less in direct sales, but this issue has several ramifications. It is not uncommon for buyers to attempt to get a concession that the livestock be kept off feed and water for a period, usually overnight, and in addition apply a "pencil shrink." Pencil shrink is an agreed-upon percentage reduction from the scale weight in arriving at a pay weight. If the scales happen to be located some distance from the producer's premises the actual shrink (from an overnight stand, loading, hauling, unloading, and weighing) can be considerable. Then, if a pencil shrink is applied, the combined shrinkage concession may be more than producers realize. Only by the installation of adequate scales on the premises will a producer know the shrinkage.

Many producers of slaughter livestock apparently have not been convinced that direct marketing is detrimental to their interests as the number and proportion of cattle and hogs marketed direct continues to increase. Table 6.6 shows direct packer purchases for recent years. While the number of sheep purchased direct by packers has declined in recent years, the proportion has increased as numbers purchased at terminals have declined (Table 6.5) and numbers purchased at auctions have not changed greatly (Table 6.4) except for a noticeable decline in direct auction purchases in 1976.

It is not the intent here to pass judgment on whether producers should do their own selling and buying. The operator of a commercial cattle feedlot who is buying and selling cattle every day undoubtedly is in a different situation than many farmer feeders who buy one lot of feeder cattle a year. Yet even in the latter case, much depends upon the qualifications of the individual.

The qualifications of a livestock marketer, whether he be a commission agent, an order buyer, a packer buyer, or a producer doing his own selling and buying, include three major attributes: (1) He must have the ability to judge livestock. In this he must be able to evaluate all the characteristics which affect price—grade (including both quality grade and yield grade), carcass yield (including fill and shrinkage factors), and weight. (2) He must know the market value of the particular class, grade, and weight of livestock under consideration—at that particular time. In other words, he must be up-to-date on the market situation. (3) He must

TABLE 6.6.   DIRECT PURCHASES BY PACKERS, INCLUDING ALL EXCEPT TERMINAL MARKET AND AUCTION PURCHASES, 1960–1976

Year	Cattle (1,000 Hd)	(%)	Calves (1,000 Hd)	(%)	Hogs (1,000 Hd)	(%)	Sheep (1,000 Hd)	(%)
1960	8,420	38.6	2,572	42.5	47,104	61.0	7,654	54.0
1965	13,455	45.1	2,351	34.3	46,613	62.9	8,127	62.4
1970	21,014	65.3	1,332	34.0	55,398	68.5	6,986	72.5
1971	22,134	68.6	1,099	32.4	62,956	69.3	7,245	74.0
1972	23,150	72.2	875	31.6	58,686	70.4	7,285	74.3
1973	22,508	73.0	680	30.9	53,462	70.3	6,530	73.0
1974	22,527	69.6	845	30.4	54,653	70.0	6,599	75.1
1975	23,671	65.9	1,447	32.4	48,726	71.6	5,742	74.4
1976	24,715	66.3	1,460	30.0	49,032	71.5	4,925	75.2

Source: USDA (1977) and earlier issues.

have the ability to bargain. In the bargaining process it is imperative to know livestock and the current market situation, but that is not enough. Bargaining involves personality traits, attitude toward the other party, knowledge of trade language, persuasiveness without antagonizing the other party, and ability to leave the impression that the other party obtained full value in the bargain.

In developing the attributes of a livestock marketer, training can be important but does not completely substitute for experience. Market evaluation of quality characteristics changes over time. Demand and supply factors change. To stay abreast of the changing situation requires continuous attention. Many professional marketers upon returning from a vacation will spend time up-dating themselves to the current situation before attempting to carry out transactions.

Analogies seldom portray a situation perfectly. However, producers in day-to-day problems find themselves faced with many decisions about whether to hire a professional or to do the job themselves. Few producers hesitate to make routine carpenter, plumbing, electrical, and mechanical repairs. On the matter of personal health and legal problems, operators almost invariably utilize services of doctors and lawyers. It would be possible under many circumstances for an individual to handle his own legal case, but few attempt it, and few would consider attempting to doctor themselves. The question of doing one's own marketing probably lies somewhere in between these extremes. Just as some individuals can do better than others in carpentering and plumbing, and some can come closer to handling their own legal problems than others, some livestock producers have the knowledge and ability to do their own marketing. Others undoubtedly would be ahead financially by hiring the services of professionals.

Outlets Used in Direct Marketing

For those who desire to use direct marketing, a number of alternative channels are available in both feeder livestock and slaughter livestock.

The types of markets within this group can be subclassified into two categories. On the one hand, operators deliver livestock to an established place of business where the price is generally on a "take it or leave it" basis. This includes direct sale to packing plants, packer buying stations and local markets (the latter includes concentration yards and locally-owned and -operated stockyards). Only limited opportunities are available for bargaining over price at these markets. On the other hand there are outlets where the operators do not have an established place of business. Country packer buyers, country dealers, and order buyers circulate among producers (usually in a given area), visit their premises, view the livestock, and transact business on the premises or elsewhere. Here price is negotiated. Weighing conditions and delivery time are negotiated. If delivery is delayed, written contracts often are drawn up specifying terms of the trade. One producer may deal with another producer in a direct marketing transaction. This often is the case with breeding stock and is a common occurrence with feeder livestock.

These market outlets were defined earlier. At this point a brief commentary is presented on characteristics of those outlets in most prevalent use.

Direct to Packing Plant.—This is a situation where producers haul their livestock to a packing plant, and sale is made upon arrival. The sale may be live grade and weight, or carcass grade and weight. Most packing plants are prepared to buy livestock on a live basis upon delivery. The price is established basically by the plant management. Producers, of course, may call or visit the plant to ascertain paying prices, but bargaining is limited.

Packer Buying Stations.—Packer buying stations simply are outlying yards owned and operated by packing firms as assembly points. Buying stations are competitive devices designed to attract receipts since they are more convenient for producers located in the vicinity of the station. Trucking expense and shrinkage usually are less for short distance hauling. Pricing arrangements at buying stations are similar to those of plant purchases, although discounts may be applied to offset the firm's absorption of trucking and shrinkage, and possible feed expense—depending upon the degree of competition encountered with other packers.

Local Markets.—Local markets include locally-owned stockyards and concentration yards. Historically, a difference existed in that concentration yards were granted certain privileges by railroads in assembling and forwarding hogs by rail.

Concentration yards developed primarily in the Corn Belt, but with the advent of truck transportation, the importance of concentration yards has declined. It is worth noting, however, that the granting of through-rail rates on livestock which are unloaded at a point, sorted, reloaded and forwarded to another point can on occasion be an important consideration. In transportation parlance this is known as "transit privilege" or feed-in-transit privilege. It is applied also to products other than livestock. Grain may be unloaded, stored for a time, and/or milled into various products, the products reloaded and shipped to another destination at an additional charge which is the balance of the through-rate. Through-rate is that which would have applied had the grain moved directly from point of original loading directly to the final destination. Without the transit privilege the sum of the two rates (from origin to point of unloading and from this point to final destination) invariably is greater than the through-rate. Transit privilege is applicable to cattle, which may be unloaded for grazing, reloaded at the end of the grazing season and forwarded to another point (Riley et al. 1952). In this case the balance of the through-rate applies just to the original weight. Gain added from grazing must pay the applicable flat rate from point of reloading to final destination. While its use is not as prevalent as in former years, this can amount to important saving in freight charges where applicable.

Locally-operated stockyards apparently are more prevalent in southern, southeastern, and eastern states than elsewhere. However, many auctions in other parts of the country act as local markets on off-auction days. For example, a day or two of the week may be specified for the purchasing of hogs. The hogs are bought outright by the auction operator upon delivery. The auction operator may or may not have a standing agreement for sale to a packer. If he has such an agreement, he is performing the function of a packer buying station. A Kansas study (Beaton and McCoy 1970) showed that local market (and dealer) operations of auction firms were just about a break-even proposition from a profit standpoint.

Country Packer Buyers.—Many, and perhaps most, of the larger packers employ buyers who travel designated areas buying livestock directly from farms and commercial feedlots. Some larger packers have well organized and equipped buying systems. Some have two-way radio communication with a head buyer's office so that instant instructions may be relayed regarding current market conditions, plant needs, relative availability of livestock as observed by various buyers and other contacts, and pricing instructions. Such a system permits a high degree of coordination in procurement objectives within a firm.

The development and growth of commercial feedlots provide a concentration of livestock which attracts packer buyers. Small producers are at a relative disadvantage not only in the reluctance of country packer buyers to visit their yards, but with only a few head their bargaining power is less than that of large operators.

Country Dealers.—Country dealers usually are one-man operations, although a few are regional and even inter-regional in scope. Livestock are bought outright, the dealer taking title and attempting to resell at a profit. Smaller dealers operate in a relatively local area without an established place of business; however, holding and sorting pens are typical. Dealers are found in all parts of the United States. In various parts of the country localized terminology may label them as truck buyers, traveling buyers, traders, speculators, scalpers, or pinhookers. In spite of the somewhat derogatory connotation to some of these labels, country dealers have survived all the changes that have occurred in the development of the U.S. marketing system. Dealers serve the function of assembling livestock—bringing together odd head and small lots into larger, more uniform, more marketable lots. The larger regional and inter-regional operators, of course, deal in numbers by the carlot.

In early U.S. history, relatively large lots were needed for droving. With the development of railroads, larger lots (carlots) could be shipped at lower rates than less-than-carlots. Dealers thrived in these circumstances. In a countermove, producers formed shipping associations whereby several producers could ship together in carlots and qualify for the lower rates. Development of truck transportation quickly changed this situation. Shipping associations waned. Dealers, however, simply changed with the times. Many purchased trucks and either went into strictly hauling operations or a combination dealer and hauling operation. With the expansion of auction markets, local dealers added auction buying and selling to their country operations. As is the case with terminal market dealers, these country dealers provide a market at all times. They tend to keep geographically separated markets in line by arbitrage operations. Dealers constantly are looking for opportunities to buy in one market and resell in another. One market will not remain out-of-line long before dealers start buying in the low-priced market and reselling in the high-priced market.

Dealers handle all classes of livestock, but their activity is substantially greater in feeder livestock (particularly feeder cattle and feeder lambs) than in slaughter livestock. All types of markets are used by dealers in disposing of their purchases. They sell in the market which appears most advantageous whether it be a producer, order buyer, local market, auction, terminal, or packer.

Many studies have been made by researchers in various states on dealer operations. Unfortunately, no comprehensive national studies have been made and most of the state reports are out-of-date. A regional study covering the North Central Region in 1956 (Newberg 1963) provides the latest information on dealer activities in the midwest. At that time there were 5401 dealers in the North Central Region—down from 9880 in 1940. This study showed that ". . . most dealers had volumes below 2000 animal units per year and half of them had volumes of less than 1000 animal units per year" (Newberg and Hart 1963).

Many dealers carry on some order buying along with their dealer activities and those with facilities mix some local market operations in the business. Newberg and Hart (1963) showed that dealers commonly had other occupational interests with farming the most common, trucking second, and auctioneering third.

Livestock dealers who handle livestock which moves in interstate trade are under regulations by the Packers and Stockyards Act. They must register with the USDA and file bond when dealing on their own account. Dealers who maintain scales for livestock which moves interstate must operate those scales and have them tested as any other market under P&S. In addition, dealers are prohibited by P&S regulations from using unfair, discriminatory, or deceptive trade practices and they must maintain records of transactions carried on for their own account (USDA 1963).

Order Buyers.—Order buyers buy livestock but do not take title. They merely act as agents in buying for another party and receive reimbursement in the form of a fee (or commission). Order buyers, therefore, provide an outlet for direct sales by producers.

Data are not available on the extent of producer sales to order buyers. In former years it probably was almost nil in slaughter livestock. In feeder cattle and feeder lambs order buyers have long played an important role in furnishing a connecting link in market channels. With the recent expansion in commercial cattle feedlots and a similar, but less extensive, trend in hog feeding, it is probable that order buyers are buying directly from these operators just as are packer buyers. The trend in size of feeding units and the tendency for these operators to market direct indicates that order buyers will of necessity go directly to producers to fill their orders.

Feeder cattle and feeder lamb order buyers for years have dealt directly with producers, especially in the range states. While data are not available on the extent of this activity, the increasing demand for feeder cattle will tend to expand the use of this outlet by producers in the future.

Other Producers.—On a local basis, direct sales among producers have always been common practice. In former years it was not a customary practice for farm cattle feeders to go beyond their local area in buying replacement stock, although some larger operators did travel to range areas and deal directly with ranchers. The increased demand for feeder cattle resulting from commercial feedlot expansion has precipitated changes in feeder cattle procurement by both commercial feedlot operators and farm feeders. Commercial feedlot operators attempt to buy directly from producers as well as through other channels. Most farm feeders depend upon order buyers, but some significant moves are under way.

The American Farm Bureau in 1960 established an affiliate—the American Agricultural Marketing Association. Then individual states at their option set up state marketing associations. The purpose is ". . . to assist producers of agricultural commodities to organize their market power, when the need and desire exists. The American Farm Bureau Federation established this affiliate because it recognizes that the marketing of farm products by negotiating contracts with the buyer of the agricultural products will be a dominant factor in the future" (Shuman 1963). Early efforts of this organization were directed at fruits and vegetables. As experience was gained it moved into some areas of livestock marketing. It is not known to have effected contracts in livestock marketing at this time, but it has organized a feeder cattle marketing program. For example, the Kansas Agricultural Marketing Association, through contracts with affiliates in southeastern states, buys and transports feeder cattle to Kansas for its members. This type of activity, and other modifications, are likely to increase as the trend toward direct marketing increases. The National Farmers Organization operates collection points and acts as bargaining agent for its members.

COOPERATIVE LIVESTOCK MARKETING

Cooperative marketing associations are recognized as a special form of business organization whereby producers own and operate the business for their mutual benefit. Two factors have been largely responsible for the development of cooperatives: dissatisfaction with services available from existing market agencies, and dissatisfaction with marketing charges of existing agencies. In 1922, the U.S. legislature enacted the Capper-Volstead Act in recognition of farm producers' disadvantage with other sectors of the economy in the marketing process. Among other things, this Act made it clear that farm producers had the right to organize cooperative marketing associations without violating existing antitrust laws.[3] A number of states also have statutes recognizing cooperatives as a special form of business activity. Cooperatives may be incorpo-

---

[3]Clodius (1962) and Nourse (1962) point out that the mere fact of being a cooperative does not protect an association from antitrust laws if the association unduly enhances price by trade restraint activities.

rated or nonincorporated as is the case with other businesses. In a cooperative, however, the profits (referred to by co-ops as savings) ordinarily are distributed back to the members on the basis of patronage. This is in contrast to other forms of business which distribute profits on the basis of ownership or investment.

Long before enactment of the statutes authorizing cooperatives, farmers had banded together in cooperative ventures, including several forms of livestock marketing cooperatives. In colonial days, co-ops were formed for importation of purebred cattle. Droving often involved informal cooperation.

Shipping Associations

The first cooperative shipping associations were formed during the early 1880's to counteract excessive margins taken by local dealers, and as a means of assembling livestock in lots large enough to gain the advantage of carlot rail freight rates. By combining shipments, scattered local producers could qualify for carlot rates. In addition, this allowed them to send their livestock to the more competitive terminal markets. This proved to be a sound educational process as producers became aware of market conditions beyond their immediate locality.

The formation of shipping associations spread throughout the heavy producing region of midwestern states, reaching a peak during the 1920's. For the most part they were concerned with assembly and forwarding stock to terminal markets. They were not engaged in buying and selling. Livestock in the shipments was consigned to established private commission firms for sale at the terminals. Opposition was encountered at some terminals as sorting and maintenance of identity for individual shippers presented some problems. But this was not a controlling factor. The movement was reversed by technological developments which largely removed the need. Truck transportation and road improvements opened the way for smaller shipments. Improved communications permitted wider and more accurate dissemination of market news. Development of artificial refrigeration and improved transportation were instrumental in decentralization of the packing industry. Cooperative shipping associations as originally organized and operated are largely a thing of the past. However, from a functional standpoint the National Farmers Organization assembly points are performing some of the original services.

Terminal Market Commission Firms

Another achievement in cooperative livestock marketing was the formation of terminal commission firms. Part of the stimulus came from the resistance of established agencies to shipping association activity. In addition, there was dissatisfaction with commission charges and distrust of commission firm allegiance. With shipping associations as a source of shipments, it was logical for producers to carry integration an additional step and organize their own terminal commission agency.

Initially, the opposition was intense and several early attempts failed. Established commission firms opposed cooperatives due to fear of losing business. Yard dealers objected because the cooperatives formed stocker and feeder divisions which completed movement of one producer's stock back to another producer without movement through a dealer. Feelings were high enough that many packer and order buyers boycotted the early cooperative commission firms.

Some success had been attained by World War I. The Equity Cooperative Exchange had been established at South St. Paul and the Farmers Union at several markets, but the breakthrough occurred during the 1920's. A conference called by the American Farm Bureau Federation in 1921 resulted in a marketing strategy which, among other things, called for the establishment of terminal market commission firms with stocker and feeder divisions, shipping associations, and a national livestock producers' organization. An outgrowth of this was formation of the National Live Stock Producers Association. Affiliated commission agencies with stocker and feeder divisions were organized at a number of terminals. Some mergers have occurred among the affiliates and some have ceased operation, but most still are in operation. Other cooperative commission firms also are in operation. While opposition has not entirely disappeared, the cooperatives generally are accepted and in many cases are the most enthusiastic advocates of terminal markets. The number of co-op commission firms reached a peak in the early 1930's. The decline since that time is associated primarily with the decline of livestock marketing at terminal, and that also has affected non-cooperative commission firms. In spite of the decline, cooperatives still are a viable factor at terminal markets. Ward et al. (1978) reported that 10% of commission firms at terminal markets in 1975 were cooperatives and they handled 20% of the livestock at those markets. In some instances, the terminal agencies have branched out into local operations. An example of this is the Producers Livestock Marketing Association, a merged association of terminal commission firms in St. Joseph, Mo., and Omaha, Neb., which also operates an auction market in central Nebraska. Another such example is the Producers Live Stock Association of Columbus, Ohio, which represents a merger of three terminal commission firms and has a number of country branches operating auctions and local markets.

Meat Packing

From time to time cooperatives have ventured into meat packing. Fox (1957) reported that of 17 attempts between 1914 and 1920 all failed.

Failures were attributed to ". . . lack of operating capital and member support, poor facilities, inadequate volume of livestock, inexperienced and unskilled management, keen competition and unsatisfactory sales outlets." While a number of later attempts failed, there have been some notable successes. Shen-Valley Meat Packers Inc. has been in continuous operation since 1949 as has Farmland Industries, Inc. since 1959. Farmland has a history of steady growth and development. It is one of the top 12 meat companies in the United States (Leith 1977). Operations of Farmland include both pork and beef slaughtering and processing facilities.

Other Cooperative Activities

A number of livestock auctions, concentration yards, buying stations, country commission and dealer operations are carried out by producer cooperative associations. Cooperatives have assumed leadership in use of livestock teleauctions in the United States. In 1976 there were 17 cooperatively operated feedlots in 7 states, and a substantial number of cooperative sow farrowing units located primarily in midwestern and plains states (Ward et al. 1978). The American Farm Bureau Federation, through its state affiliates, is engaged in feeder cattle procurement for its members. National Farmers Organization operations, already noted as a bargaining association, is another example of cooperative livestock marketing.

An area of growth in cooperative livestock marketing in recent years has been specialized feeder cattle and feeder pig sales. All parts of the country have experienced growth in this area, but southeastern states have been particularly active. In 1967, 200 southeastern auctions were engaged in cooperative livestock activities. The principal activities of 182 were marketing livestock, and 138 of these were devoted exclusively to specialized feeder cattle and feeder pig sales (Haas 1970).

Most specialized feeder livestock sales use the auction method where livestock is pooled and ownership is commingled. In pooling, animals are sorted by weight, grade, and other physical characteristics. This makes possible the sale of larger lots of uniform quality. Producers are paid for the weight of their particular animals from the pooled price received for the entire lot.

Cooperatives have a long, and somewhat turbulent, history of marketing wool in the United States. Early attempts date back to the 1840's (Ward et al. 1978). After a number of early failures a period of successful operations brought 40% of the U.S. shorn wool through cooperatives during World War II. Wool marketing is carried out primarily by local pools which in turn are affiliated with regional cooperatives most of which operate on pooling arrangements.

The Future for Cooperative Livestock Marketing

Cooperatives never have attained a dominant status in marketing of livestock on a national scope. However, in localized areas cooperative influence has been felt and the effect has been beneficial to producers. Early shipping associations enhanced the bargaining position of isolated producers in relation to local dealers and local market operators. At the same time they provided some relief in transportation costs. Upon passage of the Packers and Stockyards Act (1921) terminal market commission firms supported enforcement of provisions which alleviated unfair and discriminatory practices. These same agencies were instrumental in providing production and marketing credit, fostering improvements in market news services and grade standards, sponsoring research, and in carrying out various educational activities designed to improve quantity and quality of livestock produced.

From the standpoint of marketing efficiency, cooperatives have the dual objective of improving both operational and pricing efficiency. Results of improved operational efficiency are easier to measure and appear to have been the chief pursuit of most cooperative endeavors. This has been in the form of reduced marketing costs—coming back to producers as patronage refunds. Mehren (1965) has stated that "probably the three keys to co-op success have proved to be: keen sensitivity to changing needs of the market; willingness of members to unite in a common front and commit themselves to a common program of action; aggressive, alert, efficient farmer leadership and professional management."

Ward (1977) suggests an integrated cattle marketing system where "The cooperative performs two interrelated functions: (1) assists cattlemen who want to integrate forward as far as feeder cattle growing, cattle feeding, or cattle slaughtering, and (2) markets cattle and beef for its members." This proposal provides for flexibility on the part of the producer in deciding whether, when and how far to integrate his individual operation, but places marketing responsibility in the cooperative.

Success in the future will depend upon how well changes are anticipated and how well action programs are implemented which benefit producer members.

BIBLIOGRAPHY

ABEL, H., and BROADBENT, D. A. 1952. Trade in western livestock at auctions. Utah Agr. Expt. Sta. Bull. *352*.

ARMSTRONG, J. H. 1976. Hog marketing now and in the future. Swine Conference of the American Farm Bureau Federation, St. Louis, Mo. Jan.

BEATON, N. J., and McCOY, J. H. 1970. Economic characteristics of Kansas livestock auctions. Kansas Agr. Expt. Sta. Bull. *537*.

BLACK, W. E. and UVACEK, E. 1972. Alternative systems for marketing Texas livestock. Texas Agr. Ext. Serv. Food and Fiber Economics, Vol. 1, No. 7, Oct.

BOWEN, C. C. and THOMAS, P. R. 1970. Auction selling of slaughter cattle on a carcass basis. Ohio Coop. Ext. Serv. Bull. 510.

BROADBENT, E. E. 1970. Are we willing to adjust? *In* Long-Run Adjustments in the Livestock and Meat Industry: Implications and Alternatives, T. Stout (Editor). Ohio Agr. Res. Develop. Center, Res. Bull. *1037* Also North Central Regional Publ. *199*.

BROADBENT, E. E. and PERKISON, S. R. 1971. Operational efficiency of Illinois country hog markets. Illinois Agr. Expt. Sta. Bull. 110.

CHAMBLISS, R. L. and BELL, J. B. 1974. Selling feeder cattle in commingled lots: a pilot study of pooling in livestock auction markets. Virginia Polytechnic Institute, Agr. Econ. Res. Rept. 12.

CLODIUS, R. L. 1962. Lesson from recent anti-trust decisions. J. Farm Econ. Dec., 1603–1610.

CRAMER, C. L. 1958. Why the early week market? Missouri Agr. Expt. Sta. Bull. *712* Also North Central Regional Publ. *91*.

ECKERT, A. R. 1965. Prices, motivations and efficiency in price determination in Nebraska livestock auction markets. Univ. Nebraska Dept. Agr. Econ. Rept. *40*.

ENGELMAN, G. *et al.* 1953. Relative accuracy of pricing butcher hogs on foot and by carcass weight and grade. Minn. Agr. Expt. Sta. Tech. Bull. *208*.

FARRIS, D. E. and COUVILLION, W. C. 1975. Vertical coordination of beef in the South. Southern Cooperative Series, Bull. 192.

FARRIS, D. E. and DIETRICH, R. A. 1975. Opportunities in cattle marketing. Texas Agr. Expt. Sta. Dir. 77-1, SP-1.

FORKER, O. D. *et al.* (Editors). 1976. Marketing alternatives for agriculture, is there a better way? Cornell University, Nat'l Public Policy Educational Committee Publication No. 7, Nov.

FOWLER, S. H. 1961. The Marketing of Livestock and Meat, 2nd Edition. Interstate Printers & Publishers, Danville, Ill.

FOX, R. L. 1957. Farmers' meat packing enterprises in the United States. USDA Farmers Coop. Serv. Gen. Rept. *29*.

FOX, R. L. 1965. Livestock and wool cooperatives. *In* Farmer Cooperatives in the United States. USDA Farmer Coop. Serv. Bull. *1*.

GURSS, G. D. 1977. Annual report of public livestock markets in Kansas. Kansas Animal Health Dept. Topeka, Kan. June.

HAAS, J. T. 1970. Livestock cooperatives in the southeast. USDA Farmer Coop. Serv. Res. Rept. *13*.

HAAS, J. T. *et al.* 1977. Marketing slaughter cows and calves in the Northeast. USDA Farmer Coop. Serv. FCS Res. Rept. *36*.

HAWKINS, M. H. *et al.* 1972. Development and operation of the Alberta hog producers marketing board. Univ. of Alberta, Dept. Agr. Econ. and Rural Soc. Bull. 12.

HOLDER, D. L. 1974. A tele-o-auction for marketing lambs. USDA Farmer Coop, Serv. FCS Special Report No. 4.

HOLDER, D. L. 1977. Cooperative marketing alternatives for sheep and lamb producers. USDA. Farmers Coop. Serv. Marketing Res. Rept. No. 1081. Aug.

JOHNSON, R. D. 1972. An economic evaluation of alternative marketing methods for fed cattle. Nebraska Agr. Expt. Sta. SB 520.

LEITH, W. G. 1977. Livestock and meat marketing—a challenge for cooperatives. Kansas Cooperative Council Meeting, Topeka, Kans. March.

LOWE, J. C. 1968. Hog marketing by teletype. Manitoba Dept. of Agr. Publ. No. 471. Winnipeg. Oct.

McCOY, J. H. et al. 1975. Feeder cattle pricing at Kansas and Nebraska auctions. Kansas Agr. Expt. Sta., Bull. 582.

MEHREN, G. L. 1965. Potential for cooperatives in livestock. Proc. Stockholders' and Directors' Meeting Natl. Live Stock Producers Assoc., Chicago, March.

NAIVE, J. J. et al. 1957. Accuracy of estimating live grades and dressing percentages of slaughter hogs. Indiana Agr. Expt. Sta. Bull, 650.

NEWBERG, R. R. 1959. Livestock marketing in the North Central Region. I. Where farmers and ranchers buy and sell. Ohio Agr. Expt. Sta. Res. Bull. 846. Also North Central Regional Publ. 104.

NEWBERG, R. R. 1963. Livestock marketing in the North Central Region. II. Channels through which livestock moves from farm to final destination. Ohio Agr. Expt. Sta. Res. Bull. 932. Also North Central Regional Publ. 141.

NEWBERG, R. R., and HART, S. P. 1963. Livestock marketing in the North Central Region. IV. Livestock dealers and local markets. Ohio Agr. Expt. Sta. Res. Bull. 962. Also North Central Regional Publ. 150.

NOURSE, E. 1962. Lessons from recent anti-trust decisions. J. Farm Econ. Dec., 1614–1623.

PHILLIPS, V. B., and ENGELMAN, G. 1958. Market outlets for livestock producers. USDA Agr. Marketing Serv., Marketing Res. Rept. 216.

POWERS, M. J., and BENDT, D. R. 1968. Livestock marketing in the Upper Missouri River Basin. II. The Sioux City Stockyards—facilities and costs of operation. S. Dakota Agr. Expt. Sta. Bull. 548. Also North Central Regional Publ. 188.

RILEY, H. M. et al. 1952. Feed-in-transit privilege in marketing livestock. Kansas Agr. Expt. Sta. Circ. 288.

SHUMAN, C. B. 1963. Best interest of farmers lies in working to form own supply-management program. In Selected Papers on Marketing. Am. Meat Inst., Chicago.

STOUT, T. T., and FELTNER, R. L. 1962. A note on spatial pricing accuracy and price relationships in the market for slaughter hogs. J. Farm Econ. 44, Feb., 213–219.

USDA. 1935. The direct marketing of hogs. USDA Bur. Agr. Econ. Misc. Publ. 222.

USDA. 1963. The Packers and Stockyard Act as it applies to livestock dealers. USDA, P&S Admin., Agr. Marketing Serv. 319, Revised June 1963.

USDA. 1970. Packers and stockyards resume. USDA, P&S Admin. 8, No. 13, Dec.

USDA. 1972. Pork marketing report—a team study. USDA Farmer Coop. Serv. (unnumbered), Sept.

USDA. 1977. Packers and stockyards resume, USDA, Packers and Stockyards Administration, Vol. 15, Dec.

USDA. 1978. Meat animals production, disposition, income. USDA, Econ. Stat. and Co-op. Serv. MtAn. 1–1 (78) April.

VIA, J. E. and HAAS, J. T. 1976. Increasing efficiency in a country hog marketing system. Maryland Agr. Expt. Sta. MP 883.

WARD, C. E. 1977. Contract integrated, cooperative cattle marketing system. USDA. Farmer Co-op. Serv. Marketing Res. Rept. No. 1078.

WARD, C. E. et al. 1978. Livestock and wool cooperatives. USDA. Farmer Coop. Serv. Bull. I.

WOHLGENANT, M. K. and GREER, R. C. 1974. An evaluation of selling alternatives for Montana pork producers. Montana Agr. Expt. Sta. Bull. 675.

WILLIAMS, W. F., and STOUT, T. T. 1964. Economics of the Livestock-Meat Industry. Macmillan Co., New York.

Meat Packing and Processing

7

EARLY DEVELOPMENT

Chapter 2 reviewed briefly some developments that helped shape the structure of the meat packing industry. Among the more important were development of railroads and refrigeration. The railroad system made possible the assembly of large concentrations of livestock at terminal markets. It was logical that packing plants should also locate at these points. Refrigeration and use of refrigerated rail cars opened the way for year-round slaughtering and distribution.

Two additional early developments are worthy of note: (1) development of mechanical power, and (2) public assent to the corporate type of business organization. Mechanical power gave firms the mechanics of large-scale operation, while the corporate business structure gave them the economic power to organize, finance, and perpetuate large-scale operations.

Thus, it is not surprising that in the development period meat packing evolved as a highly concentrated industry. Many other early American industries developed along essentially the same pattern. The "captains of industry" were adept at both technological and economic innovations. Among the latter were trusts, combinations, mergers, pools, and holding companies. While the basic U.S. business and legal philosophies were oriented toward a highly competitive free enterprise economy, individual interests were directed toward developing imperfections in the competitive model. Individual objectives centered on gaining an advantage over competitors, i.e., to compete the competitors out of existence. This was (and still is) construed by the general public as a trend toward monopoly, and hence, as inherently undesirable. A considerable body of literature exists on these early developments and public reaction as expressed by laws, regulations, and various "movements" (Dewey 1959; Dirlam and Kahn 1954; Kaysen and Turner 1959; Watkins 1927; Wilcox 1960). Among the more important public enactments were the Interstate Commerce Act, 1887; Sherman Antitrust Act, 1890; Clayton Act, 1914; Federal Trade Commission Act, 1914; Packers and Stockyards Act, 1921; Robinson-Patman Act, 1936, and others.

The meat packing industry was subjected to criticism and investigation along with others. The U.S. Senate in 1888 authorized a study to determine possible monopolistic elements in the transportation, production, and distribution of meat. This investigation confirmed alleged collusion and other monopolistic practices. However, no direct corrective action was taken. In 1904, Congress authorized an investigation to determine whether monopolistic practices were holding cattle prices down relative to beef prices. In general, the results of this study indicated no undue profits accruing to packers. No action was taken in antimonopoly aspects, but a by-product of this investigation was the revelation that unsanitary conditions existed in some packer operations. As a result, Congress enacted the Federal Meat Inspection Act in 1906.[1]

With the onset of World War I, food prices rose relative to livestock prices, and packer profits increased. Dissatisfaction was intense on the part of both consumers and livestock producers. President Wilson authorized the Federal Trade Commission in 1917 to investigate the food industry, including meat packing. The report, issued in 1919, revealed that the so-called "big five" packers (Armour, Cudahy, Morris, Swift, and Wilson) held a dominant position in meat packing. Engelman (1975) states, "In 1918 the 'Big Five' accounted for about three-fourths of the cattle slaughtered at Federally-inspected plants, over 72 percent of the sheep, and 53 percent of the hogs." That was construed by the Commission to be prima facie evidence of monopolization of the industry. The Commission recommended that the monopoly be broken up, and proposed various steps for doing this.

As a result of this study, the U.S. Department of Justice in 1919 began proceedings for prosecution of the "big five" under provisions of the Sherman Act. While the Department of Justice was preparing the case, the packers, in conference with the Attorney General, settled the case out of court in 1920 in what is known as the Packers Consent Decree. In arriving at the consent decree, the packers did not admit guilt to monopolistic practices but, nevertheless, agreed to certain provisions designed to prevent monopolistic control of packing and distribution of meat. The provisions were comprehensive and restrictive, including, among other things, that these packers would: (1) divest themselves of physical facilities in public stockyards, stockyard railroads, market newspapers, and public cold storage facilities; (2) not engage in the retailing of meat and a large number of additional specified commodities; and (3) submit to perpetual jurisdiction of the U.S. District Court whereby the court could add provisions, if necessary, to carry out the intent of the decree in abolishing monopolistic control of these packers.

Another outgrowth of the investigation was enactment by Congress of the Packers and Stockyards Act in 1921. Subsequently the packers have, from time to time, sought relief from consent decree provisions. Minor

---

[1]This was not the first indication of public interest in sanitation in the meat industry. The Meat Inspection Act of 1890 provided for inspection of meat products entering export trade but was inadequate for comprehensive inspection.

relief was initiated in 1971 when the Department of Justice granted freedom from restrictions on manufacturing and wholesaling certain items previously prohibited.

CLASSIFICATION AND NUMBER OF MEAT PACKING AND PROCESSING PLANTS

Classification

Meat packing and processing plants are classified in a number of ways. For some purposes they are classified as "Federally inspected" (FI), or "non-Federally inspected" (non-FI). Non-FI packing plants are further subclassified according to size as large, medium or small. Livestock slaughtered by farmers is classed as "farm slaughter." All other is classed as "commercial." Packers also are classified, for some purposes, on the geographical extent of plant location. National packers are firms with plants located generally throughout the country or at least in several regions. Regional packers are those with plants generally in a given region. It is obvious that all classifications are not mutually exclusive. National packers may also be large *and* FI. Commercial slaughter includes FI *and* large, medium, and small non-FI plants. Prior to 1966 custom slaughtering done for farmers was included in farm slaughter; since 1966 it has been classed as commercial slaughter.

Type of Inspection.—Federally inspected plants are those at which federal employees make various checks and tests designed to provide the consuming public with wholesome meat and meat products. In the interests of general welfare, the U.S. Congress, in 1906, enacted the Meat Inspection Act providing for federal inspection.[2] All packers who ship meat in interstate or international trade must be under federal inspection. All meat purchased by the federal government for military personnel and for welfare programs must be from FI plants. Non-FI plants are limited in meat shipments to trade within the borders of the state where located. Individual states are responsible for inspection of meat moving in intrastate trade. Some municipalities have separate inspection.

Under provisions of the U.S. Wholesome Meat Act of December 1967 each state was required to either set up inspection specifications equivalent to federal standards or submit all plants to federal inspection. Prior to this act, a considerable variation existed in state inspection standards and their enforcement among states.

---

[2]The first Meat Inspection Act was passed in 1890. It was aimed primarily at meat for export and did not compare in effectiveness with the Act of 1906.

Data on FI plants have always been more complete and uniform than on non-FI plants due to centralized assembly (in Washington, D.C.) of required reports. This probably will continue to be the case. However, the new law necessitates more complete information on state inspected plants than formerly.

Subclassification by Size.—Most FI plants are relatively large, but a substantial variation exists in size of non-FI plants. Subclassification of non-FI plants by size is based on an arbitrarily selected set of size intervals, or ranges, or total live weight slaughtered annually. The intervals are as follows: more than 2,000,000 lb—large; 300,000 to 2,000,000 lb— medium; less than 300,000 lb—small.

Information currently available on small plants is less complete than for the others. While small plants consitutue ½ of all plants in number, they slaughter only 1% of the livestock. Many of them are locker plants.

In previous years the term "wholesale" packer was used where "large" packer now is used, and "local" packer was used where "medium" packer now is used. The term "small" has been used consistently for a number of years.

Number of Slaughter Plants

Table 7.1 is a tabulation of the number of slaughter establishments in the United States from 1965 to 1978. In that period the total number of establishments (FI plus non-FI) declined 20%. FI plants more than doubled. It is likely that at least part of this change resulted from non-FI plants switching to federal inspection following enactment of the U.S.

TABLE 7.1.   NUMBER OF U.S. SLAUGHTER ESTABLISHMENTS, 1965, 1970, ANNUAL 1975–1978

Year	Federally Inspected	Non-Federally Inspected			Total Non-Federally Inspected	All Plants
		Large	Medium	Small		
1965	571	855	1522	4750	7157	7728
1970	726	567	2596	3845	7008	7734
1975	1485	na	na	na	4602	6087
1976	1741	na	na	na	4515	6255
1977	1687	na	na	na	4440	6127
1978	1750	na	na	na	4434	6184
Change 1965 to 1978 in number	1179	—	—	—	−2723	−1544
Change 1965 to 1978 in percentage	206.5	—	—	—	−38.0	−20.0

Source: USDA (1978A) and earlier issues.

Wholesome Meat Act of December 1967. This act, among other things, tended to tighten inspection standards on non-FI plants.

Location of Slaughter Plants

Slaughter plants are distributed roughly in proportion to the density of livestock production. However, an exception exists in the location of lamb slaughter plants. The heaviest lamb production is in range areas, but the bulk of slaughter facilities is in eastern regions. One reason is that to comply with Kosher requirements lambs must be slaughtered relatively close to the point of consumption.

Volume of Livestock Slaughtered

Tables 7.2 and 7.3 give the number of head slaughtered of various species. Attention is called to the substantial increase (amounting to 95%) in cattle slaughter. Calf slaughter declined 70% in the 12 yr period 1961–1973, then increased sharply as grain finishing declined. This reversal is not expected to continue. Sheep and lamb slaughter has declined persistently, while hog slaughter has held to a fairly steady trend—aside from cyclical ups and downs.

Table 7.4 shows that FI packers' slaughter was 93.9% of all livestock slaughtered for the commercial trade in 1977. Federally inspected slaugh-

TABLE 7.2.   CATTLE AND CALF SLAUGHTER: NUMBER BY CLASS OF SLAUGHTER, 48 STATES, 1961–1976

	Cattle					Calves				
	Commercial					Commercial				
Year	FI (1000 Hd)	Other Non-FI (1000 Hd)	Total [1] (1000 Hd)	Farm (1000 Hd)	Total (1000 Hd)	FI (1000 Hd)	Other Non-FI (1000 Hd)	Total [1] (1000 Hd)	Farm (1000 Hd)	Total (1000 Hd)
1961	19,968	5,666	25,635	836	26,471	5,005	2,696	7,701	379	8,080
1962	20,339	5,745	26,083	828	26,911	4,980	2,515	7,494	363	7,857
1963	21,662	5,570	27,232	838	28,070	4,535	2,298	6,833	371	7,204
1964	25,133	5,685	30,818	860	31,678	4,820	2,434	7,254	378	7,632
1965	26,614	5,733	32,347	824	33,171	5,076	2,344	7,420	368	7,788
1966	27,319	6,408	33,727	444	34,171	4,432	2,215	6,647	214	6,861
1967	27,780	6,089	33,869	426	34,295	4,002	1,917	5,919	188	6,107
1968	29,592	5,434	35,026	388	35,414	3,876	1,567	5,443	170	5,613
1969	30,537	4,700	35,237	339	35,576	3,637	1,226	4,863	146	5,009
1970	30,793	4,232	35,025	329	35,353	3,024	1,048	4,072	131	4,203
1971	31,419	4,166	35,585	310	35,895	2,806	883	3,689	132	3,821
1972	32,267	3,512	35,779	304	36,083	2,421	632	3,053	131	3,184
1973	30,521	3,166	33,687	340	34,027	1,808	441	2,249	127	2,376
1974	33,319	3,493	36,812	515	37,327	2,355	633	2,987	185	3,172
1975	36,904	4,008	40,911	553	41,464	3,894	1,316	5,209	197	5,406
1976	38,992	3,663	42,654	542	43,196	4,438	912	5,350	178	5,528

Source: USDA (1977).
[1]Totals based on unrounded data.

TABLE 7.3.   HOG AND SHEEP AND LAMB SLAUGHTER: NUMBER BY CLASS OF SLAUGHTER, 48 STATES, 1961–1976

| | Hogs | | | | | Sheep and Lambs | | | | |
| | Commercial | | | | | Commercial | | | | |
Year	FI (1000 Hd)	Other Non-FI (1000 Hd)	Total [1] (1000 Hd)	Farm (1000 Hd)	Total (1000 Hd)	FI (1000 Hd)	Other Non-FI (1000 Hd)	Total [1] (1000 Hd)	Farm (1000 Hd)	Total (1000 Hd)
1961	65,632	11,702	77,335	4,635	81,970	15,036	2,154	17,190	347	17,537
1962	67,770	11,564	79,334	4,090	83,424	14,692	2,145	16,837	331	17,168
1963	71,577	11,747	83,324	3,793	87,117	13,955	1,867	15,822	325	16,147
1964	71,667	11,352	83,018	3,266	86,284	12,947	1,647	14,595	300	14,895
1965	63,708	10,076	73,784	2,610	76,394	11,710	1,297	13,006	294	13,300
1966	63,729	10,282	74,011	1,314	75,325	11,553	1,184	12,737	267	13,003
1967	70,915	11,209	82,124	1,297	83,421	11,516	1,275	12,791	244	13,034
1968	74,789	10,371	85,160	1,241	86,401	10,888	996	11,884	235	12,119
1969	75,682	8,156	83,838	1,120	84,958	10,070	621	10,691	232	10,923
1970	78,187	7,630	85,817	1,107	86,924	10,010	542	10,522	248	10,800
1971	86,667	7,771	94,438	1,089	95,527	10,256	473	10,729	237	10,966
1972	78,759	5,948	84,707	1,050	85,757	9,905	396	10,301	224	10,525
1973	72,264	4,531	76,795	1,051	77,846	9,234	363	9,597	201	9,798
1974	77,071	4,691	81,762	1,323	83,085	8,556	291	8,847	226	9,073
1975	64,926	3,761	68,687	1,137	69,824	7,552	283	7,835	222	8,057
1976	70,454	3,330	73,784	1,166	74,950	6,474	240	6,714	201	6,915

Source: USDA (1977).
[1]Totals based on unrounded data.

ter would approach this proportion even if farm slaughter were excluded since in 1977 the latter amounted to only 1.4% of total slaughter. In the 13 yr interval, the proportion of all species slaughtered in FI plants increased from 84 to 93.9%. The percentage of calves slaughtered in FI plants (85.1%) is considerably less than other species—cattle 92.5%, hogs 95.8%, and sheep and lambs 96.5%. However, veal constituted only 1.6% of total red meat production in 1977. Data on FI slaughter (since it includes more than $9/10$ of all livestock slaughtered) give a fairly adequate picture of the total slaughter industry for many aspects of market analysis. Due to mandatory reporting requirements and collection of the data in a central office, this information becomes available to market analysts earlier than data on non-FI plants.

TABLE 7.4.   SLAUGHTER BY CLASS OF PLANT AS PERCENTAGE OF TOTAL COMMERCIAL SLAUGHTER, BY SPECIES, 48 STATES, 1964, 1969 AND 1977

| | FI Plants | | | Non-FI Plants | | | Total Commercial | | |
Species	1964 (%)	1969 (%)	1977 (%)	1964 (%)	1969 (%)	1977 (%)	1964 (%)	1969 (%)	1977 (%)
Cattle	83	88	92.5	17	12	7.5	100	100	100
Calves	59	63	85.1	41	37	14.9	100	100	100
Hogs	87	91	95.8	13	9	4.2	100	100	100
Sheep and lambs	89	95	96.5	11	5	3.5	100	100	100
All species	84	89	93.9	16	11	6.1	100	100	100

Source: USDA (1978A) and earlier issues.

DECENTRALIZATION AND DECLINE IN CONCENTRATION

Following World War II a series of developments commenced which led to an alteration of the structural organization of the meat packing industry. A reversal began in the degree of concentration. Simultaneously, a trend began in decentralization. Decentralization and lessening of concentration has taken place by a gradual, but not uniform, dilution of the position held by the big five. The total number of firms (and plants) increased in a roughly cyclical pattern. Most of the increase was by firms other than the big five—leading to a lessening of concentration.

Reasons for Decrease in Concentration

It is possible that the consent decree had some effect on the lessening of concentration. The big five became the big four in 1923 when Armour acquired Morris. But the big four probably became less aggressive in further expansion.

Development of the motor truck and construction of all-weather highways probably were the greatest factors in decentralization of the packing industry. As mentioned in earlier chapters, livestock producers became skeptical of marketing practices at terminals during hard times of the 1920's and especially during the depression years of the 1930's. Packers found that many producers were willing to market direct under these circumstances and with packing plants reasonably close, trucks provided the means of delivery where railroads were previously the only feasible transportation.

Other factors also influenced packers to construct plants at interior points. Approximately ⅓ of the live weight is dropped during slaughtering. Thus, total transportation costs can be reduced by slaughtering close to areas of production. Some other costs also can be reduced. Labor wage rates typically are lower in the more rural areas than in large cities. Eventual unionization usually raises wages at interior points, but until that occurs a plant may realize a labor cost advantage over competitors. Property taxes and utility costs often are less in smaller towns than in large cities, and cost of living often is less for plant employees.

The acceptance of federal grades, especially for fresh beef, has tended to encourage smaller firms which located at interior points. Since branding of fresh meat has not been accomplished to any material extent, smaller new packers were able to compete favorably with larger established firms. Fresh beef is traded largely on U.S. federal grades. Fresh pork is seldom sold on U.S. grades. However, a substantial portion of fresh pork is sold by interior packers to processors located close to consumption areas.

From the standpoint of packing plant technology, the new, smaller plants could compete very well with older, larger plants. Many of the latter were approaching obsolescence, but the big companies were reluctant to modernize. A modern large cattle slaughter plant has some economies of size as compared with a modern small plant, primarily in labor utilization (Logan 1966), but these economies are relatively small beyond a capacity of around 100 head per hour. In actual operation they might be offset by less than optimum achievement in other costs.

Measures of Concentration Trends

As derived from census data, the 4 largest packers, by 1947, accounted for 41% of the value added by all meat packing firms (see Table 7.5). This 41% is not exactly comparable to the 70% determined by the FTC for the big five in a 1918 study (based on 1916 data). Nevertheless, it is close enough in concept to indicate a substantial reduction in concentration.[3]

Table 7.5 shows a continued decline in the degree of concentration since 1947. This holds whether one considers the 4 largest, the 8 largest, or the 20 largest firms. The fraction of value added by the 4 largest packers declined from 41% in 1947 to 19% in 1972. During this period the number of firms engaged in meat packing increased from 1999 to 2833 in 1963, then declined to 1207. Some firms operate more than one establishment (or plant), but for the years for which data are available, the trend in number of establishments roughly parallels that of number of firms at about 1.05 establishments per firm.

Little change occurred in the degree of concentration in meat processing between 1954 and 1972 (see Table 7.5). "Meat processors" by census definition include firms primarily engaged in the manufacturing of processed meat from purchased meat. These firms do little slaughtering. The 4 largest processors accounted for 16% of value added in 1947 and 19% in 1972. This, by most standards, would be classified as a relatively low degree of concentration. When the 20 largest processors are grouped, the same general trend emerges.

Data presented in Table 7.6 show the trend in degree of concentration in the meat packing industry by specie of livestock handled. The declining trend in concentration has been most striking for cattle slaughtering. The largest 4 firms slaughtered 49% of the cattle in 1920 and 19% in 1975. The

---

[3]The FTC study determined percentages on number of livestock, while data in Table 7.5 are percentages of value added in the slaughtering process. Obviously, the two sets of data would be closely correlated.

TABLE 7.5.   CENSUS CONCENTRATION RATIOS FOR MEAT SLAUGHTERING AND PROCESSING FIRMS, INDUSTRY BASIS, VARIOUS YEARS, 1947–1972

Year	No. of Firms	Value Added by Manufacture (Million Dollars)	Percentage of Value Accounted for by Firms Ranking:			
			1–4	1–8	1–20	All Firms
Meat Packing (SIC 2011)						
1947	1999	977	41	54	63	100
1954	2228	1397	39	51	60	100
1958	2646	1749	34	46	57	100
1963	2833	1908	31	42	54	100
1967	2529	2220	26	38	50	100
1972	1207	1099	19	26	38	100
Meat Processing (SIC 2013)						
1954	1254	334	16	24	35	100
1958	1430	442	17	25	36	100
1963	1273	563	16	23	35	100
1972	1207	1099	19	26	38	100

Source: U.S. Dept. of Commerce (1975).

TABLE 7.6.   PERCENT OF U.S. COMMERCIAL LIVESTOCK SLAUGHTER BY THE FOUR RANKING FIRMS IN EACH SPECIES, 1920, 1930, 1940, ANNUAL 1950–1975

Year	Cattle	Calves	Hogs	Sheep
		Percent		
1920[1]	49.0	34.4	43.8	61.8
1930	48.5	45.5	37.5	68.1
1940	43.1	45.6	44.3	66.1
1950	36.4	35.4*	40.9	63.6
1951	32.0	34.6*	40.5	62.9
1952	34.3	36.0*	39.3	63.5
1953	34.4	39.0*	37.9	62.4*
1954	32.4	37.5*	38.7	61.4*
1955	30.8	36.6*	40.6*	61.0*
1956	29.8	37.4*	40.2*	61.5*
1957	29.3*	35.4*	38.7*	58.4*
1958	27.4	32.4*	35.9*	56.6*
1959	24.7	29.8*	33.5*	54.4*
1960	23.5*	29.0*	34.9*	54.7*
1961	24.2*	30.1*	33.7*	54.7*
1962	23.7*	28.2*	34.4*	55.4*
1963	22.9*	29.1*	33.8*	54.5*
1964	22.6*	32.1*	34.9*	56.8*
1965	23.0*	32.4*	35.2*	57.8*
1966	22.4*	30.4*	31.7*	59.0*
1967	22.2*	30.2*	29.8*	58.1*
1968	21.5*	29.0*	30.1*	54.2*
1969	23.0*	27.3*	33.5*	60.4*
1970	21.3*	23.8*	31.5*	53.1*
1971	21.4*	21.6*	31.8*	53.2*
1972	22.3*	21.8*	31.6*	54.7*
1973	22.8*	23.7*	32.9*	51.8*
1974	20.9*	23.5*	32.7*	55.7*
1975	19.3*	24.3*	33.1*	57.5*

Source: USDA (1978C).
[1]Data for 1920 includes the "Big Five" (Armour, Cudahy, Morris, Swift, and Wilson) which became the "Big Four" in 1923 when Armour acquired Morris.
*Includes one or more firms other than the original "Big Four" (Armour, Cudahy, Morris, Swift, and Wilson).

decline in calf slaughter was from 34% to 24%, for hogs from 44% to 33%, and for sheep from 62% to 52%. Concentration in sheep slaughter traditionally has been substantially higher than in other species, and remains relatively high.

The evidence clearly indicates a decrease in concentration at the national level. However, Engelman (1975) points out that, "The market for livestock is quite circumscribed inasmuch as most slaughter livestock is sold out of first hands by the producer to a packer or other buyer located within 50 to 100 miles." Aspelin and Engelman (1976) state that, "The meat packing industry tends to be highly oligopsonistic (oligopolistic) at the state level—much more so than it is nationally. Four ranking firms account for 65 percent or more of slaughter for different species at the state level in most cases . . . Of the concentration shown for the 40 individual states, only in 12 cases did the four-firm ratio drop below 65 percent in steers and heifers, 11 states for cows and bulls, 21 states for all cattle, 5 states for calves, 4 states for hogs, and sheep and lambs—no states. These levels of concentration describe highly oligopsonistic markets in most states."

SPECIALIZATION

There are two aspects to packing plant specialization: horizontal specialization refers to the number of different species slaughtered in a plant, while vertical specialization refers to the extent to which meat is processed in the plant where slaughtering takes place. A plant in which only one specie is slaughtered is considered highly specialized horizontally. If no processing is done, the plant is considered to be highly specialized vertically. A plant which slaughters only hogs and sells the pork only in carcass form is highly specialized in both ways—this is the so-called "kill and chill" pork packing plant.

Horizontal Specialization

In early days it was customary to construct plants in which all species were slaughtered. However, as decentralization of the packing industry progressed there were apparent economies to single specie plants. Anthony (1966) reported that "Horizontal specialization of FI slaughter plants is increasing." While no recent studies have confirmed a continuation of this trend, it is common knowledge that most of the recently constructed plants have been single specie. Economies in construction costs, labor specialization, and sales specialization make it unlikely that the trend will change.

Vertical Specialization

The concept of vertical specialization is similar to that of vertical integration—i.e., the control by ownership or contract over successive stages of production or marketing. Slaughtering and processing[4] are successive stages in meat production. Meat can be processed by a firm in the plant where slaughtering takes place, it can be shipped to a specialized processing plant owned by the same firm, or it can be sold to another firm for processing. The less processing done by a plant, the more specialized is that plant.

Empirical analysis of vertical specialization is limited. In one of the most comprehensive studies, though not recent, Anthony and Egertson (1966) reported that, ". . . a rather high proportion of the slaughter plants were doing only a limited amount of processing. Half of the slaughter plants which engage in meat processing activities processed only an average of 14% of their slaughter production." In a related study Anthony (1966) stated that, "Fifty percent of the plants processed less than ½ of their slaughter output. Ten percent of the slaughter plants processed 50–100% of output. A fairly large number of FI slaughter plants (32%) processed more than they slaughtered; that is, they acquired meat for processing from other plants. The average processing of these plants was 333% of their slaughter."

Studies are not available showing the trend in specialization over a period of time. However, there are economies to be gained in breaking, fabricating, and boning at point of slaughter. Significant increases have occurred in production of boxed beef. A number of pork plants which started as kill and chill operations are now engaged in further processing. The processing of fresh meat in connection with slaughter operations is expected to increase, which will tend to reduce specialization in slaughter plants. There appear to be no economies, however, in manufacturing processes (or prepared meats such as cured, bologna, sausages, etc.) at point of slaughter. This aspect of processing probably will continue to be oriented to locations near heavy consumption areas and will tend to foster a higher degree of specialization in that aspect of processing which we call manufacturing (as contrasted to cutting, breaking, and fabricating of fresh meat).

---

[4]Processing is defined here to include boning, carcass breaking (fabrication), boxing, and other such fresh meat operations. For other purposes processed meat is defined to include only such meat as bologna, luncheon meats, sausages, cured meat, smoked meat, etc., and, in addition, ground (hamburger) and comminuted meat—with only ground and comminuted being in the fresh state. A more explicit comment is given on the definitions of processed meat and processing in Chapter 8.

DIFFERENTIATION OF PRODUCT

Branding and differentiating fresh meat presents a difficult problem. Only limited success has been achieved in this area. Some packers have established cutting and trimming characteristics in wholesale cuts as a basis for differentiating their product. Swift has an enzimatic tenderizing process which it promotes under the Proten trademark. However, the widespread use of federal grades in wholesaling and retailing fresh beef and lamb has largely forestalled the differentiation of these products by individual firms. More than ½ of all beef and lamb slaughtered under federal inspection is officially graded. Only a relatively small proportion of pork is retailed in the fresh state, and at the wholesale level individual specifications are used in preference to U.S. grades.

A different situation exists with processed meats. Here, processors have been able to develop individualized characteristics such as flavor, color, texture, brand-name, and packaging. By advertising and other promotional efforts processors have succeeded in differentiating their products.

Table 7.7 presents data from the National Commission on Food Marketing (1966) study on the extent of branding by packers. The report stated, "Processed items—such as bacon, hams, sausage, and luncheon meats—are labeled with the packer or retailer name. A number of variations exist in product composition, packaging and branding, and these frequently differ among firms. Some packers have promoted their brands enough to influence consumers. As a result, in some sections of the country highly advertised brands sell at higher prices than unadvertised brands.

". . . Two thirds or more of fresh beef, veal, lamb and pork was sold unbranded by packers. But less than 10% of cured pork items and processed meats was sold unbranded by most groups of packers.

"On branded items, the packers' brands predominated. Less than 5% of the cured and processed meat sold by the largest 4 packers carried the customers' labels. However, the second largest 4 packers and other reporting packers sold around 10% of their cured and processed meats under custom labels. Customer branded meats were sold chiefly to large retail chains." There is no evidence of change at the packer level since that study was made.

CONDITIONS OF ENTRY INTO THE MEAT PACKING AND PROCESSING INDUSTRY

Entry into an industry is defined as the initiation of operations by a new firm. Exit from an industry is the discontinuance of operations by an

TABLE 7.7.   PROPORTIONS OF SPECIFIED MEAT ITEMS SOLD BY TYPE OF BRAND, 1954–1965[1]

Firm Group	Seller's Brand	Customer's Brand	Un-branded	Not Indicated
Beef and veal				
4 largest meat packing companies[2]	26.6	3.8	69.6	...
Second 4 largest meat packing companies[3]	34.6	3.2	62.2	....
All other firms reporting (131 firms)	17.8	2.1	69.1	11.0
Lamb and mutton				
4 largest meat packing companies[2]	28.6	...	71.4	...
Second 4 largest meat packing companies[3]	5.6	...	94.4	...
All other firms reporting (19 firms)	6.8	...	93.2	...
Fresh and frozen pork				
4 largest meat packing companies[2]	...	...	100.0	...
Second 4 largest meat packing companies[3]	34.7	...	65.3	...
All other firms reporting (75 firms)	13.3	...	64.9	21.8
Cured hams, picnics, and bacon				
4 largest meat packing companies[2]	86.4	4.9	8.7	...
Second 4 largest meat packing companies[3]	76.2	10.9	12.9	...
All other firms reporting (62 firms)	75.1	11.8	10.1	3.0
Other processed meats				
4 largest meat packing companies[2]	85.8	4.8	9.4	...
Second 4 largest meat packing companies[3]	81.2	9.8	9.0	...
All other firms reporting (77 firms)	77.7	9.0	4.5	8.8

Source: National Commission on Food Marketing (1966).
[1] Firms were requested to report total sales by customer class and brand type for a period of at least 2 months between Nov. 1, 1964, and Apr. 30, 1965. Data furnished were converted to monthly averages. Percentages were computed from totals of firm groups.
[2] Armour, Morrell, Swift, and Wilson.
[3] Hormel, Hygrade, Oscar Mayer, and Rath.

established firm. Firms already established and operating in an industry which is difficult to enter (has high barriers to entry), are hypothesized to behave differently (to have some degree of control over prices and profits) than firms in an industry which is easy to enter. High barriers to entry do not in themselves furnish proof of adverse economic behavior. However, if an industry (meat packing or processing) has high barriers to entry, this would at least provide some grounds for further investigation of behavior and performance of firms in that industry. Livestock producers and consumers both have a stake in this issue just as they do in any other structural characteristic which affects relative bargaining power, prices of livestock, and retail prices of meat.

Conditions of entry and exit in the meat packing and processing industry have not been documented recently. Anthony (1966) investigated the situation for the 1950–1962 period and the National Commission on Food

Marketing (1966) made a similar study for a shorter, but later, and partially overlapping period, 1958–1964. Both studies showed a relatively high rate of exit and entry—indicating relative ease of entry. It is apparent in more recent years that conditions of entry have tightened. A combination of circumstances work in that direction. Costs of construction and operation have increased substantially with the persistent inflationary trend. Packer and processor relative bargaining power with national retail chains has declined. Red tape and costs of complying with recent environmental requirements have provided a disincentive to entry and hastened the exit of some firms.

Data developed in the studies mentioned above indicate that entry into processing is not as easy as entry into meat packing. There are several possible explanations. Processors have been able to brand and differentiate their products more successfully than packers have differentiated fresh meat. A newcomer in processing, therefore, has the problem of winning customers who may have preferences for a particular brand. Another possible factor is that processors in general have a more stable geographic orientation. Many of them are located near areas of heavy consumption which have been relatively stable. Livestock production, on the other hand, and particularly finished cattle production, has shifted considerably and new packing plants have tended to follow.

It should be noted that these analyses included only inspected plants. As was apparent in an earlier section on packing plant classification, the FI plants are considerably larger in average size and slaughter a substantially larger percentage of livestock than do the non-FI plants. In commenting on possible implications of results to the entire industry, Anthony (1966) states, "Far more complete data are available for meat packers operating under federal inspection than for those not receiving this service. As a result, most evidence presented in this report pertains directly to this sector of the meat packing industry and only indirectly to the industry as a whole as it is represented by the federally inspected sector. Because of the importance of the federally inspected sector, the necessary assumption that it is representative is not especially damaging. Nevertheless, the conclusions possible from the evidence presented are limited." A comparable qualification is appropriate for conclusions regarding the meat processing industry.

INTEGRATION IN THE PACKING INDUSTRY

Horizontal Integration by Packers

A number of packers are integrated, both horizontally and vertically. National packers and many regional packers have a number of establishments (plants) at different locations. This is horizontal integration. While the original big five packers have declined in degree of horizontal integration, several have expanded by merger, acquisition, and internal growth. Empirical evidence is lacking on reasons why firms expanded horizontally, but possible motives are (1) attempts to gain economies in purchasing supplies or in selling products, (2) possible forestalling of competitors from gaining a foothold in particular locations, and (3) attempts to increase total profits by increasing volume of meat handled.

The public has been concerned in previous times with the amount of horizontal integration, to the extent that it was related to concentration in the industry. With the decline in concentration, as discussed in a previous section, concern with horizontal integration has abated. However, horizontal integration can expand while concentration in the industry is decreasing. Data reported by the National Commission on Food Marketing from 200 responding firms indicated that horizontal type acquisitions by meat packing firms exceeded horizontal type disposals during the period 1947–1964. More acquisitions were made by the 4 largest firms than any other 1 size of group, but total acquisitions by other (smaller) size firms exceeded that of the largest 4.

Conglomerate Integration

Conglomeration in business means the collection of diverse business endeavors under one firm's management. During the 1950's and 1960's conglomerates expanded at a rapid rate. Some ran into financial difficulties in the 1970's and the pace of development changed rather abruptly.

During the period of rapid development, however, meat packing firms were involved in conglomerate mergers which might be called conglomerate integration. One of the most widely publicized conglomerate mergers was the take-over of Wilson & Company by Ling Tempco Vaught—a conglomerate with many diverse interests. Others are the merging of Armour & Company with Greyhound Corporation, Cudahy with General Host, and John Morrell with United Brands. Swift and Company more or less expanded into the conglomerate—Esmark. None of the original big five remains as a separate independent packing operation.

It should be noted that several of the national packers (e.g., Wilson and Armour), over the years, voluntarily diversified into a number of lines unrelated to meat packing. In a sense they expanded into conglomerate ventures. However, these generally did not attract the attention which accompanied recent conglomerate merger activity.

Vertical Integration by Packers

As mentioned earlier, the carrying out of processing activities by slaughterers involves integration of successive stages in meat production.

This has never been a matter of serious public concern. Processing is considered to be in the domain of packers. However, concern has been voiced in recent years by segments of the cattle industry over packer and food chain backward vertical integration into cattle feeding. From time to time, public concern also has been expressed over lamb and hog feeding by packers. Federal legislation has been proposed to curb cattle feeding by packers but has not been enacted.

Data are not available on packer feeding prior to 1952, however since then the Packers and Stockyards Administration of the USDA has reported livestock feeding activities of meat packers and retail food chains. The National Commission on Food Marketing also investigated certain aspects of vertical integration by packers.

Data in Table 7.8 indicate that major packer cattle feeding tends to follow the cattle cycle. Feeding of calves by this group generally has declined, while the feeding of hogs has increased. Lamb feeding, after reaching a peak in the early 1970's, has declined. Packers other than the 10 major packers have generally increased the feeding of all species. Retailers are only lightly involved in cattle feeding.

Special studies conducted by USDA in 1961 and 1965 show that packer associated interests (such as packing plant owners, directors, officers, employees, nonreporting subsidiaries, and affiliates) feed additional cattle which are not included in the regular annual report. This is shown in Table 7.9. It also may be seen in Table 7.9 that the rate of increase in feeding by packers has been at about the same rate at which total fed cattle marketings increased during the past 10 yr. Packer feeding has been 6–7% of fed cattle marketing since about 1960—if feeding by packer-associated interests are excluded. When feeding by associated interests is included, the proportion was 8.8% in 1961 and 11.5% in 1965. Integration into feeding by packers is accomplished by ownership of feedlots, ownership of cattle which are fed on a custom basis, contractual arrangements for feeding, or various combinations of these alternatives.

Aspelin and Engelman (1966) investigated the effect on prices of packer feeding by one slaughterer at a leading terminal market in 1962. They concluded that "packer feeding was about 3% of cattle slaughter by packers located near the market. The primary packer-feeder in the area fed about 10% of its cattle slaughter. Data on movement of cattle by weeks throughout the year did not show evidence that shipments of packer-fed cattle consistently offset fluctuations in the receipts of cattle on the local (terminal) market. Apparently, packer-fed shipments did not serve to stabilize the market supply to any great extent.

"The statistical analysis indicated that packer feeding had significant depressive effects on weekly average prices at the terminal market, relative to prices at other terminals. Price effects were within the range of $0.25–$0.50 per cwt on the weekly average price for choice steers at the

TABLE 7.8. LIVESTOCK FEEDING ACTIVITIES OF MEAT PACKERS, 1962, 1965, ANNUAL 1970-1976[1]

Type of Packer and Year	Cattle		Calves		Hogs		Sheep and Lambs	
	No. of Firms	1000 Head	No. of Firms	1000 Head	No. of Firms	1000 Head	No. of Firms	1000 Head
10 Major packers								
1962	8	296.4	2	19.8	1	1.1	6	510.4
1965	8	308.6	1	30.8	2	4.0	6	676.9
1970	8	503.0	1	7.0	2	43.4	5	827.1
1971	7	294.4	1	2.0	3	53.7	5	577.4
1972	9	292.0	2	1.9	3	95.4	5	884.0
1973	9	351.0	—	—	3	69.0	5	806.3
1974	9	434.9	1	1.5	3	52.4	5	814.7
1975	8	387.7	1	0.6	3	57.6	4	588.4
1976	8	569.6	—	—	2	69.9	3	563.3
Retail food chains								
1962	2	25.5	—	—	—	—	—	—
1965	3	58.8	—	—	—	—	—	—
1970	3	38.0	—	—	—	—	1	8.9
1971	2	45.4	—	—	—	—	1	8.5
1972	2	49.2	—	—	—	—	1	30.7
1973	2	40.5	—	—	—	—	1	155.4
1974	1	25.2	—	—	—	—	1	104.3
1975	1	23.2	—	—	—	—	—	—
1976	1	20.3	—	—	—	—	—	—
Other packers								
1962	197	614.4	47	25.3	42	26.3	20	40.3
1965	184	862.1	41	31.1	21	39.1	14	149.0

MEAT PACKING AND PROCESSING 189

TABLE 7.8. *(Continued)*

Type of Packer and Year	Cattle No. of Firms	Cattle 1000 Head	Calves No. of Firms	Calves 1000 Head	Hogs No. of Firms	Hogs 1000 Head	Sheep and Lambs No. of Firms	Sheep and Lambs 1000 Head
1970	118	837.6	25	68.9	15	21.7	7	397.5
1971	104	1246.2	21	64.0	10	21.3	6	333.3
1972	96	1506.6	15	30.7	8	19.0	4	488.5
1973	95	1185.1	9	18.5	9	37.2	3	583.2
1974	102	1106.2	6	16.2	10	31.2	3	410.3
1975	99	1000.0	9	60.6	10	54.6	3	426.4
1976	92	1161.8	15	51.8	10	86.4	4	484.2
Total								
1962	207	936.3	49	45.1	43	27.4	26	550.7
1965	195	1229.5	42	61.9	23	43.1	20	825.9
1970	129	1378.6	26	75.9	17	65.1	13	1233.5
1971	113	1586.0	22	66.0	13	75.0	12	919.2
1972	107	1847.8	17	32.6	11	114.5	10	1403.2
1973	106	1576.6	9	18.5	12	106.2	9	1544.9
1974	112	1566.3	7	17.6	13	83.6	9	1329.2
1975	108	1410.8	10	61.2	13	112.2	7	1014.8
1976	101	1751.8	15	51.8	12	156.3	7	1047.4

Source: USDA (1978B) and earlier issues.
[1]Summarized from annual reports of packers filed with Admin. USDA. Includes livestock fed by or for meat packers and removed from feedlot for slaughter or sale during the reporting period. Separate feeding activities by owners, or employees of meat packers or nonreporting subsidiaries or affiliates are not included.

TABLE 7.9.   NUMBER OF PACKERS FEEDING CATTLE AND NUMBER OF HEAD FED COMPARED WITH TOTAL FED MARKETINGS OF CATTLE IN 39 STATES, 1954–1976

Year	No. of Packers Feeding	No. of Cattle Fed By Packers[1] (1000 Hd)	Fed Cattle Marketings in 39 States[2] (1000 Hd)	Packer Feedings as a Percentage of Fed Marketings (%)
1954	165	564.9	9,482	6.0
1955	161	545.8	10,762	5.1
1956	157	520.8	11,331	4.6
1957	151	557.6	11,285	4.9
1958	176	729.1	11,787	6.2
1959	157	617.0	12,843	4.8
1960	165	856.7	13,621	6.3
1961	206	919.2 (1251.2)	14,561	6.3 (8.8)
1962	215	981.4	15,434	6.4
1963	211	1175.6	16,808	7.0
1964	190	1126.8	18,319	6.2
1965	204	1291.4 (2059.2)	18,936	6.8 (11.5)
1966	202	1473.7	20,597	7.2
1967	198	1547.8	21,920	7.1
1968	174	1397.3	23,286	6.0
1969	154	1632.8	24,930	6.6
1970	141	1739.5	25,830	6.7
1971	123	1652.0	26,312	6.3
1972	113	1880.4	27,780	6.8
1973	109	1620.9	26,091	6.2
1974	115	1623.5	24,125	6.7
1975	114	1410.8	21,370	6.9
1976	104	1803.6	25,168	7.2

Source: USDA (1978B) and earlier issues.

[1]The number of cattle (including calves) fed by meat packers was summarized from annual reports of packers filed with the Packers and Stockyards Administration, USDA. Feeding by packer-associated interests is not included. Most packer feeding is reported on a fiscal year basis.

[2]USDA Econ. Res. Serv. *Livestock and Meat Situation*, May 1967 and October 1965 issues and *Livestock and Meat Statistics*. Estimates for 39 states are on a calendar year basis and account for most fed cattle (and calf) marketings in the United States. The figure for 1954 was based on earlier data from Breimyer (1961).

[3]Numbers and percentages enclosed by brackets include feeding by packer-associated interests. Data not available for other years.

market. According to the mathematical analysis, a given increase in packer-fed supplies transferred to plant had more than ten times as much effect upon the local price for choice steers as did the same increase in market supplies of choice steers."

A major criticism leveled at this study was the limited sample upon which it was based, i.e., 1 packer and 1 market. Nevertheless, it called attention to a public concern over the expansion in packer feeding. While this is the only attempt at quantitative analysis of the effect on prices of packer feeding, it was followed by a number of articles of a subjective nature. The following summary draws heavily from one such article.

Various motives undoubtedly exist for packer feeding. Farris (1967) suggests the following:

(1) A more even supply of cattle may increase plant operating effi-

ciency. Receipts fluctuate from seasonal factors and erratically from day to day and week to week. If livestock are not available, plants are shut down. This is an expensive proposition due to high fixed costs in plant and equipment. Packers can reduce fixed cost per unit of output by continuous operation. Labor also has a minimum fixed cost in that unions are guaranteed a minimum week's pay whether livestock is available or not.

(2) Livestock procurement costs may be lowered. Fewer livestock buyers are needed and transportation costs can be reduced if cattle are fed near the plant by the packer.

(3) Risks may be reduced. The packer can control not only numbers, but also quality of livestock and thereby meet specifications of meat purchasers. In addition, price risk may be reduced by strengthening the packer's bargaining position in buying livestock—especially during periods when supplies of particular grades are scarce.

(4) New organizational and technological innovations may be discovered. By being actively engaged in feeding, a packer may increase his knowledge of that aspect of the industry and participate in development of new technology.

(5) Implementation of new developments may occur rapidly. Access to funds may permit packers to implement new developments more rapidly than farmers or commercial feeders.

(6) Public policy generally sanctions freedom of enterprise. Livestock producers have integrated forward into meat packing, even though the extent is limited. This in itself could be grounds for arguing the case for packer backward integration. There is merit in consistency in public policy.

Farris (1967) also recognizes some potential problems associated with packer feeding:

(1) Market price manipulation may be possible. Vertical integration may provide a means of altering the market structure and in turn allow a business conduct not possible under purely competitive conditions. A packing firm able to dominate purchases at a particular location could exercise some control over prices. According to Farris (1967), "The packer practice of negotiating for larger numbers of cattle several days in advance of shipment from the feedlot may also give packers opportunities to use such cattle in affecting paying prices."

(2) The pricing process may be affected by reduced proportion of cattle bought and sold in the marketing channel. Prices in established markets are widely publicized and used. Reduction in transactions at these markets erodes the base upon which they are established.

(3) The organizational structure of agricultural production may be adversely affected. Since packer feeding generally is large scale, its expansion may adversely affect family farm type feeding operations. The same

observation can be made on large-scale commercial feedlots in general.

(4) Packer feeding may bring unfair competition to specialized producers. This is an extension of (3) above. Since packers have earnings from other interests which could be used to cover losses in feeding operations, it has been suggested that losses could be sustained for extended periods in a given area, to the detriment of specialized feeders. It also has been suggested that packers could destabilize supplies and enhance price swings. This point, as well as several others mentioned above, prompted the National Commission on Food Marketing to propose the public reporting of activities of large, diverse firms engaged in feeding.

(5) The total volume of cattle fed may be higher if packers feed—leading to depressed producer prices. Historically, variations in production have been associated with profit levels. It seems likely this will continue to be the case, whether feeding is done by packers or others.

In summing up public policy considerations Farris (1967) commented, "In considering policy alternatives toward packer feeding, it may be possible to adopt a course which would avoid the undesirable possibilities associated with the practice, yet preserve its efficiency, flexibility, and potential for technological gain. One alternative would be to permit packers to feed or have custom fed for them a certain proportion of their slaughter capacity. Analysis of various considerations would be necessary to ascertain what prohibition would best minimize adverse side effects while retaining efficiency advantages of packer feeding. Without better factual knowledge, it may be desirable to place increased emphasis on supervising market practices of integrated firms more closely, while determining whether it would be in the public interest to restrict vertical integration."

EARNINGS IN THE PACKING INDUSTRY

Historically, the meat packing industry has been noted for low earning rates compared to other U.S. industries. This is apparent in Fig. 7.1, 7.2, and 7.3 where net income after taxes as a percentage of net worth is plotted for a number of industries from 1945 to 1976. Earnings in packing were the lowest of any industry shown from the beginning of that period until the 1970's. During some years of the 1970's meat packing earnings have exceeded average earnings of chain stores and the baking industry, but have been consistently below earnings of other industries. These figures represent the average performance for the entire packing industry. Undoubtedly variation exists in earnings among firms within the industry. Recent data are not available, but ranges shown in Table 7.10

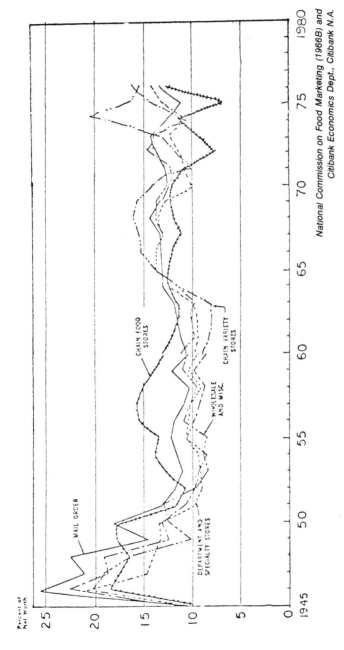

FIG. 7.1.   NET INCOME AFTER TAXES AS A PERCENTAGE OF NET WORTH 1945–1976

National Commission on Food Marketing (1966B) and
Citibank Economics Dept., Citibank N.A.

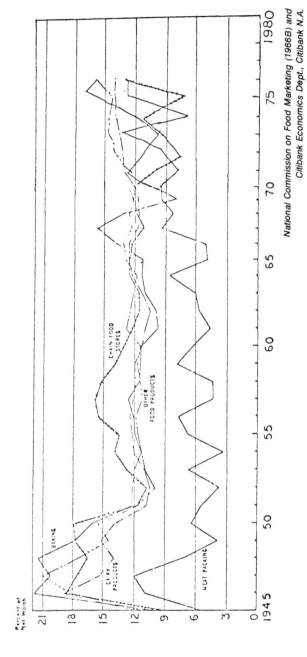

FIG. 7.2.   NET INCOME AFTER TAXES AS A PERCENTAGE OF NET WORTH 1945–1976

National Commission on Food Marketing (1966B) and
Citibank Economics Dept., Citibank N.A.

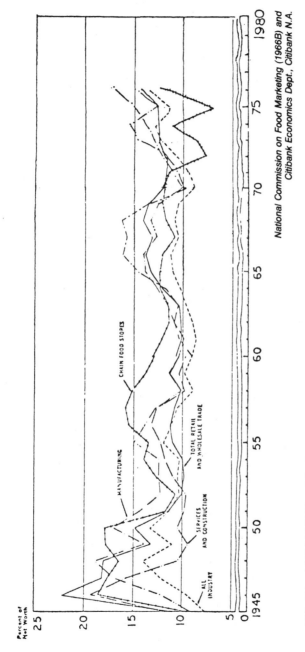

FIG. 7.3.   NET INCOME AFTER TAXES AS A PERCENTAGE OF NET WORTH 1945–1976

National Commission on Food Marketing (1966B) and
Citibank Economics Dept., Citibank N.A.

TABLE 7.10.   EARNING RATES FOR LARGE MEAT PACKING COMPANIES, 1947–1964[1]

	Net Income after Taxes as Percentage of:					
	Total Sales		Total Assets		Net Worth	
			Firms Ranking			
	1–4	5–8	1–4	5–8	1–4	5–8
Year	(%)	(%)	(%)	(%)	(%)	(%)
1947	1.70	1.12	10.2	9.2	15.5	14.9
1948	0.68	1.21	3.6	10.2	6.0	15.7
1949	0.63	0.52	2.7	3.5	5.0	5.4
1950	0.80	0.99	3.6	6.3	6.3	10.4
1951	0.58	0.87	3.0	6.1	5.1	10.6
1952	0.56	0.68	2.7	4.4	4.8	8.0
1953	0.84	0.62	4.0	4.3	7.0	7.7
1954	0.44	0.60	2.0	4.3	3.4	7.5
1955	0.75	1.17	3.5	7.6	5.6	13.7
1956	0.70	1.43	3.2	8.5	5.6	15.4
1957	0.42	0.80	1.9	4.7	3.3	8.2
1958	0.46	0.71	2.2	4.4	3.6	7.4
1959	0.90	1.23	4.1	7.5	6.8	12.3
1960	0.76	1.05	3.4	5.8	5.7	9.8
1961	0.64	0.43	3.0	2.4	4.9	3.8
1962	0.73	0.51	3.3	2.9	5.7	4.5
1963	0.77	0.78	3.4	4.4	5.8	6.6
1964	1.10	1.16	5.0	6.5	8.6	9.9
Avg 1947-1964	...	...	...	...	6.0	9.5

Source: National Commission on Food Marketing (1966).
[1] Ranked according to red meat sales in 1963. Largest 4 companies include Armour, Morrell, Swift, and Wilson. Companies in second group of 4 include Hormel, Hygrade, Oscar Mayer, and Rath. The data are for fiscal years ending approximately Oct. 31.

illustrate such differences. Earnings for the largest 4 ranged from 3.3 to 15.5% of net worth with an averege of 6% for the period. Earnings for the second largest 4 firms were higher, averaging 9.5%. (Figure 7.2 shows average industry earning on net worth in recent years at higher levels—the 1973–1976 average was 11.5%.) Earnings as a percentage of sales are notoriously low—running less than 1¢ on the dollar in most years; but this statistic is not very revealing of firm and industry profits without also citing sales volume.

Earnings of specialized cattle slaughtering firms at 16.0% and specialized meat processing firms at 13.1% were considerably higher than average industry rates or earnings of the largest firms. Specialized hog slaughtering firms earned about ½ the rate of cattle slaughterers, but still were above the industry average.

Relatively low earnings in the meat packing industry probably result from a combination of factors. To some degree it reflects the highly competitive nature of the industry. Also, it probably is at least partially due to operational inefficiencies. The variation in receipts due to cyclical and seasonal factors has resulted in a chronic overcapacity and unused facilities at times other than peak receipts. This tends to increase average costs and reduce profits. Under historical supply conditions, overcapacity has been virtually inherent in the industry. During cyclical and seasonal downswings in receipts, packers compete with each other for available livestock in an attempt to maintain volume. This tends to increase procurement costs and reduce profits. As has been noted earlier, structural changes are taking place which are leading to a greater degree of stability in supply. Both cyclical and seasonal trends are becoming less pronounced.

Although empirical evidence is unavailable on the subject, specialization in slaughtering probably improves operational efficiency and enhances profits. Many of the newer, specialized plants have incorporated technologically-improved equipment, more efficient layouts, and sophisticated management techniques, all of which enhance operational efficiency.

BIBLIOGRAPHY

ANTHONY, W. E. 1966. Structural changes in the federally inspected livestock slaughter industry 1950–62. USDA Econ. Res. Serv., Agr. Econ. Rept. *83*, Revised Feb.

ANTHONY, W. E., and EGERTSON, K. E. 1966. Decentralization in the livestock slaughter industry. USDA Econ. Res. Serv., Suppl. Agr. Econ. Rept. *83*.

ASPELIN, A., and ENGELMAN, G. 1966. Packer feeding of cattle, its volume and significance. USDA, C&MS Marketing Res. Rept. *776*.

ASPELIN, A. and ENGELMAN, G. 1976. National oligolopy and local oligopsony in the meat packing industry. USDA, Packers and Stockyards Administration (Unpublished).

BUTZ, D. E., and BAKER, G. L. 1960. The Changing Structure of the Meat Economy. Harvard Univ. Graduate School Business Admin., Boston.

COX, R. W., and TAYLOR, F. R. 1965. Feasibility of cooperatively owned slaughter plants. N. Dakota Agr. Expt. Sta. Agr. Econ. Rept. *39*.

DEWEY, D. Monopoly in Economics and Law. Rand McNally & Co., Chicago.

DIRLAM, J. B., and KAHN, A. E. 1954. Fair Competition. Cornell University Press, Ithaca, N.Y.

ENGELMAN, G. 1975. Trends in livestock marketing before and after the Consent Decree of 1920 and the Packers and Stockyards Act of 1921. USDA, Packers and Stockyards Administration. Statement to the Subcommittee on SBA-SBIC Legislation, House Small Business Committee. June 23.

FARRIS, P. L. 1967. Economic evaluation of cattle feeding by meat packers. Econ. Marketing Inform. Indiana Farmers, May 4–5.

IVES, J. R. 1966. The Livestock and Meat Economy of the United States. Am. Meat Inst., Chicago.

KAYSEN, C., and TURNER, D. F. 1959. Antitrust Policy: A Legal and Economic Analysis. Harvard University Press, Boston.

LOGAN, S. H. 1966. Economies of scale in cattle slaughter plants. Natl. Comm. Food Marketing Tech. Study *1*, Suppl. *2*.

LOGAN, S. H., and KING, G. A. 1962. Economies of scale in beef slaughter plants. Calif. Agr. Expt. Sta., Gianni Found. Res. Rept. *260*.

NATIONAL COMMISSION ON FOOD MARKETING. 1966. Organization and competition in the livestock and meat industry. Natl. Comm. Food Marketing Tech. Study *1*.

STAROBA, A. R. *et al.* 1977. Costs and returns for small slaughter plants in North Dakota. North Dakota State University, Dept. of Agr. Econ. Agr. Econ. Rept. No. 119.

USDA. 1977. Livestock and meat statistics. USDA, Econ. Res. Serv., Stat. Rept. Serv., Agr. Mktg. Serv. Stat. Bull. No. 522.

USDA. 1978A. Livestock slaughter, annual summary 1977. USDA, Econ. Stat. and Coop. Serv. MtAn 1-2-1 (78), March.

USDA. 1978B. Packers and stockyards résumé, USDA, Packers and Stockyards Admin. Vol. XV, No. 3, Dec.

USDA. 1978C. The future role of cooperatives in the red meats industry. USDA, Econ. Stat. and Coop. Serv. Mktg. Res. Rept. 1089. April.

U.S. DEPARTMENT OF COMMERCE (1975). Census of manufactures, 1972. U.S. Dept. of Comm., Bur. of Census, Special Report Series: Concentration Ratios in Manufacturing Industries. Washington, D.C. Oct.

WATKINS, M. W. 1927. Industry Combinations and Public Policies. Houghton Mifflin Co., Boston.

WILCOX, C. 1960. Public Policies Toward Business. Richard Irwin, Homewood, Ill.

WILLIAMS, W. F., and STOUT, T. T. 1964. Economics of the Livestock-Meat Industry. Macmillan Co., New York.

8

Meat Marketing—Wholesale

The previous chapters examined routes and means by which live animals are moved to the point of slaughter. But meat in the packing plant must reach the ultimate user to attain its full economic value. Between the packing house and the consumer lies a vast and complex system of market channels. Three things complicate the situation: (1) the quantities involved are enormous, (2) the distances of movement are great, and (3) the products have a relatively high degree of perishability.

U.S. consumers in 1977 were provided 18.8 billion kilograms (41.4 billion lb) of red meat (including imports)—an average of approximately 87 kilograms (192 lb) for every man, woman, and child in the country. But the bulk of meat is not produced in close proximity to areas of heavy population. Consumers are concentrated on the east coast, around the Great Lakes, on the west coast, and more recently in the so-called Sun Belt. Meat production is concentrated in the Corn Belt and in Central and Southern Plains states.

Under present economic conditions and with the present state of technology, fresh meat is not stored for an appreciable length of time. It is moved rapidly through market channels into the hands of consumers. While great strides have been made in meat preservation (refrigeration revolutionized the distribution and merchandizing of meat), refrigerated meat must be moved into consumption in a matter of days. Freezing, freeze drying, drying, and various methods of curing increase the storability of meat in varying degrees. For certain cuts of pork (hams, bellies and shoulders) freezing and curing are used extensively—partly to take advantage of seasonal price trends. Vacuum bagged, boxed beef can be held three to four weeks, but in terms of total meat production, and particularly with beef and lamb, storage is not a major factor. Market channels move the relatively perishable product from packers, to processors, to retailers and food service outlets. Significant changes have occurred and are occurring, in this system.

Refrigeration, transportation, and communication improvements probably would have to be classed as the major factors affecting meat distribution. But development of chain supermarkets, structural changes within the packing industry, and technological developments in processing and packaging also are important factors. The big chain retailers purchase wholesale meat by the carlot, often directly from packers: the smaller neighborhood type retailer, of necessity, purchases smaller lots from a more local wholesale type of distributor. Smaller packers often have problems making contact with purchasers in distant markets. They find an increasing need for intermediaries, independent wholesalers, brokers, and jobbers. A rapidly expanding hotel, restaurant, and institution (HRI) trade (often referred to as the food service industry) has been another recent development of importance in meat marketing.

Figure 8.1 is a generalized diagram of the major channels of meat distribution. A cell is shown for by-products simply to call attention to the fact that the output of a packing plant consists of more than meat. The actual distribution of meat is considerably more complex than indicated. For example, the cell labeled independent wholesalers, brokers, etc., does not do justice to the many types of wholesalers nor to the functions they perform. Involved here is the handling of fresh, cured, and processed meat; the handling of carcasses, primal, subprimal and portion control cuts; boning; buying outright versus simply acting as agent; peddling; etc. Some retail establishments (and some restaurant chains) also operate slaughtering and processing facilities, while some packers also operate retail establishments. There is trade among packers and among processors which may go direct or through agents. Sometimes more than one agent may handle a given lot of meat (either physically or by title) as it is transferred from one point in the distribution system to another.

In general, the transfer of title, i.e., the buying and selling, can be classified as either a wholesale or retail operation. To a large extent the cuts of meat involved in these transfers also can be classified as wholesale and retail, but this distinction gradually is losing the universality it once had. Where, at one time, all of the cutting into retail cuts was done in the retail store, a strong trend is underway to move that function to earlier points in market channels. This will be discussed in greater detail in a later section. Here it will only be mentioned that a major factor preventing its movement all the way back to the packer and processor level is lack of an acceptable method of preservation for the retail cut. A limited quantity of retail cuts already is moving, primarily as frozen meat, through wholesale channels.

WHOLESALING DEFINED

Wholesaling operations are defined here as movements from point of slaughter (packer) to retail outlets, HRI outlets, and export. Under some circumstances, the definition of a wholesaler or wholesaling is made contingent upon the breaking of a carcass into wholesale cuts. In this sense some difficulty is encountered in differentiating between manufacturing and wholesaling. In the above definition it makes no difference who does

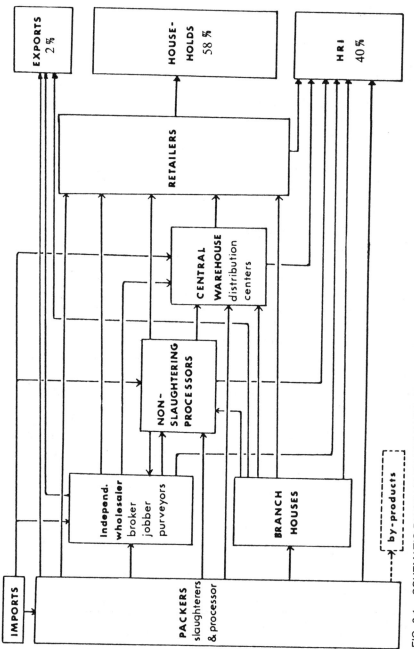

FIG. 8.1. SCHEMATIC DIAGRAM OF MEAT MARKETING CHANNELS

the breaking or the extent of breaking; any movement through distribution channels to retail, HRI, or export points is classed as wholesaling.

Products included are processed meats[1] as well as fresh meat. The National Livestock and Meat Board has prepared charts of wholesale (and retail) cuts of beef (Fig. 8.2), veal (Fig. 8.3), pork (Fig. 8.4), and lamb (Fig. 8.5). These wholesale cuts correspond generally to primal cuts, in trade terminology. However, standardized cutting has not been accepted nationwide. Wholesale trade also is carried on in sides and quarters of beef, and in entire carcasses (unsplit) of veal and lamb. Wholesale shipments of carcass pork may be either "packer style" (jowls attached, head and leaf fat removed, and carcass divided into halves) or "shipper style" (head and leaf fat attached and carcass not split). The latter constitutes only a minor fraction of pork shipments. A considerable quantity of portion control cuts is traded on a wholesale basis to the HRI trade and, to a limited extent, retail cuts are being wholesaled to retailers. This type of activity is carried on by wholesalers who specialize in breaking down to portion cuts and retail cuts.

In earlier days, larger packers built "full line" plants—plants in which all species and classes of livestock were slaughtered. Many also had a "sausage kitchen" in the same establishment where smoking, curing, and manufacturing of processed meats were done. Others did nothing but slaughtering. For some years the trend has been away from full line plants and away from combination slaughtering and manufacturing of processed meats. Now, many simply are kill-and-chill plants with shipment of sides or quarters, although, as mentioned, a trend is well underway to more cutting, boxing, and further processing at the packer/ processor level.

Figure 8.1 indicates wholesale movement from packers (who may process as well as slaughter) to the following: branch houses, independent wholesalers and brokers, processors, HRI trade, export, and direct to retail outlets. Many additional linkages are indicated and some are not shown. No attempt is made in this discussion to give a completely de-

---

[1]Processed meat includes such items as bologna, luncheon meats, sausages, cured meat, smoked meat, and ground and comminuted meats even though the latter may be in the fresh state. The term sausage carries a double meaning. It is used to indicate fresh ground pork, but also is used as a general term to include all prepared meats of the bologna, luncheon meat types, cured and smoked sausage, etc. The word "processing," when used in the term "central processing unit," also has another meaning from that which might be implied in the above definition of processed meat. A central processing unit is one in which meat is broken down by a central unit of a larger operating unit, e.g., of a chain of retail stores or restaurants. To make the situation even more confusing, wholesalers who engage in the breaking down of carcasses to wholesale cuts or retail cuts are known as "fabricators." This is a misnomer, as the word fabricate ordinarily refers to an assembly process whereas the cutting of meat is disassembly. However, there is no confusion in the meat trade where these terms are commonly used and well understood.

tailed description of the wholesaling of meat. Those who are interested will find more detail by consulting Fowler (1961), Ives (1966), and Williams and Stout (1964).

What goes on in this complex network of marketing may seem rather remote to the average livestock producer. In terms of distance, many of the things which happen are far removed from the feedlot scene. Nevertheless, they are vitally important. It is in this framework that wholesale prices are generated. What the packer gets at wholesale largely determines what he can pay for slaughter livestock. The producer who intends to do his own marketing must have an acquaintance with the wholesale market—its methods and procedures, changes that are occurring, and sources of price quotations. Livestock sales which are made on the carcass grade and weight method are related directly to wholesale carcass prices. Sales made on live grade and weight basis are only one step removed from the wholesale carcass market.

PACKER BRANCH HOUSES

Packer branch houses are satellite warehouse-processing-distribution centers owned and operated by a parent packing firm, and located in relatively large population centers away from the packing plant(s). Basically, the branch house is a sales and distribution activity with a considerable amount of carcass breaking, processing, smoking, and curing. In the hey-day of big packers (before decentralization of the packing industry got underway in the 1920's and 1930's), this was the major channel of distribution to larger cities. Meat was moved from packing plants to branch houses in car and truck lots, then from branch houses to retailers, independent wholesalers, hotels and restaurants, processors, and minor quantities to export.

The U.S. Census of Manufactures (1967) provides the latest data available on branch house, jobber, and broker operations. The number of branch houses, according to U.S. census reports, dropped from 1157 in 1929 to 522 in 1958. A slight increase occurred in the following decade, then a further decline took place (Table 8.1). In dollar amount, shipments and sales were substantially higher in 1972 than in 1929. Higher meat prices have held up the dollar volume. The National Commission of Food Marketing (1966) reported that "only about 14% of meat was handled by branch houses in 1963, compared with 30% in 1939." The decline in branch house use is attributed largely to increases in direct sales by packers to large chain retailers. These large operators by-pass intermediate steps between packer and retailer. The greater adaptability and flexibility of truck transportation makes direct deliveries more feasible and economi-

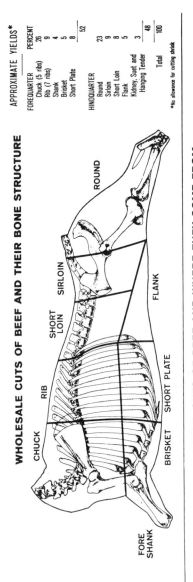

WHOLESALE CUTS OF BEEF AND THEIR BONE STRUCTURE

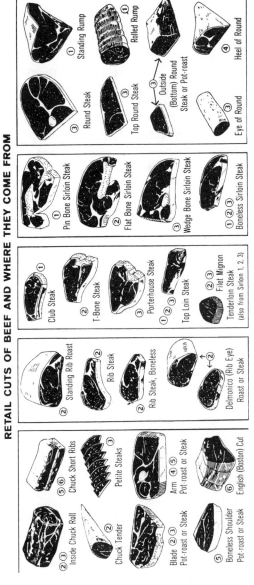

RETAIL CUTS OF BEEF AND WHERE THEY COME FROM

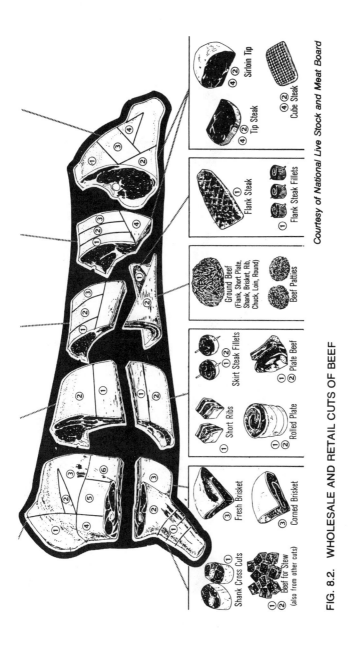

Courtesy of National Live Stock and Meat Board

FIG. 8.2.   WHOLESALE AND RETAIL CUTS OF BEEF

WHOLESALE CUTS OF VEAL AND THEIR BONE STRUCTURE

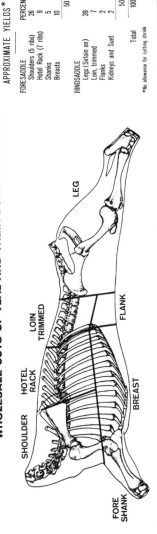

SHOULDER HOTEL RACK LOIN TRIMMED LEG

FLANK

BREAST

FORE SHANK

APPROXIMATE YIELDS*	PERCENT
FORESADDLE	
Shoulders (5 ribs)	26
Hotel Rack (7 ribs)	9
Shanks	5
Breasts	10
	50
HINDSADDLE	
Legs (Sirloin on)	39
Loin, trimmed	7
Flanks	2
Kidneys and Suet	2
	50
Total	100

*No allowance for cutting shrink

RETAIL CUTS OF VEAL AND WHERE THEY COME FROM

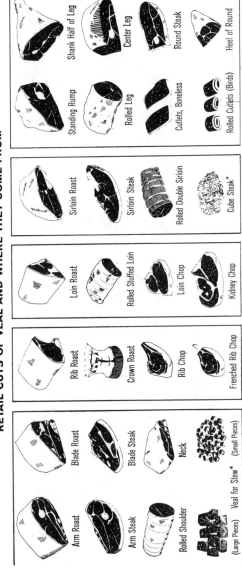

Arm Roast   Blade Roast   Rib Roast   Loin Roast   Sirloin Roast   Standing Rump   Shank Half of Leg

Arm Steak   Blade Steak   Crown Roast   Rolled Stuffed Loin   Sirloin Steak   Rolled Leg   Center Leg

Rolled Shoulder   Neck   Rib Chop   Loin Chop   Rolled Double Sirloin   Cutlets, Boneless   Round Steak

(Large Pieces)   (Small Pieces)   Frenched Rib Chop   Kidney Chop   Cube Steak*   Rolled Cutlets (Birds)   Heel of Round

Veal for Stew*

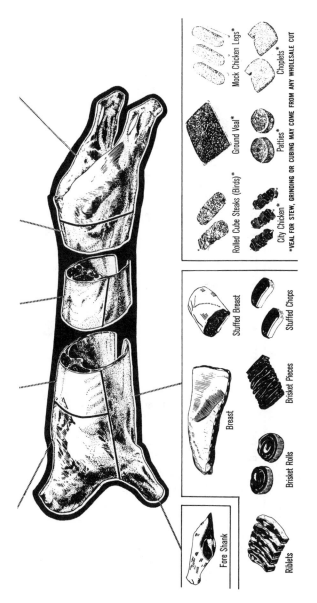

FIG. 8.3.   WHOLESALE AND RETAIL CUTS OF VEAL

Mock Chicken Legs*

Choplets*

Ground Veal*

Patties*

Rolled Cube Steaks (Birds)*

City Chicken*

*VEAL FOR STEW, GRINDING OR CUBING MAY COME FROM ANY WHOLESALE CUT

Stuffed Breast

Stuffed Chops

Breast

Brisket Pieces

Brisket Rolls

Fore Shank

Riblets

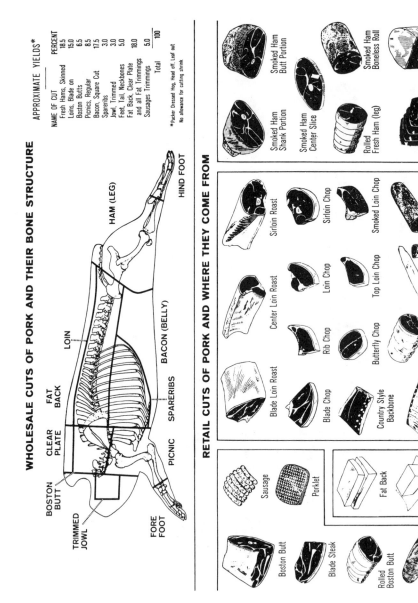

WHOLESALE CUTS OF PORK AND THEIR BONE STRUCTURE

APPROXIMATE YIELDS*

NAME OF CUT	PERCENT
Fresh Hams, Skinned	18.5
Loins, Blade on	15.0
Boston Butts	6.5
Picnics, Regular	8.5
Bacon, Square Cut	17.5
Spareribs	3.0
Jowl, Trimmed	3.0
Feet, Tail, Neckbones	5.0
Fat Back, Clear Plate and all Fat Trimmings	18.0
Sausages Trimmings	5.0
Total	100

*Packer Dressed Hog. Head off. Leaf out
No allowance for cutting shrink

BOSTON BUTT
CLEAR PLATE
FAT BACK
LOIN
HAM (LEG)
TRIMMED JOWL
PICNIC
SPARERIBS
BACON (BELLY)
FORE FOOT
HIND FOOT

RETAIL CUTS OF PORK AND WHERE THEY COME FROM

Smoked Ham Shank Portion
Smoked Ham Butt Portion
Smoked Ham Center Slice
Smoked Ham Boneless Roll
Rolled Fresh Ham (leg)
Canned Ham
Sliced Cooked "Boiled" Ham

Blade Loin Roast
Center Loin Roast
Sirloin Roast
Blade Chop
Rib Chop
Loin Chop
Sirloin Chop
Country Style Backbone
Butterfly Chop
Top Loin Chop
Smoked Loin Chop
Back Ribs
Rolled Loin Roast
Tenderloin
Canadian Style Bacon

Boston Butt
Blade Steak
Rolled Boston Butt
Smoked Shoulder Butt
Sausage
Porklet
Fat Back
Lard

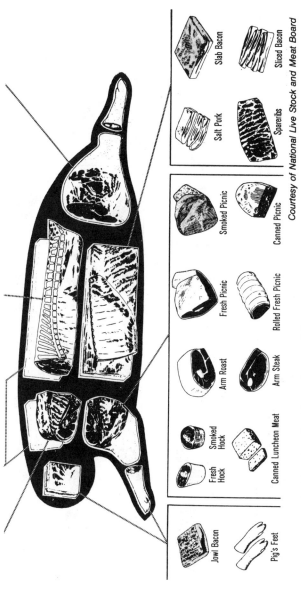

Courtesy of National Live Stock and Meat Board

Slab Bacon

Sliced Bacon

Salt Pork

Spareribs

Smoked Picnic

Canned Picnic

Fresh Picnic

Rolled Fresh Picnic

Arm Roast

Arm Steak

Fresh Hock

Smoked Hock

Canned Luncheon Meat

Jowl Bacon

Pig's Feet

FIG. 8.4.   WHOLESALE AND RETAIL CUTS OF PORK

WHOLESALE CUTS OF LAMB AND THEIR BONE STRUCTURE

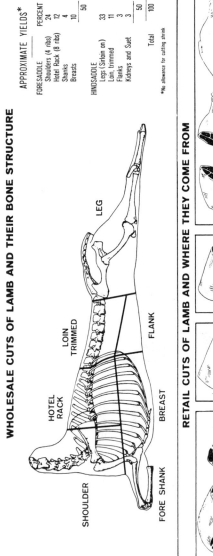

APPROXIMATE YIELDS*

FORESADDLE	PERCENT
Shoulders (4 ribs)	24
Hotel Rack (8 ribs)	12
Shanks	4
Breasts	10
	50
HINDSADDLE	
Legs (Sirloin on)	33
Loin, trimmed	11
Flanks	3
Kidneys and Suet	3
	50
Total	100

*No allowance for cutting shrink

SHOULDER

FORE SHANK

HOTEL RACK

LOIN TRIMMED

LEG

FLANK

BREAST

RETAIL CUTS OF LAMB AND WHERE THEY COME FROM

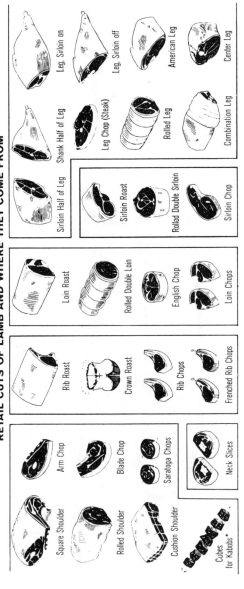

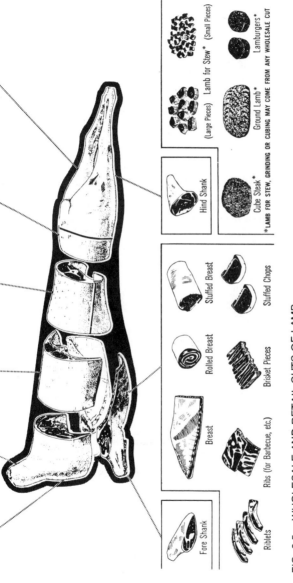

Fore Shank

Riblets

Breast

Ribs (for Barbecue, etc.)

Rolled Breast

Brisket Pieces

Stuffed Breast

Stuffed Chops

Hind Shank

Cube Steak*

(Large Pieces) Lamb for Stew* (Small Pieces)

Ground Lamb*

Lamburgers*

*LAMB FOR STEW, GRINDING OR CUBING MAY COME FROM ANY WHOLESALE CUT

FIG. 8.5.   WHOLESALE AND RETAIL CUTS OF LAMB

cal. Increasing operational costs, particularly labor, in large cities also have been factors in the decline of branch houses.

PACKER DIRECT SALES

Concurrent with the decline in branch house operations has been an increase in wholesale selling at the packing plant level. Part of this is due to the desire and ability of large retail chains to deal directly with sources of supply of wholesale meat, part is due to packers' desire to cut out costly

TABLE 8.1.   MEAT SLAUGHTERING PLANTS, PROCESSING PLANTS, PACKER SALES BRANCHES, MEAT MERCHANT WHOLESALERS, MERCHANDISE AGENTS: NUMBER OF ESTABLISHMENTS AND DOLLAR VOLUME OF SHIPMENTS AND SALES, CENSUS YEARS 1929–1972

Census Years	Meat Packing[2] Plants	Processing[2] Plants	Packer Sales[3] Branch Offices	Meat Merchant[3] Wholesalers (Jobbers)	Merchandise[3] Agents (Brokers)
		No. of Establishments			
1929	1277	—	1157	2225	130
1939	1392	1197	924	2552	84
1948	2154[1]	1264[1]	731	3200	58
1954	2367	1316	665	4357	97
1958	2810	1494	522	4482	154
1963	2992	1341	577	5170	134
1967	2697	1374	616	5041	163
1972	2475	1311	464	4847	245
		Shipments and Sales (Million $)			
1929	3,435	—	—	690	145
1939	2,540	—	1076	520	116
1948	8,970[1]	1601[1]	2717	1,977	577
1954	10,265	1541	2702	2,866	521
1958	11,972	2066	2263	3,891	606
1963	12,436	2130	2446	5,371	810
1967	15,576	3008	2811	7,395	853
1972	23,024	4632	4250	12,611	1470

Source: U.S. Department of Commerce (1975) and earlier issues.
[1]Meat packing and processing figures are for 1947.
[2]Census of Manufacture: 1972 Statistics by Industry.
[3]Census of Business Wholesale Trade—Summary Statistics 1972.

branch house operations, and part is due to increased use of federal grades and specification buying which facilitates buying by telephone. In regard to the latter point, some large buyers still buy by inspection (rather than description) in order to take advantage of certain specifications not pinpointed by federal grade standards.

Direct plant sales are accomplished by plant sales departments, car and truck routes, and the maintenance of sales offices in cities away from the location of packing plants.

Plant Sales

Practically every large packer and many medium sized ones maintain a carlot sales department, and, in addition, practically all handle less than carlot sales. Carlot sales departments deal directly with large buyers—primarily retail chains. The shipments move by refrigerated boxcar or truck directly to the buyer's warehouse or processing center. Local sales of less than carlot quantities are handled either by buyers visiting the plant to make selection or salesmen taking local orders.

Car and Truck Routes

In that range of territory outside of local sales and not covered by either branch houses or sales offices away from the plant, less than carlot sales are handled by car and truck routes branching out from the plant. Usually orders are taken by route salesmen who make rounds of retail stores, restaurants, hotels, and institutions. Orders are relayed back to the plant and delivery made periodically—possibly several times a week by truck or rail. Trucks have the decided advantage of store door delivery. Trucks and refrigerated cars are loaded so that the last order in will be the first to come out, and so on around the route. The optimization of routing a fleet of trucks in a manner to minimize delivery costs is a problem that lends itself well to mathematical programming. Significant savings often can be made by access to solutions of this type.

Packer Sales Offices

Many packers are too small to maintain branch houses. Some larger ones prefer not to operate branch houses. At the same time some in both categories want to handle sales without going through an intermediary. One way of doing this is to establish sales offices in cities away from the plant. These offices take orders and solicit orders, catering primarily to larger customers. They perform the selling function of branch houses, but do not physically handle the meat. Orders are relayed back to the plant, and shipments are made directly from packing plant to purchaser.

INDEPENDENT WHOLESALERS

Many packers by choice or necessity engage the services of an intermediary in selling their products. Small independent packers located in production areas great distances from high consumption areas often do not have the connections or facilities for direct sales. Others prefer to emphasize plant production aspects and hire experts, so-to-speak, to do the marketing associated with the buying and selling of carcasses and wholesale cuts. A multitude of other activities is performed by independent operators who function in channels between the packers and retailers, HRI buyers, and the export trade. Among these activities are breaking (or fabricating), boning, freezing, processing, etc.

A major distinction can be made within this group of intermediaries on whether the intermediary buys the meat outright (takes title) or simply acts as an agent on a commission basis. By following a strict definition, the former can be classified as jobbers and the latter as brokers. In the meat trade, the term jobber ordinarily is used in a narrower sense; however, there is some ambiguity in its meaning. In general, the term jobber is used to indicate the type of outlet of the wholesaler rather than method of operation. In this discussion, jobber will be used in the broader sense—i.e., independent wholesalers who buy (take title) and resell, hopefully at a profit. Within this general classification, reference will be made to various specialities which are related to the activities performed. By the same token, it is possible to indicate specialities, or subclassifications, within the broker class.

Wholesaling operations have grown substantially in recent years, as shown in Table 8.1. The number of meat wholesalers more than doubled between 1929 and 1967. During the same time the value of shipments and sales by wholesalers increased more than 19 times. This was a period of relatively rapid fragmentation of the packing industry and rapid expansion in both chain retail operations and the food service trade (HRI). The latter is directly related to enormous growth in franchised motel and restaurant chains. Important also has been the growth in institutional feeding, e.g., educational dormitories, other state institutions, and factory and store cafeterias. These types of buyers not only purchase large quantities, but want regular delivery of uniform quality meat. Many of them are discriminating buyers with substantial bargaining power.

A series of cold and hot wars following World War II necessitated large meat purchases for military installations. Military requirements have tapered off since termination of U.S. involvement in Vietnam, but that outlet still utilizes large quantities of meat. Military purchases generally are on contract bids as are governmental school lunch and welfare purchases.

Under these circumstances many packers find it advantageous either to sell their meat to a jobber and let him take it from there, or employ the services of a broker who acts as their agent for a fee. The expansion in independent wholesaling activities in recent years seems to attest to the fact that wholesaling is a complicated, competitive aspect of the meat business and that there is a place for specialists in the performance of wholesaling functions.

Jobbers

As mentioned, the term jobber is used here to mean a meat wholesaler who takes title (he buys the products and resells) hoping to make a profit. Mention already has been made that some of these operators deal in carcass meat, some in wholesale cuts, and some in retail and HRI cuts. Some specialize in breaking (fabricating), others in boning, freezing, etc. Data in Table 8.1 indicates that jobbers handle a significant volume of meat. One type of specialist within this group deserving of particular mention is the "purveyor." This term generally is restricted to meat wholesalers who deal rather exclusively with the food service industry. Particular purveyors often limit their clientele to a specific segment of that industry. For example, some handle only the very highest quality meat and deal exclusively with high class hotels, night clubs, and steamship lines. Others cater to franchised motel chains, etc. Purveyors with long-established reputations are among the most discriminating buyers. Ives (1966) reports that "this specialized type of wholesaling has had its greatest growth during and since World War II, coincident with the growth of food service establishments in these years."

Boners specialize in the preparation of boneless meat. This, primarily, is lower grade beef. This meat moves on through channels to processors, the military, and retailers.

The meat "peddler" is another independent jobber. Other terms used for this operator are "wagon or truck jobber" and "truck distributor." The peddler is a small operator, often consisting of only one man who does the buying, selling, and delivering. The importance of peddlers has declined along with the decline in neighborhood meat retailers and neighborhood grocery stores.

Brokers

The meat broker and live cattle broker perform similar functions at different levels in the marketing system. By definition a broker acts as agent for the principals to a transaction. He facilitates transfer of title by bringing the seller and buyer together on terms of trade, but does not take title in his own name. His reimbursement is a fee (a commission) usually paid by the seller.

Some packers without direct sales organizations prefer the use of brokers rather than selling to jobbers. Under any type of wholesaling arrangement, it is not unusual for shifting demand and supply conditions to leave packers and processors in distress situations at times. In such cases brokers may be called upon to assist in moving the meat.

While brokers are not involved in the movement of a major fraction of the total meat supply (see Table 8.1) they are an important link in the system for certain packers, and under certain conditions. Motts (1959) reported that brokers packer agents handle less than 5% of the total meat supply. The National Commission on Food Marketing (1966), in a sample of packers and processors, found a range from 9 to 27%, depending upon type of product and type of firm.

WHOLESALE PRICING

Meat pricing at the wholesale level long has been somewhat of a mystery to producers. Unlike live animal markets, where the producer is on more or less familiar ground, there are no public markets for wholesale meat² such as auctions or terminals where transactions can be observed. Public and private sources have published quotations of wholesale prices for some time, but the quotations themselves give no insight into the pricing mechanism, and many smaller producers are not even aware of the sources of such quotations.

Basically, pricing at wholesale is done either by formula or negotiation. There are, of course, variations in both of these approaches.

Negotiated Prices

This simply is private bargaining between seller and buyer. It is analogous to private treaty in live animal marketing in that principals to the transaction rely on knowledge of the product, familiarity with current market conditions, and expertise in bargaining. The negotiation may take place by telephone based on stipulated specifications, or purchasers (or their representatives) may personally inspect and select meat in the wholesaler's cooler.

The prices retailers are willing to pay for wholesale meat is a reflection of consumer demand. In other words, retailer demand is a derived demand—derived from consumer demand. The prices at which wholesal-

---

²In early days there were public markets; e.g., Faneuil Hall was erected in Boston in 1742 for trading in provisions. Other large cities followed with provisions markets. Even when operating, however, quotations from these markets were largely unavailable to producers. In those days packers were concerned primarily with packing operations and left merchandising of meat up to wholesalers. As the packers established branch houses, public meat markets largely disappeared.

ers are willing to sell is a reflection of the costs they have incurred in obtaining the meat. In the case of a packer-wholesaler this is reflected in the cost of live animals and is a derived supply. Under competitive conditions, trading prices are expected to tend toward an equilibrium of the supply-demand situation. It is not surprising that stable equilibrium conditions are not found in the real world as supply and demand conditions are constantly changing and assumptions of a perfectly competitive market are constantly upset. The crux of this approach to pricing is well-informed buyers and sellers with at least approximate equality in bargaining power.

Prior to the 1920's and 1930's, the packing industry was highly concentrated in the hands of a relatively few big packers, while meat retailing was highly decentralized in a large number of relatively small firms. Under these conditions there is little question but what the balance of bargaining power was with the packer-wholesalers. That situation has changed. Development of large retail chains and cooperative and voluntary wholesale buying arrangements among independent retailers has shifted the balance of bargaining power. There has been a great deal of conjecture about the implications of the present situation, but a study by the National Commission on Food Marketing in reports issued in 1966 failed to develop evidence of undue market power of any sector in the food economy.

One variation of negotiated pricing that usually merits special attention is "offer and acceptance pricing." Large volume buyers under this arrangement invite wholesalers (packers, independent wholesalers, processors, etc.) to make offers on orders in which the purchaser sets forth rather rigid specifications. Specifications include such things as grade (may be both quality and yield), sex, weight (range per cut), and terms of delivery. Sales by specification normally carry the agreement or understanding that any product which does not meet the specifications, upon delivery, may be rejected by the purchaser. This can leave the supplier in a serious situation. However, unofficial reports indicate that only a small fraction of shipments is rejected.

Upon receipt of offers from several prospective suppliers the purchaser compares offering prices. Other things being equal, the lowest price will be accepted. Some would argue that such an arrangement can scarcely be classed as negotiation. This undoubtedly is the case at times. However, often there is discussion and some negotiation, particularly on deviations from the specifications.

Offer and acceptance pricing ranks high in operational efficiency. Very little cost is involved in a transaction. When buying is accomplished on specification the transaction can be consummated by telephone. Salary and travel expense of personnel who otherwise would personally select carcasses or cuts is averted. Sellers are relieved of some expense in attending these buyers.

From the standpoint of the mechanics of operation, this method also has the potential for a high degree of pricing efficiency. Specifications can be quite rigid. The crucial point is how well price differentials among meats of various specifications reflect the real differences to final users. Sellers whose offers are rejected are left somewhat in the dark in not knowing by how much their offer exceeded the acceptance price. For the system to work effectively over a period of time sellers need to have either invitations from alternative buyers, or alternative outlets. The degree of competition in this respect has a direct bearing on price level of offers made. If enough time is allowed in delivery of the product, suppliers conceivably could push equivalent acceptance prices back to live animal prices. This, of course, could work either way on the level of livestock prices.

Formula Pricing

In formula pricing the transaction price is determined by applying a formula to some predetermined base price for the particular product being traded. The base prices most generally used are from *The National Provisioner's Daily Market and News Service*—commonly called the "Yellow Sheet." The National Provisioner is a Chicago-based, privately-operated firm which assembles and publishes, on a subscription fee basis, daily closing prices for wholesale meat and meat products. Prices are obtained by telephone from sellers, buyers, and brokers in the meat trade. National Provisioner claims that only freely competitive transactions are used in the quotations and that each is verified. Another source of wholesale price quotations is the *Meat Market Research and Reporting Service* Meat Sheet which is known as the "Pink Sheet." This publication originates at Elmhurst, Ill., and is privately owned. The Meat Sheet is of more recent origin than the Yellow Sheet and, as yet, has not attained as widespread acceptance as the Yellow Sheet. The USDA also assembles and publishes wholesale meat prices, but at this stage, Yellow Sheet quotations are the dominant base for meat formula pricing.

One common carlot carcass beef quotation is for the River Markets (i.e., Omaha, Sioux City, St. Paul, St. Joseph, Kansas City). For example, using the Yellow Sheet price as the base, traders adjust by formula to arrive at prices applicable to, say, New York, Boston, Omaha, Kansas City, Denver, etc. A typical example of formula pricing would be the case of a Kansas City packer agreeing to deliver a carlot of carcass beef (meeting certain specifications) to Boston on a continuing basis at the Yellow Sheet quotation (or some agreed upon amount under or over the Yellow

Sheet) for beef of agreed specifications—plus transportation cost from Kansas City to Boston. Formula pricing also is used for pricing cattle sold on carcass grade. One typical arrangement here is for a packer (say a Great Plains packer) and nearby commercial feedlot manager to agree that cattle delivered two weeks hence will be priced "in the meat" at, say, $2 under the Yellow Sheet quotation on the day the cattle are delivered and slaughtered. Feedlot managers, as a rule, are not enthusiastic about that approach to pricing cattle, but during periods of over-supply and draggy cattle markets packers sometimes insist on it.

Williams (1970) states that "The formula works fairly well within the territory under the influence of the northeastern market." He further indicates that "Numerous other examples could be presented, however, in which the implied Chicago price is $0.50–$1.00 higher or lower when based on delivered prices as compared with FOB prices or on alternative delivery point prices. The implied price at Chicago on a shipment from Denver to Atlanta often is $0.75 under the Chicago price of the same or a similar shipment to New York . . . The practice of adjusting prices at other markets to a Chicago base price leads to other problems. For example, the transportation cost employed does not include all transfer costs. Then, too, there is the question of which transportation cost shall be employed. The truck load rate from Chicago to New York usually is quoted as $1.50, but this is an average figure. The piggyback rate is $1.25. Which should be used in adjusting New York prices? Transportation costs often are not the same in both directions between two markets. It is also well to remember that while the system may operate satisfactorily within the trading area defined, the Yellow Sheet is used by all sectors of the trade everywhere."

Formula pricing is widely used. The National Commission on Food Marketing found in a 1965 survey that of the meat packers and processors responding, 41% of beef and veal, 24% of lamb and mutton, 41% of fresh and frozen pork, 29% of cured hams, picnics and bacon, and 20% of other processed meat transactions (with most important customers) were determined by formula pricing. "In nearly all cases the quotation source was the Yellow Sheet" (National Commission of Food Marketing 1966). Stout et al. (1968), in an Ohio study, found that, "All 24 firms employed formula prices in purchasing some or all of their fresh meats." The Food Commission concluded that use of formula pricing was increasing. Unofficial estimates now indicate as much as 75% of wholesale beef is priced by formula. The Food Commission also mentioned several important implications of formula pricing. These include: (1) whether formula pricing tends to perpetuate geographic price patterns unrepresentative of changing supply and demand conditions, (2) whether prices used as base prices are subject to manipulation by trade interests, and (3) whether base prices accurately reflect equilibrium supply-demand conditions.

Does Formula Pricing Perpetuate Geographic Patterns?—The reasoning here is that if prices today are based on yesterday's closing prices, and if yesterday's prices had been based on the previous day's closing prices, and so on—it is conceivable that the pattern of prices would remain fixed over an area even though supply and demand conditions changed. Furthermore, if all, or most, meat were formula priced day after day, the continuing day by day quotation of price reporting services (i.e., Yellow Sheet, Meat Sheet, USDA) would not adequately reflect changing supply and demand conditions. The reasoning is logical. It would appear that the price level would remain fixed under these circumstances. The record shows, however, that geographic price relationships have not remained fixed and price level has varied. There are several apparent explanations. Not all meat is traded on a formula basis. Some of the prices collected in assembling Yellow Sheet quotations undoubtedly come from negotiated sales. Furthermore, some negotiation takes place even where the base price may be used as a point at which bargaining starts. There is no evidence at this time that formula pricing has perpetuated geographic patterns, although this cannot be ruled out as a possibility if greater use were made of it on a strict formula basis.

Can Formula Pricing Be Manipulated?—In a discussion of the beef marketing system, DeGraff (1960) stated "no one organization involved in processing or distributing beef is big enough or powerful enough to dominate the market or to dictate prices in the market. The reasons why this is true center on two points. First, beef, like any major food, has a nationwide market. Prices among the regions and localities of the country are closely tied together by competitive forces, by highly developed transportation, and by effective market news services, both public and private . . .

"Second, in order to control such a market, or prices in such a market, it would be necessary to accomplish what is impossible for any food industry firm, or indeed any group of firms. It would be necessary to have a large measure of control over the following: (a) the wants and preferences of consumers; (b) the availability and price of substitute products; (c) the level and distribution of consumer incomes; and (d) a substantial part of the supply of the product involved . . .

"There are, of course, always possibilities of monopoly attempts or of collusion in any one market area. They seldom get far or last long . . ."

Comments by Williams (1970) indicate that the Yellow Sheet is vulnerable to manipulation. He states that the National Provisioner's ". . . sample of open trades selected as closing necessarily is smaller than the total number of trades for a given weight and grade. In short, closing prices require more judgment and are more vulnerable to effects of a few 'planted' sales . . .

"Several brokers have stated in informal interviews that they had, and within limits could, manipulate the Yellow Sheet. It seems that techniques have been developed which induce Yellow Sheet editors to use price reports of 'private sales' which cannot be checked . . .

"To the extent that it exists, the practice of formula pricing on a fixed forward basis increases incentives for manipulating Yellow Sheet prices. In some degree USDA reports, as well as the Yellow Sheet, are vulnerable. It must be nearly impossible to fully verify even a high percentage of the eligible trades in an area which includes more than half of the nation."

While some conditions and incentives may exist for manipulation, there apparently is a lack of any clear-cut evidence that this has been done. The extent to which the Yellow Sheet is used seems to indicate a substantial degree of confidence in its reliability.

Do Formula-Generated Prices Accurately Reflect Equilibrium Prices?—As long as base prices are derived from prices determined in a freely competitive market, and as long as the price reporting agency, whether it be private or public, obtains an adequate sample of prices, then they could be said to reflect equilibrium prices. Both the National Provisioner and USDA contend that a sufficient volume of competitively determined sales are made to accurately reflect the market (Williams 1970). Current trends in market structure, however, indicate that the number and volume of trades meeting the requirements indicated above, is becoming smaller. If this trend continues, and there is reason to believe that it will, it is probable that this approach will not represent an equilibrium situation and may not adequately reflect the major movement of meat in wholesale channels.

Implications with Respect to Market Efficiency.—There is no question that formula pricing is very efficient from an operational standpoint. It utilizes a minimum of time and manpower. One or two persons by use of the telephone could purchase all of the meat requirements for a national retail food chain. The same situation applies at the packer-processor level. That probably is a major reason retailers and wholesalers alike appear to be satisifed with formula pricing.

From the standpoint of pricing efficiency the system depends upon competition to establish prices which reflect true supply and demand conditions. That requires bona fide negotiated prices as the source of price reporting services quotations. While questions have been raised about the adequacy of reported prices there is, as yet, no clear cut evidence to the contrary.

BIBLIOGRAPHY

ARMSTRONG, J. H. 1968. Cattle and beef buying, selling and pricing handbook. Indiana Coop. Ext. Serv. May.

BUTZ, D. E., and BAKER, G. L. 1960. The changing structure of the meat economy. Res. Div. Harvard Business School.

DE GRAFF, H. 1960. Beef Production and Distribution. University of Oklahoma Press, Norman.

DIETRICH, R. A., and WILLIAMS, W. F. 1959. Meat distribution in the Los Angeles Area. USDA Agr. Marketing Res. Rept. 347.

FOWLER, S. H. 1961. The Marketing of Livestock and Meat, 2nd Edition. Interstate Printers & Publishers, Danville, Ill.

IVES, J. R. 1966. The Livestock and Meat Economy of the United States. Am. Meat Inst., Chicago.

KOLMER, L. et al. 1959. Consumer marketing handbook. I. Meat. Iowa Coop. Ext. Serv. Nov.

MOTTS, G. N. 1959. Marketing handbook for Michigan livestock, meat and wool. Mich. Agr. Expt. Sta. Spec. Bull. 426.

NATIONAL COMMISSION ON FOOD MARKETING. 1966. Organization and competition in the livestock and meat industry. Natl. Comm. Food Marketing Tech. Study 1. Govt. Printing Office, Washington, D.C.

PRICE, J. F., and SNELL, J. G. 1965. Meat processing handbook. Mich. Coop. Ext. Serv. Sept.

STOUT, T. T., and HAWKINS, M. H. 1968. Implications of changes in the methods of wholesaling meat products. Am. J. Agr. Econ. 50, 660–675.

STOUT, T. T., HAWKINS, M. H., and MARION, B. W. 1968. Meat procurement and distribution by Ohio grocery chains and affiliated wholesalers. Ohio Agr. Res. Develop. Center Res. Bull. 1014.

USDA. 1969. Feasibility of a physical distribution system model for evaluating improvements in the cattle and fresh beef industry. USDA Agr. Res. Serv. 52–36.

U.S. DEPARTMENT OF COMMERCE. 1975. Census of Manufactures 1972. U.S. Dept. of Comm., Bur. of Census, Vol. 1.

WILLIAMS, W. F. 1958. Structural changes in the meat wholesaling industry. J. Farm Econ. 40, 315–329.

WILLIAMS, W. F. 1970. Implications of developments in the pricing structure of the livestock-meat economy. In Long-Run Adjustments in the Livestock and Meat Industry: Implications and Alternatives. Ohio Agr. Res. Develop. Center Res. Bull. 1037. Also North Central Regional Publ. 199.

WILLIAMS, W. F., and STOUT, T. T. 1964. Economics of the Livestock-Meat Industry. Macmillan Co., New York.

9

Meat Marketing—Retail

Production, slaughter, processing, wholesaling—these functions would serve no purpose and the product would have no value without the final step—that is, getting the meat into the hands of consumers. In the United States, some 98% of total meat production is consumed domestically—the remaining 2% is exported. Most domestically consumed meat is channelled through retail outlets, but an increasing fraction goes through HRI outlets. It is estimated that HRI trade handles approximately 40% of final meat sales and that share is increasing each year. In evaluating the retail trade it should be noted that smaller units of the HRI trade obtain meat supplies from retail stores. It has been estimated that some ¼ of all HRI purchases are supplied by retailers. Farm slaughter, once an important source of meat, now amounts to about 1.4% of the total. From the retailer's standpoint meat is a most important item. Meat generates approximately 25% of the total gross retail sales.

DEVELOPMENTS IN FOOD RETAILING

Significant changes have occurred in food retailing during the past 60 yr, and particularly since the 1930's. Two major developments are closely associated with these changes—development of the chain system,[1] and development of the supermarket. These were not simultaneous. The chain system came first, resulting in a rapid increase in number of stores. This was followed by development of the supermarket, which resulted in consolidation and reduction in store numbers. The total volume of business expanded greatly during this period. Figure 9.1 illustrates the trend in number of stores and sales for the period 1940 to 1977.

Retail Chains

The origin of corporate retail chains dates back to 1859 with organization of the A&P Tea Company (DeGraff 1960). Early development of chains was almost entirely a matter of horizontal integration; and the emphasis was on number of stores rather than size of individual units. "Between 1910 and 1930, chains grew from 2000 to 45,000 stores and

---

[1]Chains are defined as firms with 11 or more stores.

encompassed ⅓ of one of the nation's largest industries. The speed and scope of this development was alarming to businessmen as well as the public" (National Commission on Food Marketing 1966).

Early chain activity also centered on the economics of obtaining supplies. Primarily, it was one of integrating wholesaling operations and retailing under one management. This was vertical integration—the control of successive stages in production and marketing channels. Some

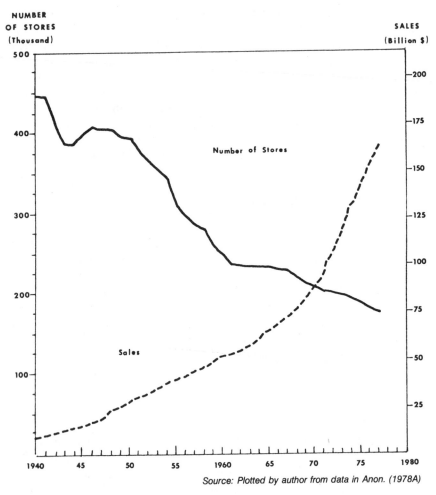

Source: Plotted by author from data in Anon. (1978A)

FIG. 9.1.   NUMBER OF GROCERY STORES AND ANNUAL SALES, 1940–1977

chains rather quickly went one step further and brought processing operations under their management.[2] Thus, significant gains were made in
operational efficiency which were reflected in prices below competing
nonchain stores. Management was able to standardize, coordinate, and
control activities that previously were under widely scattered, independent operations. However, operational efficiency was not the only source
of economic strength. Perhaps of even greater significance was the gain
in bargaining power that came with mass purchasing. This enabled the
chains to obtain price concessions from suppliers that small, independent
operators could not obtain. Chain purchasing departments dealt directly
with packers for meat supplies, by-passing entirely established wholesale
agencies.

Rapid expansion of chain activity placed small, independent meat markets and grocery stores under severe competition. The situation was
aggravated by widespread economic depression during the 1930's. Many
were absorbed by the chains. The small business operators did not give
up without a struggle. They agitated for, and obtained, legislation designed to thwart unfair competition. The Robinson-Patman Act passed in
1936 reflected the failure of the earlier Clayton Act in restraining discriminatory practices. The Robinson-Patman Act contained provisions
aimed at prevention of price concession to some (the large) buyers and not
to others, and other discriminatory trade practices which allegedly gave
advantages to large purchasers. Many states also imposed taxes and
regulatory strictures on chain operations. Nevertheless, expansion of
chains continued.

Supermarkets

While early chain activity centered on economies in procuring supplies,
it soon became apparent to grocers that economies also were available by
possible changes in selling. A major development here was the supermarket. The supermarket idea—large volume, low cost, and mass
retailing—actually originated among independent operators, but it was
adopted almost immediately by the chains. Where previously the emphasis was on number of outlets, a shift began during the 1930's to fewer
and larger outlets. The emphasis shifted to volume—not just for volume
per se, but for economies of size that came with volume. In addition, such
innovations as self-service, cash-and-carry, and standardized accounting
and operating procedures came into prominence.

---

[2]As will be noted later, meat processing by retailers has been on only a limited scale. Most
extensive processing is in dairy products, fruits, vegetables, and bakery products.

Economies in both purchasing and sales allowed large volume operators to reduce margins. The small, independent specialized meat markets, as well as grocery stores, were at a serious competitive disadvantage, and numbers declined sharply. The small independent meat market all but vanished. However, independent grocers (including meat departments) did not give up.

To counteract corporate chains, the independents made two major adaptations: (1) independent wholesalers sponsored the development of voluntary chain arrangements which gave independent retailers advantages of large-scale purchasing and many associated services, and (2) retailers formed cooperative wholesale purchasing and servicing organizations. In the former, a large wholesaler takes on many of the purchasing and servicing functions of the centralized purchasing department of a corporate chain. This may include many kinds of assistance for its members in operation and management of retail stores. The cooperative chain differs in that the independent retailers organize, own, and operate the wholesale agency for the benefit of its members. It would be a mistake to visualize all independent retailers as small operations. Many of them are large supermarkets—just as large as corporate supermarkets but with fewer stores. Many of the independents, however, own and operate more than one supermarket. There also are many superettes in the independent system, both voluntary and cooperative. In marketing terminology, supermarkets are defined as stores with annual sales of $1,000,000 or more. Those with annual sales below $1,000,000 are classed small stores.

Development of chain operations and expansion of supermarkets were restricted during World War II. However, immediately upon termination of the war, expansion continued at an accelerated rate. "The rapid postwar adoption of the supermarket came at the expense of many thousands of small grocery stores. The number of grocery stores operated by single-store firms dropped by more than 130,000 between 1948 and 1963. Stores with annual sales less than $5000 dropped 86%" (National Commission on Food Marketing 1966).

Recent developments in food retailing which are relevant to the meat industry are (1) the discount store, (2) the convenience store, (3) central cutting and packaging, and (4) "fast foods."

Discount Houses

The discount house in a sense is a throw-back to the notion of early supermarket operation. Basically they feature high volume, limited service, low cost, and low price merchandising. They usually carry a wide variety of goods. Some carry no food at all. Some carry food but only a limited selection of meat items, while others carry a full line of meat. Many discounters are organized as corporate chains.

Perhaps the best evidence that discounting has made inroads on established food retailers is found in the fact that established firms responded by adopting discount tactics themselves. It was reported that in 1969, 47% of the chain supermarkets and 27% of independent supermarkets were discounting—and further, that the meat department in 19% of the supermarkets and 22% of the superettes were discount pricing meat (Anon. 1971). In 1977 the extent of discounting among chain supermarkets remained at 47% while 31% of independent supermarkets were using a discount policy (Anon. 1978B).

The practice of discounting created some repercussions throughout the retail industry. In 1972 A&P initiated a discount program known as WEO ("Where Economy Originates"). Retaliatory price cuts by competitors were, at least in part, associated with reduced earnings for the industry for several years (Parker 1975). The severity of price cuts apparently then was relaxed somewhat. Wide adoption of discounting, if that were continued, probably would defeat its purpose. The price elasticity of demand for food in total is relatively low—as indicated in an earlier chapter.

Discounting is not limited to retailing. In 1969 it was reported that 33% of grocery wholesalers discounted prices (Anon. 1971B). In 1977 the proportion was 39% (Anon. 1978C).

Convenience Stores

Convenience stores are relatively small installations in carefully selected locations with potential for a large volume of traffic. They feature long hours of operation and a reasonably wide range of items with a limited number of lines of a particular item. Attempts are made to hold operating costs as low as possible, but prices generally run higher than in discount or supermarket operation. The emphasis is on convenience in location, service, and items carried.

The neighborhood grocery stores that survived the chain-supermarket revolution are, in essence, convenience stores. However, the recent movement is not a revival of neighborhood stores. Some are independently owned and operated, but corporate chain systems dominate the field.

The sharp growth in convenience stores is shown in Table 9.1. In a little over 20 yr the movement has grown from virtually nothing to 30,000 stores with 4.5% of all grocery sales. There appears to be no doubt of continued growth. The extent of growth probably will be closely associated with trends in affluence of consumers. Additional convenience

TABLE 9.1.   CONVENIENCE STORE GROWTH

Year	No. of Stores	Share of U.S. Total Grocery Sales (%)
1957	500	0.2
1960	2,500	0.7
1962	3,500	0.9
1965	5,000	1.1
1967	8,000	1.7
1968	9,600	2.1
1969	11,620	2.6
1970	13,250	2.7
1971	15,075	3.1
1972	17,600	3.5
1973	20,300	3.8
1974	22,700	4.1
1975	25,000	3.8
1976	27,400	4.1
1977	30,000	4.5

Source: 1957–1969 Anon. (1971), 1970–1977 Anon. (1978D)

stores will be built if patronage is forthcoming. Capital investments per unit are small compared to supermarkets.

In 1970, processed meat was handled by 97% of convenience stores and fresh meat by 20%. However, the selection of cuts was limited. Under present methods of operation, convenience stores are not major outlets for meat, when considered on a per store basis. Nevertheless, with continued expansion in number of stores, the aggregate quantity of meat can become increasingly important. Current research underway at Kansas State University in the technology of production and marketing of frozen fresh meat is particularly relevant to convenience stores. Wider consumer acceptance of frozen meat would enhance convenience store sales.

Central Cutting and Packaging

Central cutting and packing is the use of a central unit consisting of cold storage, cutting, packaging, and delivery facilities for the servicing of a number of retail outlets. Centralized warehousing and breaking has been customary for many years for large chain metropolitan operations. The extension to retail cutting and wrapping is of more recent origin. Centralized cutting and packaging for retail appears to be increasing, but at this time there are several factors retarding its acceptance. Under present packaging techniques it is difficult to maintain the bloom of fresh beef. The irregular sizes of packaged cuts are difficult to pack and handle without breaking the packages—necessitating some rewrapping. Lack of standardization in cutting is a bottleneck to extensive centralization. This is not a problem in a given metropolitan area, but is a problem as the area under consideration widens. Certain areas have become accustomed to particular methods of cutting and to the use of names for particular cuts that would not readily move in other areas. Centralized cutting permits a greater degree of specialization in use of labor and more efficient techniques that reduce labor costs. Widespread adoption probably would result in a decrease in the number of meat cutters needed for a given volume of meat, resulting in labor union reluctance.

The farther centralized cutting and wrapping can be carried back in meat marketing channels, the greater are the savings in transportation. Elimination of bone and waste fat would reduce the tonnage of shipments by some 20–25%, and hence, an approximately equivalent reduction in freight costs.

As noted in the previous section, the freezing of fresh meat appears also to be particularly adapted to centralized cutting and packaging—if techniques can be perfected to maintain the bloom and acceptable appearance in the packaged meat. The Kansas State University research project mentioned earlier indicates that this can be done.

Centralized cutting and wrapping has been used successfully by some retailers for a number of years. Yet its use has not expanded significantly. Apparently a break-through will be required in some of the restraining factors if its growth is to be stimulated.

"Warehouse" Type Retail Food Stores

This type of outlet may approach the ultimate in no-frills retailing. With an objective of minimizing costs and associated margins, warehouse type outlets generally operate out of an unpretentious appearing building. Aisle space is relatively crowded. Shelf space for canned products may consist of stacked containers in which the goods were received. In some instances customers mark prices on individual items as they are placed in the shopping cart, bring paper bags and boxes, and do their own bagging and/or boxing at the check-out counter. Many warehouse retail stores feature generic labelled products (i.e., cans and packages without brand name). In general, the selection of meat cuts, the consistency of meat quality, its packaging, and display are on somewhat less rigid standards than in top level supermarkets, but there are exceptions. Computer assisted check-out is growing in acceptance at this type store, as well as in other types. The emphasis in warehouse retail stores is on economy.

Fast Foods

"Fast foods" is one of the latest developments in food retailing and has experienced phenomenal growth. Fast foods are fully prepared, eat-in or carry-out foods ready for consumption without additional preparation. For years larger grocery stores have operated delicatessen departments in which such meat items as luncheon meats, bacon and canned hams are carried. In some instances the delicatessen may feature ribs and chicken cooked in the store and sold warm. However, generally it remains for the consumer to cut, serve, and furnish accessory items.

Retail food operators recently have noted, with some envy, the acceptance and expansion of franchised, chain operated, ready-to-eat food establishments (such as Kentucky Fried Chicken, McDonalds, etc). Currently there is discussion and some experimentation with adapting this type of outlet to retail grocery stores, possibly as a separate department. No data are available on the volume of meat moving through current franchised establishments, but there is no question that it is significant. It appears likely that retail grocers will attempt to capture some of this market. If this happens, it will put the retailers more directly into the HRI trade, but is not likely to change meat marketing channels or methods. Added outlets with promotional activity, however, could increase the demand for meat.

STRUCTURAL CHARACTERISTICS

In previous sections it was noted that food retailing changed from an industry of predominantly small, independent operators to one typified by chain operations with large supermarkets during the past 50–60 yr. Mergers, purchases, and extensive new construction were used by chains in their expansion activities. Integration, both horizontal and vertical, was used extensively in expansion and market control. Many firms developed their own brand names and through advertising and promotion attempted to differentiate their products from those of competitors. Capital requirements for establishing a new store rose to heights that could not be attained by small firms.

There are characteristics associated with the structure of an industry which are presumed to have some effect on the conduct or behavior of firms in setting price and product policies. These, in turn, are presumed to bear some relationship to the performance of firms and the industry. In other words, under structural conditions which permit the attainment of a relatively high degree of bargaining power it might be hypothesized that firms would be able to buy inputs at prices relatively low compared to costs of production. If the same conditions existed in the selling arena, selling prices might be set relatively high in comparison to costs. Under these circumstances, the resulting performance would be profits in excess of those under more highly competitive conditions.

As evidenced by legislation designed to restrict monopolistic tendencies and unfair and discriminatory practices (e.g., Sherman Act, Clayton Act, Robinson-Patman Act) and investigations by such bodies as the Federal Trade Commission and Justice Department, the trend in these characteristics of the food industry was a matter of serious social concern. This concern extends not only to small, independent retailers but also to farmers and ranchers and to consumers in general. The Federal Trade Commission has investigated various aspects of food retailing from time to time. Congressional staff and Congressional Commissions also carry out studies. The U.S. Congress in 1964 set up a National Commission on Food Marketing. This was the most exhaustive study of recent times.

Degree of Concentration in Food Retailing

The degree of concentration in an industry is indicated by the size distribution of firms in that industry. One measurement of concentration is the market share. This is determined by calculating the percentage of an industry's business done by the 4 (or 8, or 12, or ?) largest firms. The larger the proportion of business done by a given number of firms, the greater is the degree of concentration in that industry. It generally is hypothesized that the greater the degree of concentration the greater is the possibility of firms enhancing their economic or market power. Industries dominated by a relatively few firms (e.g., U.S. automobile or steel manufacturers) presumably have a greater degree of control over price and product policy than an industry of many firms with none large enough to claim any dominance (e.g., farming). It may be noted that the mere existence of a high degree of concentration does not in itself provide proof that firms in fact utilize the economic power they may have, but the calculation of market shares usually is a first step in structural analysis. If it can be shown that concentration is relatively low, then this factor generally can be eliminated as a source of undue market power.

Earlier discussion in this chapter indicated a tendency toward concentration in the food retailing industry. It was pointed out that due to the chain and supermarket movements the trend was toward fewer and bigger firms. "The rate of expansion of the chain store movement reached its peak in the early 1930's when stores belonging to chains totaled about 80,000. By this time the chain organizations had encompassed ⅓ of the nation's retail food business" (National Commission on Food Marketing 1966). Chain store movement was followed by expansion in supermarkets. "Between the late 1940's and the late 1950's, retail chain organizations and groups of affiliated independents expanded rapidly . . . The supermarket movement progressed from 28% of the grocery store business going to supermarkets in 1948, to 69% in 1963." (The proportion increased to 81% by 1972—see Table 9.2.)

TABLE 9.2.   SHARE OF U.S. GROCERY STORE BUSINESS ACCOUNTED FOR BY SUPER-
MARKETS (sales in million $)

Size of Store in Annual Sales	1948	1954	1958	1963	1967	1972
$1,000,000 or more						
Sales	2,757	10,753	18,757	26,405	38,474	63,140
Share of grocery store sales	11.9	32.6	45.5	52.7	61.0	72.0
Increase in grocery stores						
sales over 1948 (%)	—	174	282	343	512	600
$500,000 or more						
Sales	6,437	16,017	25,202	34,460	47,046	70,730
Share of grocery store sales	27.8	48.7	61.2	68.8	75.0	81.1
Increase in grocery store						
sales over 1948 (%)	—	75	120	147	169	292

Source: National Commission on Food Marketing (1966B) as compiled from Bureau of Census.
1967 and 1972 added by author as compiled from U.S. Dept. of Commerce (1971) and U.S. Dept.
of Commerce (1975).

"The late 1950's and early 1960's marked a change in behavior and
growth patterns in most markets in the United States. The replacement
of small stores by supermarkets in most areas slowed considerably . . . Thus on the whole, a saturation point was reached, with supermarkets doing ⅔ of the grocery business. The other ⅓ went through
convenience stores and other small stores" (National Commission on
Food Marketing 1966). The number of convenience stores has expanded
rapidly in recent years; this has slowed considerably the rate of decline in
total stores numbers. Table 9.3 indicates that the number of stores was
virtually stabilized from 1964 to 1968, but the downward trend has continued since then.

TABLE 9.3.   NUMBER OF U.S. GROCERY STORES, TOTAL SALES, AND AVERAGE
SALES PER STORE, 1960–1977

Year	No. of stores	Total sales (Billion $)	Avg sales per store ($)
1960	260,050	51.700	198,808
1961	248,800	53.450	214,831
1962	234,870	56.200	239,281
1963	231,000	58.200	251,948
1964	228,600	61.600	269,466
1965	227,050	64.925	285,950
1966	227,005	67.850	298,892
1967	226,170	72.435	320,268
1968	226,700	75.935	334,958
1969	219,330	81.825	373,068
1970	208,300	88.415	424,460
1971	204,900	94.470	461,054
1972	201,050	101.700	505,844
1973	199,560	113.130	566,897
1974	198,130	130.835	660,349
1975	191,810	142.530	743,079
1976	183,700	151.980	827,327
1977	175,820	162.800	925,947

Source: Anon. (1978A)

Figure 9.2 shows that the growth of affiliated independent grocers (affiliated in wholesale purchasing under either voluntary or cooperative arrangements) has roughly paralleled that of the chains since 1948. This, together with expansion of convenience stores, has tended to neutralize the impact of growth in established chains.

The market share, at the national level, of the 4 largest grocery chains remained relatively consistent at 20% to 22% from the late 1940's to the 1970's. There are no absolute standards for evaluating whether a concentration ratio of that magnitude is excessive. Bain (1959) suggested that a ratio of less than 35% would be considered unconcentrated. It is generally agreed, however, that measures of concentration at local levels are more meaningful than those at the national level. Shoppers of consumer goods are more or less limited in access to retail outlets, and grocery shoppers frequently are the most limited.

A report by the Federal Trade Commission (U.S. Department of Commerce 1975) reported that in metropolitan areas the market share of the 4 largest food retailers averaged more than 50%, and that food chains holding the largest market shares had higher gross margins and higher net profits than stores with lower market shares. Bain (1959) classified sellers with 50% market share as moderately concentrated.

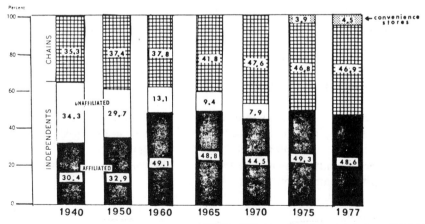

Source: Plotted by author from data in Anon. (1978A)

FIG. 9.2.   SHARE OF TOTAL GROCERY STORE SALES BY CHAINS, AFFILIATED INDEPENDENTS, UNAFFILIATED INDEPENDENTS, AND CONVENIENCE STORES

Vertical Integration

Vertical integration is the control of successive stages of production or marketing—usually by ownership or contracts. Vertical integration in meat marketing by retailers would be classed as backward integration. Successive backward stages, if carried all the way, would include wholesaling, processing, slaughtering, feedlot finishing, growing, and primary cow-calf operations. Vertical integration by retailers is not a major factor in the meat marketing system. Studies by the National Commission on Food Marketing (1966), while indicating a slightly increasing trend, showed that purchases of hogs for slaughter amounted to less than ½ of 1% of commercial slaughter, that purchases of cattle and calves usually were less than 2½%, and purchases of sheep were 3% to 3½%. A USDA (1977) study showed only 1 retailer feeding cattle, and none feeding calves, hogs, or sheep in 1976.

During a period of uncertain and unstable economic conditions in 1973 some retailers purchased livestock and engaged packers to slaughter and process them on a custom basis. That was done primarily out of necessity rather than choice, as price ceilings on meat, consumer boycotts, and rapid inflation had disrupted normal market channels. Although not classed as vertical integration, it may be noted that direct transactions between retailer buyers and packers (and processors) in effect integrate the wholesaling stage by eliminating it.

Among the motives for backward integration is the possibility of establishing a brand name and thereby effecting product differentiation. This, of course, is possible in processed meats, but to date, has extremely limited application to fresh meat. As will be noted in a later section, contracting for production under "private label," in effect, is vertical integration.

Differentiation of Product

In a perfectly competitive market situation it is impossible for firms to establish and capitalize on brand names. This is an axiom of competitive theory. That corporate chains and affiliated independents have been able to establish brand names attests to the fact that competition is something less than the competitive norm.

Food retailers, as in many other industries, have several major options in distributing branded products. For example: (1) They can own outright the manufacturing and processing facilities which enable them to establish their own brand names, manufacture to their own specifications, and carry out their own advertising and promotional activities. (2) They can contract with an established manufacturer for production under their (the retailer's) own "private label." In this case the retailer lays out the specifications of the product and carries out the advertising and promotion. (3) They can handle products manufactured and branded by an independent manufacturer or processor. Larger, well-established manufacturers carry out extensive promotional and advertising activities. Retailers, ordinarily, simply handle these products without advertising. (4) A fourth category is the same as (3) above, except the manufacturer does no advertising, and ordinarily neither does the retailer.

While large retail chains have successfully differentiated many products under their own labels, success with differentiating meats has been limited. In the case of fresh meats a common approach is the selection of a word label, other than USDA's official grade names (i.e., prime, choice, good, etc.), but carrying some connotation of quality. Distinctive packaging has not been a major factor in differentiating fresh meats. Processed meats are more adapted to branding, as this opens the possibility of distinctive flavor, color, texture, packaging, etc. No recent studies have been made in this area, but the National Commission on Food Marketing showed substantial differentiation of such products as bacon and wieners by retailers.

Barriers to Entry

Market structure theory holds that established firms in an industry which has relatively high barriers to the entry of new firms may behave differently (in price and product policies) from firms in an industry with low barriers to entry. Technically, barrier to entry is defined as price or cost advantage held by established firms in an industry relative to potential new entrants (Bain 1956).

Prior to chain and supermarket developments, food retailing was considered an industry with relatively low barriers to entry. This accounted for the numerous neighborhood "pa and ma" grocery stores. The Food Commission states, "There are strong indications, however, that post-World War II developments in food retailing have brought about a significant change in entry conditions. First, there have been dramatic organizational changes in food retailing, which themselves suggest alterations in the condition of entry. Second, various parts of this study have developed evidence which sheds light directly on the changing conditions of entry."

Organizational changes have resulted in the reduction in number of stores, increasing average size of stores, increasing capital requirements for starting new stores, and declining profits for small volume stores relative to supermarkets. Small retailers are at a disadvantage in obtaining desirable locations in competition with chains and affiliated groups.

Most of the desirable locations are in shopping centers, and developers give preference to established firms. Advertising rates favor large-scale operators and trading stamp companies prefer to have exclusive agreements with larger retailers. As already discussed, the concentration associated with corporate chain and affiliated groups give them more bargaining power and price concessions in their buying programs. Vertically integrated buying arrangements and advantages gained in conglomerate aspects of geographically large coverage of chain operations give large-scale operators advantages which are, in effect, barriers to entry by new firms.

The relevance of indicated barriers to entry in food retailing is the same for meats as most other food products. Relatively high barriers to entry give firms in large chain operations the ability to exercise some control over pricing the product. Retail meat prices are considerably more rigid than either wholesale meat prices or live animal prices. The points discussed in this section on structural characteristics (degree of concentration, vertical integration, differentiation of products, and barriers to entry) at least partially explain the basis upon which retailers derive the market power to exercise price control. These actions affect marketing margins and profit rates of large retailers.

RETAIL PRICING

Retailers have one common problem, whether they be chain or independent operator, supermarket or small store. Each has operating expenses which must be covered to stay in business. In addition, retailers hope to make a profit.

The pricing of processed meats differs little from nonmeat merchandise—the product sold can be identified with the one bought at wholesale. A predetermined margin can be applied to the wholesale cost to determine a target retail selling price. In practice, this is not as simple as it may appear if the price of each item is to cover expenses to particular items, or even to particular departments within a store.

The pricing of fresh meat, however, is considerably more complicated than that of processed meat. Most fresh meat is bought in wholesale cuts which must be fabricated (broken down) into retail cuts. In this process some shrinkage (weight loss) inevitably occurs. This may be called "cutting shrinkage." In addition, a so-called "store shrinkage" must be expected due to weight loss in rewrapping damaged packages, pilferage, spoilage, and returns by dissatisfied customers. These shrinkages are costs that must be covered, and it makes little difference whether they are included in operating expenses or listed separately.

In the cutting process, bone may be removed and fat may be trimmed away. The salvage value of bone and fat is negligible.

Figures 8.2, 8.3, 8.4, and 8.5 show the numerous retail cuts derived from various wholesale cuts. Retailers may buy sides, quarters, primal or subprimal wholesale cuts. The extent of fabrication by the retailer, the extent of cutting shrinkage, and the amount of salvage will vary with the type of wholesale cut. But, probably of greater importance, is the variation in quantity of saleable meat from carcass to carcass.

Thus, in retailing fresh meat, the individual items (cuts) sold at retail are different from the item purchased at wholesale and the quantity sold at retail is different from the quantity purchased at wholesale.

It is well known that some retail cuts are more preferred than others. Yet all must move into consumption at a fairly uniform rate. This means that individual cuts must be priced at differential rates per pound—the more desirable cuts at higher prices than less desirable cuts. The aggregate weighted average price per pound for *all* retail cuts derived from the wholesale cut must cover all expenses (including shrinkage) and yield a profit, if a profit is to be attained.

Table 9.4 illustrates several aspects of the retailer's problem in merchandising meat. It should be immediately apparent that the total value of the carcass to the retailer is dependent, not only upon retail price per pound, but also upon the cut-out. Cut-out is defined as the quantity of saleable meat obtained from a wholesale cut. In this example, the wholesale cut is an entire carcass. Cut-out long has been recognized as an extremely important factor in meat marketing. Two carcasses of identical weight and quality grade can yield significantly different quantities of saleable meat. This characteristic prompted the adoption of official USDA yield grades in June, 1965. Other things being equal, carcasses with higher cut-out have higher value. The example shown in Table 9.4 should not be construed as representative for all 600-lb choice steer carcasses.[3] Among the variables which affect cut-out are: conformation of the animal, degree of finish, weight of the carcass, sex, method of cutting, and degree of trim. By rigid specifications, careful selection of wholesale cuts, and standardization of cutting procedures, a retailer can narrow the range of cut-out obtained. But under present conditions he cannot hold this factor completely constant.

Table 9.4 also illustrates differential pricing among various cuts. Prices are shown to range from 49¢ to $3.45 per lb. The average retail price per pound of meat purchased by this retailer was $752.66 ÷ 600 = $1.25. His wholesale cost was 98.4¢ per lb. In this case, the gross margin was $162.57 which amounted to ($162.57 ÷ 752.66)100 = 21.6% of the retail

---

[3]Actually, a carcass can, and usually does, yield a greater variety of retail cuts than shown in Table 9.4, which makes the pricing problem more complicated than indicated.

TABLE 9.4. EXAMPLE: CUT-OUT, RETAIL PRICES AND RETAIL VALUE OF CHOICE 600 LB STEER CARCASS

	Cut-out per 600 Lb Carcass (Lb)	Retail price ($)	Retail value ($)
Steak			
Sirloin	34	2.35	79.90
Porterhouse	11	2.90	31.90
T-bone	13	2.75	35.75
Club	9	3.45	31.05
Round	36	1.85	66.60
Flank	2	2.45	4.90
Roasts			
Standing rib	29	2.25	65.25
Rolled rib	10	3.45	34.50
Rump roast	28	1.95	54.60
Arm chuck	28	1.75	49.00
Blade chuck	70	1.05	73.50
Boston pot roast	16	2.15	34.40
Heel of round	16	1.75	28.00
Other			
Brisket	11	1.65	18.15
Short ribs	15	1.15	17.25
Plate	18	1.15	20.70
Boneless neck	14	1.05	14.70
Stew meat	10	1.75	17.50
Shank	7	1.25	8.75
Hamburger	69	1.19	82.11
Kidney	2	0.49	0.98
Total retail wt, avg price, and retail value	448	1.72	769.49
Salvage (bone and fat)	130	0.065	8.45
Cutting shrink	22	—	—
Total weight and value	600	—	777.94
Store shrink (3.25%)			25.28
Realized gross			752.66
Cost (wholesale)			590.09
Gross margin			162.57
Percentage gross margin			21.6

value. The percentage gross margin shown is about the national average for retail meat departments, although better managers are reported to set a target at 25% (Anon. 1969).

If detailed research information were available on the demand for each retail cut, retailers would be able to adjust prices in a manner to maximize profits. Since this information is not available, retailers have had to discover market clearing prices through trial and error pricing.

Retailers, operating on a continuous basis, can observe those cuts which move slowly and adjust prices downward to speed up the movement before spoilage occurs or before those cuts have to be converted to a different form. Steaks and roasts can be converted to ground beef, for example, but this is not the way to stay in business. If certain cuts move out rapidly, this is a signal that prices can be raised.

Retailers to some extent are constrained in pricing policies by their competitors' actions. This applies particularly to price raises, but also to a lesser degree to price cuts. A retailer who raises prices above those of his competitors on cuts of a given quality is likely to lose customers. A retailer who cuts prices may expect retaliation by competitors and his actions could precipitate a price war. The latter is not likely, however, in the normal course of price adjustments designed merely to keep the usual volume of meat moving into consumption.

Approaches to Retail Pricing

Several different approaches are used by retailers in establishing selling prices. (1) A method used by many large operators is "percentage gross margin."[4] Here the retailer calculates an average weighted selling price such that the gross margin over wholesale cost is some predetermined percentage of the selling price. In the example provided in Table 9.4, the gross margin was 21.6%. As mentioned earlier, the retailer may set up a target gross margin of, say, 25% but the margin actually realized will depend upon the cut-out and prices obtained for the various cuts. The calculation of a predetermined percentage gross margin is simple if the retailer knows (a) the cost of meat laid into his retail outlet and (b) the cutout of retail product. Let us say the delivered cost of a side of beef is *wholesale* $0.90 per lb and 1.41 lb carcass beef yields 1 lb of retail beef. The average retail price which would give a 25% gross margin is obtained by the algebraic expression:

$$x = (\$0.90 \times 1.41) + .25x$$
$$x = \$1.692$$
$$\text{where } x = \text{retail price.}$$

Of course, the retailer still would be faced with the problem of pricing individual cuts to yield an average of $1.692 per lb.

(2) Some retailers use a "percentage mark-up" over wholesale cost. One study indicated this to be the most widely used method among retailers in the North Central States (Forstad 1955). The National Commission on Food Marketing (1966) stated that "of the many operating measures utilized in food retailing, (percentage) gross margin is probably the most frequently used." It is obvious that any given absolute gross margin would comprise a higher percentage of wholesale cost than of retail value. Assume the cost and cutout identical with the example above and that the

---

[4]Percentage gross margin is synonymous with percentage gross profit. It also follows that absolute gross margin is synonymous with absolute gross profit. Absolute gross margin or absolute gross profit is expressed in cents per pound rather than as a percentage.

retailer desires a percentage mark-up of 33%. Retail price can be calculated from the simple arithmetic expression:

$$\text{retail price} = (\$0.90 \times 1.41) + .33 \,(\$0.90 \times 1.41)$$
$$\text{retail price} = \$1.688$$

(3) A third approach to retail pricing is "cents-per-pound mark-up" over wholesale cost. A *Food Topics* report indicated a trend toward this method with a further comment that 25% gross margin works out to be about 10–12¢ per lb on the costs-per-pound approach. (Anon. 1969).

(4) Smaller retailers sometimes simply use a prepared "retail meat pricing chart." These charts are available from various agencies. The format generally is similar. The chart is a Table set up in such a way that the retailer locates a column corresponding to his wholesale cost (per side of beef, for example) or he may use a column corresponding to his average desired selling price. The retailer then simply follows down that column to successive rows which show him suggested retail prices for the various retail cuts that may be derived from the side. The suggested retail prices are such that their weighted average will equal either the average wholesale cost or average desired selling price as selected at the column heading. Such charts are based on a standardized cut-out of the various cuts of meat. Earlier comments indicate the weakness of accepting an average cut-out without knowing specifications of the carcass and the cutting and trimming specifications. Even established retailers who work hard at maintaining such specifications periodically run cut-out tests to keep their operations in line. The chart prices, while possibly appropriate for a particular area at a particular time, may not agree with consumer preferences in another location or at a different time. In spite of the weaknesses, price charts may serve as guides and if up-dated and made specific to particular locations, may be very helpful.

(5) A fifth approach, one which cannot be ignored by any retailer, is to be guided by the prices of competitors. As with the pricing chart, this method ignores possible differences in cut-out resulting from differences in specifications of wholesale cuts ordered by competitors, differences in cutting methods, and in trimming specifications. Undoubtedly most retailers take note of prices charged by their competitors. Stout *et al.* (1968), in a study restricted to a sample of Ohio retailers, reported, "Product pricing procedures reflected the intensely competitive nature of meat retailing. All firms, with costs and margin goals firmly in mind, priced products within the restrictive framework permitted by competitors' prices. Although net profit goals, product mix, cut-out test results, volume and turnover, brand loyalty, and other factors entered into the pricing decisions, competitors' prices were the dominant consideration."

Meat Price Specials

The complications of retail meat pricing are further compounded by the use of "special" sales—short-term (week-end or early week) substantial price reductions, usually accompanied by intensive advertising and in-store promotion. Price cuts in a "special" may amount to 20% off regular prices (Tongue 1963). The evidence indicates that consumers do in fact respond to specials. William Tongue (1963) presented data showing that a beef special raised the tonnage of beef from about 20% of weekly store meat sales to 60–65%, smoked meats from about 1% to 41% and poultry from about 9% to 29%. DeGraff (1960) reported data from the National Association of Food Chains which indicated a smaller but nevertheless significant response. The retailer knows that a price reduction will cut his gross returns per unit, but as long as the margin still yields a positive profit he hopes the increased volume will maintain, if not improve, his net earnings.

Livestock producers have questioned the effect of meat price specials on livestock prices. No definitive research studies are available on this subject. Tongue (1963) and DeGraff (1960) both attempt to show that, logically, meat price specials may be expected to benefit the livestock industry. The reasoning, in part, is that specials result in increased volume of sales. This necessitates larger purchases at wholesale, which, in turn, would be expected to add strength to the wholesale market. Somewhat contrary to this line of reasoning is an opinion that packers, eager to obtain the large volume meat sales that go with a large chain special, may be willing to grant price concessions rather than hold for higher prices. Neither of these hypotheses has been verified by empirical analysis.

FROZEN FOOD LOCKERS AND PROVISIONERS

Sales by frozen food purveyors and the use of lockers, including home freezers, are classed as retail activity in that patron and owners are final consumers. The widespread use of home freezers is an important stimulant to the demand for meat.

The commercial frozen food locker industry, including combination slaughter-locker plants, began in the early 1900's, had its greatest period of growth during the 1930's and 1940's, and reached a peak during the early 1950's. Numbers have declined since that time, coincident with an increase in use of home freezers. Most refrigerators currently on the market have a freezing compartment which adds greatly to total home freezing capacity. Many households also have a separate freezer unit. The availability of home freezers is closely associated with consumer response to meat price specials. Housewives often stock-up on specials, using home freezers for storage.

As noted above, many small locker plants have gone out of business since the early 1950's. It is reported that many of those remaining are experiencing difficulties in complying with the U.S. Wholesome Meat Act of 1967 and associated state requirements. As a result, an acceleration is expected in the rate of decline of this class of frozen food locker plant.

Of relatively recent origin is the freezer provisioner—a specialized food service firm which contracts with households to supply frozen foods, including meat. This type of operator is oriented directly toward supplying owners of home freezers. Some firms handle freezers as well as frozen food, while others handle only food. Various services are provided including cutting, wrapping, sharp freezing, house delivery, and financing. At present, only a relatively small fraction of total meat consumed is obtained in this manner. Future growth in this segment of the meat industry will depend largely on the degree to which the quality of meat delivered conforms to expectations of consumers and the level of prices as compared to supermarket prices.

BIBLIOGRAPHY

ANON. 1969. Meat marketing trends and practices. Food Topics Jan., 64–67.

ANON. 1971. Grocery business annual report—1971. Progressive Grocer Apr., 68.

ANON. 1978A. Grocery industry report for 1977. Progressive Grocer. Apr., 55.

ANON. 1978B. Grocery industry report for 1977. Progressive Grocer. Apr., 82.

ANON. 1978C. Grocery industry report for 1977. Progressive Grocer. Apr., 84.

ANON. 1978D. Grocery industry report for 1977. Progressive Grocer. Apr., 144.

BAIN, J. S. 1956. Barriers to New Competition. Harvard University Press, Cambridge, Mass.

BAIN, J. S. 1959. Industrial Organization. John Wiley & Sons, Inc. New York. 124–133.

BUTZ, D., and BAKER, G. L. 1960. The changing structure of the meat economy. Harvard Univ., Graduate School Business Admin., Div. Res.

DeGRAFF, H. 1960. Beef Production and Distribution. University of Oklahoma Press, Norman.

DeLOACH, D. P. 1960. Changes in food retailing. Washington Agr. Expt. Sta. Bull. *619*.

FEDERAL TRADE COMMISSION. 1960. Economic inquiry into food marketing. I. Concentration and integration in retailing. Federal Trade Comm. Staff Rept. U.S. Govt. Printing Office, Washington, D.C.

FORSTAD, E. C. 1955. Retailing meat in the North Central States. Indiana Agr. Expt. Sta. Bull. *622*.

HARRISON, T. G. 1962. How we developed our meat program. Food Merchandising *38*, Feb.

LEIMAN, M. 1967. Food retailing by discount houses. USDA Agr. Marketing Serv., Marketing Res. Rept. *785*.

LEIMAN, M., and KRIESBERG, M. 1962. Food retailing by discount houses. *In* Marketing and Transportation Situation. USDA Econ. Res. Serv. *MTS-114*.

MUELLER, W. F., and GAROIAN, L. 1960. Changes in market structure of grocery retailing, 1940–58. Wisconsin Agr. Expt. Sta. Res. Rept. *5*.

MUELLER, W. F., and GAROIAN, L. 1961. Changes in Market Structure of Grocery Retailing. University of Wisconsin Press, Madison.

NATIONAL COMMISSION ON FOOD MARKETING. 1966. Organization and competition in food retailing. Natl. Comm. Food Marketing Tech. Study *7*. Govt. Printing Office, Washington, D.C.

NIX, J. E. 1978. Retail meat prices in perspective. USDA. Econ. Stat. and Coop. Serv. ESCS-23, May.

PARKER, R. C. 1975. Economic report on food chain profits. Federal Trade Commission. Staff Report (unnumbered). 19.

STOUT, T. T., HAWKINS, M. H., and MARION, B. W. 1968. Meat procurement and distribution by Ohio grocery chains and affiliated wholesalers. Ohio Agr. Res. Develop. Center Res. Bull. *1014*.

TONGUE, W. W. 1963. Week-end specials pay off at retail level and increase overall consumption of meat. Proc. 58th Annual Meeting Am. Meat Inst., Chicago, Sept. 22–25.

U.S. DEPARTMENT OF COMMERCE. 1971. 1967 Census of retail trade. U.S. Dept. of Comm. Summary and Subject Statistics. Vol. 1, Gov't. Printing Office. Washington, D.C.

U.S. DEPARTMENT OF COMMERCE. 1975. 1972 Census of retail trade. U.S. Dept. of Comm., Summary and Subject Statistics. Vol. 1, Gov't. Printing Office, Washington, D.C.

USDA. 1977. Packers and stockyards resumé, USDA. Packers and Stockyards Administration, Vol. XV, No. 3. Dec.

10

Futures Markets

Interest in, and use of, livestock futures markets increased substantially during the 1970's as farmers and ranchers sought ways of improving their risk management strategy. This is a direct reflection of concern over increasing risk which has arisen from persistently accelerating capital and operating costs and extreme product price fluctuations associated with increased but erratic exports of agricultural products and weather-induced as well as cyclical variability in domestic production. These conditions are likely to continue. Farmers and ranchers will have increasing need for means of shifting price risk. The futures market is one of several available means which can be used for that purpose.

WHAT IS A FUTURES MARKET?

A futures market is a market in which contracts are bought and sold. These are contracts in which a seller agrees to deliver and a buyer agrees to accept a specified commodity (e.g., live steers) at a future time. Terms of the contract specify: (1) commodity being traded, (2) price, (3) quantity, (4) quality, (5) place of delivery, and (6) time of delivery.

Futures trading in livestock is relatively new (it started November 1964), but there is another type of livestock contract known as forward cash contract which has been used for many years, and it is important to distinguish between the two. Many examples could be given of forward cash contracts, but the following is illustrative. It is not unusual for a rancher to enter into agreement with a buyer in the spring in which he sells calves to be delivered in the fall. This contract also would specify the price, time of delivery, place of delivery, etc. Obviously, this is a contract for delivery at a future date. The buyer in this contract may, if he chooses, sell the contract (his obligation to accept) to another buyer. Presumably he would do this if the second buyer were willing to pay him something above the stated contract price in order to obtain the contract rights. The original buyer might, sometime after entering the contract, be willing to sell to a second buyer at no premium if he had reason to believe prices would move down by the specified time of delivery.

The second buyer might be willing to pay a premium for the contract if he wanted the cattle and thought the stipulated contract price was a good buy, or if he wanted the privilege of reselling the contract to still another buyer. That can happen on an advancing market. The original seller—the rancher—conceivably could buy back the contract (though he, too, might have to pay a premium to get it) and thus cancel out his obligation to deliver. This illustration involves a contract for future delivery but, technically, it is not a "futures market contract." This is commonly known as a forward cash contract. It was this sort of trading in cash contracts for the delivery of grain that prompted the organization more than a century ago of a specialized market to facilitate the trading in cash grain contracts. From this emerged the concept and techniques for trading in futures contracts.

The term "futures market" is reserved for a specialized market in which contracts are bought and sold under formal, and regulated, conditions. Trading is done only by specified people, at a specified place, during specified hours, under specified rules and regulations.

Trades can be made only by members of the organized exchange or employees of a firm which holds membership (owns a seat) in the exchange. Memberships are limited in two ways: (a) only a stated number are made available, and (b) ethical and financial standards must be met. Limitation of membership is no handicap to anyone who wishes to trade. It simply means that a nonmember must have a member make the trade for him, and memberships are held by brokerage firms which are in the business of performing this service for a fee.

Trades can be made only on the "floor" of the designated exchange. Again, this is no handicap to a nonmember who wishes to trade, as the nonmember can place his trading order through a broker by telephone. Distance from the exchange itself is no factor.

Terms of the contracts are standardized with respect to quantity (per contract), quality, allowable tolerances in delivery of quality other than that specified in the contract, premiums and discounts for delivery of quality other than the specified contract quality, place of delivery, time of delivery, and other factors which affect the value of the commodity. The market exchanges have comprehensive self-regulating safeguards. In addition, they are supervised and regulated by the Commodity Futures Trading Commission, an agency of the federal government. Thus it is apparent that a futures market is an "organized" market.

The rancher and cattle buyer used in the illustration could meet anywhere, each acts as his own agent, draws up, and signs a contract. This may be satisfactory when only a relatively few contracts are made and when the major objective is actual transfer of the physical commodity (e.g., cattle) from seller to buyer. An active futures market will complete hundreds of trades each day. This type of market facilitates transfer of the physical commodity, but only a minor fraction of futures contracts are fulfilled by actual delivery of the commodity. The futures market serves other functions which will become apparent later.

A number of futures markets are now in operation. Markets exist for trading in live slaughter cattle, live feeder cattle, live hogs, frozen pork bellies, skinned hams, boneless beef, carcass beef, wool, eggs, iced broilers, most grains, orange juice, sugar, coffee, onions, potatoes, cocoa, plywood, lumber, propane, copper, mercury, tin, several precious metals, foreign currencies, government securities, interest rates, and others.

Some markets specialize in only one or a few commodities; others handle trades in a number of commodities. Trading in livestock and/or meat is available at the Chicago Mercantile Exchange (commonly referred to as CME, the Mercantile Exchange, or simply as the "Merc"), Mid-America Commodity Exchange, Chicago Board of Trade, Minneapolis Grain Exchange, Pacific Commodity Exchange, New York Mercantile Exchange, and the Winnepeg Commodity Exchange. Wool was traded on Wool Associates of the New York Cotton Exchange until suspended in 1977. Each of these markets also handles trades in other commodities.

Like cash contracts, futures contracts specify a time for delivery (and acceptance)—a time at which the contract matures. However, where many cash contracts specify an exact date of maturity, a futures contract specifies a period within a particular month (contract month) during which delivery may be made. Each exchange, through its organization, determines which months shall be specified as contract months, and the period within the month during which delivery shall be permitted. When the Chicago Mercantile Exchange initiated trading in live cattle in November, 1964 each of the 12 months was designated as a contract month. Traders, however, centered activity primarily in contracts maturing in February, April, June, August, October and December. In 1977 trading was reopened in the January contract. At any given time of the year trades may be made in any one of these contract months. For example, in January of a given year, traders have a choice of making trades in any of the above contract months for that year and/or contracts which mature in the following year as trade activity dictates.

Trading is done by open, verbal (as well as hand signal) bids and offers between brokers, or their representatives, on the floor of the exchange as these brokers vie with each other to fill their clients' orders to the best advantage. Floor traders who trade for their own accounts also contribute to the volume of business. At any given time, on an active market, some clients are attempting to sell and others are attempting to buy. Brokers with "buy" orders will, in the interest of their clients, attempt to make purchases at the lowest possible price. Likewise, brokers with "sell" orders will attempt to make sales at the highest possible price. This, with standardized contracts, standardized marketing conditions, and a well-organized system of market information, results in a highly competitive market. Commodity futures markets, along with securities markets, are considered to conform more closely to the perfectly competitive economic model than any other type of market.

In the process of "opening an account" with a broker, a client is required to sign a margin agreement, show evidence of financial responsibility, and deposit prescribed margin funds with his broker. "Margin" is the amount of money a trader is required to put up in order to make a futures transaction. In a sense the margin is equivalent to a down payment or "earnest" money. A trader in the futures market does not put up full value of the contract—only the margin, which amounts to a relatively small fraction of the full value. The same thing is done in many cash contracts upon signing of the contract. Full payment is made upon delivery of title. Similarly in futures trading, if a contract is held until delivery of the commodity, full payment is required.

Minimum initial and maintenance margin requirements are specified by commodity exchanges. Margin requirements normally are higher for speculative accounts than for hedging accounts. If they choose, brokerage firms may require a higher margin than specified by the exchange. If, after having made a trade, the market price moves favorably for a trader, his margin fund simply stays on deposit with the broker and is returned when the trade is closed out. However, if the market price moves unfavorably, the margin is used to cover losses, and before the margin fund is exhausted, the broker will call for additional margin, or if not forthcoming, the account will be closed out by the broker.

A trader never knows who takes the opposite end of his transaction. As indicated above, the transaction is executed by two brokers (or their representatives), one acting as agent for the seller, the other for the buyer. Each commodity exchange has a clearing association (clearing house) which performs a function similar to a bank clearing house. Brokerage firms hold membership in the clearing house and all transactions are cleared through it. For every sale cleared by one broker, an identical purchase is cleared by the broker who took the opposite end of that transaction on the floor of the exchange. A client who has made a trade has an obligation to his broker and the broker has an obligation to the clearing house. This contractual obligation is just as legal and binding as any other type of contract. An important difference, however, is that futures markets are organized in such a way that traders can easily liquidate their original position prior to maturity of the contract. This can be done simply by making an offsetting trade (if the initial trade were a purchase it can be offset by a sale, or vice versa) for the same number of contracts in the same contract month as the original trade.

Traders rarely enter the market with intentions of making or accepting delivery of the actual commodity. They liquidate their position prior to termination of trading in a given contract month. It should be emphasized, nevertheless, that however rarely it may happen, it is possible to make delivery and get delivery on a futures contract. The market would not function properly without this provision.

FUNCTIONS OF FUTURES MARKETS

Future markets are considered to perform two basic functions: price setting through speculative activity (Tomek and Gray 1970), and hedging (Working 1962). Working and others have shown that the existence of futures markets depends basically upon hedging, which is the use of futures transactions in such a way as to offset the effects of an adverse price change in the cash market.

Any person who owns a commodity automatically assumes risks, one of the most important being that market prices may drop before the commodity is sold. Cattle feeders and hog feeders have seen potential profits vanish as market prices decline. A packer who has a commitment to supply meat to a wholesaler or retail buyer for a stipulated price may see his potential profits vanish as market prices move up. Hedging in the futures market is the transference of such price risks to other parties who usually are speculators.

At any given time prices generated in the futures market represent an aggregate consensus, primarily of speculators' analyses and opinions of future prices based upon current and foreseeable circumstances. As conditions change with the passing of time, the consensus may, and usually does, change. It will be mentioned later in this chapter that historically the futures market has not been an accurate forecaster of future prices, but this does not detract from its performance in hedging and pricesetting.

Speculation in the futures market, as implied above, is the voluntary assumption of price risk. Hedging in the futures market could not adequately be accomplished without speculation. While the interests of the livestock and meat industry primarily are centered in hedging price risk, it is important to understand at least the mechanics of speculation.

Speculation

Speculation in the futures market—the voluntary assumption of price risk—is not done for the fun of it. Speculators hope to make a profit. The concept of speculation in futures is deceptively simple, but consistent and long-run profits are difficult to come by. No attempt is made here to diagnose or prescribe speculative techniques or programs. Only the mechanics of speculation and its relation to hedging will be discussed.

At any given time a speculator may, by whatever means are at his disposal, make a market analysis in which he concludes that during the next "X" days, weeks, or months the futures price for a particular commodity will move up, or down, or remain unchanged. For illustrative purposes, assume that his analysis indicates the price will move up. It is obvious that he could make a profit by buying a futures contract now and reselling it later if the price does in fact rise—providing the selling price exceeded the buying price by more than the amount of charges involved in making the trades. If the speculator's analysis had indicated that the futures market price would move down, it is equally apparent that he could make a profit by selling a futures contract now and buying it back later—again providing the buying price had dropped below the selling price by an amount exceeding the charges involved in making the trades. The futures market provides the mechanism by which speculators can make such trades, but there is nothing inherent in the market which guarantees the speculator a profit. It should be perfectly obvious that if the speculator's analysis is in error—if the market moves in the opposite direction from what his analysis showed—then he would sustain a loss.

A futures market which remains relatively stable over time is of little interest to speculators, since profit potentials arise from price changes. Commodity prices (i.e., livestock, grains, etc.), historically have fluctuated rather sharply. The price of choice 900–1100 lb slaughter steers at Kansas City was $34.00 per cwt during the spring of 1970; by fall of that same year the price was $27.00. During the same period hog prices dropped from $30.00 to $15.00. In a four month period beginning August 1973 slaughter steer prices dropped nearly $20 per cwt. Grain prices also fluctuate sharply, although the government price support programs tend to dampen the variations. Over a period of time, futures prices tend to move in the same general direction as market prices for the actual, physical commodity (the latter market is called the "cash" or "spot" market to differentiate it from the futures market).

Following are hypothetical examples of two speculative ventures—one (Example A) in which prices move in the expected direction, the other (Example B) in which prices move opposite from the expected.

Example A.—Assume a speculator has opened an account with a broker and that on a January 12 his analysis indicates the price of October live cattle futures will rise during the following three months. The speculator on January 12 places an order with his broker to buy two units of CME October cattle "at the market." "At the market" leaves it up to the broker to get the best buy possible at that time.[1] Let us say the order is executed at $48.50 per cwt. At this point (January 12) the speculator is "long" two units of October cattle. Each contract unit, by CME specification, is 40,000 lb liveweight, so the total value would be

$$2 \times \frac{40,000}{100} \times \$48.50 = \$38,800.00$$

Now assume that three months later (April 12) the speculator closes out his position. He will do this by placing an order with his broker to sell two units of CME October cattle. The broker executes the order at, we shall say, $51.00 per cwt. This liquidates the speculator's position in October cattle. The value of the sale would be

$$2 \times \frac{40,000}{100} \times \$51.00 = \$40,800.00$$

Gross profits would be $40,000.00 − 38,800.00 = $2,000.00. The broker's commission and interest on margin funds must be deducted from gross profits. Futures markets suggest commission rates and margin requirements, but individual brokerage firms have flexibility in setting the levels. We shall assume a commission of $50 per contract. That covers both ends of the transaction, i.e., the original purchase and subsequent sale. Commission on 2 contracts would be $2 \times \$50 = \$100$. Margin requirements are assumed to be $1200 per contract or a total of $2400. Interest at 9% for 3 months would be $0.08 \times \$2400 \times \frac{3}{12} = \$48$. Net profits then would be $2000 − $100 − $48 = $1852.

Example B.—In this case assume everything identical with Example A up to the point of liquidating the original position. At this point we assume prices have declined so that the liquidation sale of two units of October cattle is made at $46.50 per cwt. Under these conditions the sale value would be

$$2 \times \frac{40,000}{100} \times \$46.50 = \$37,200.00$$

Gross profits would be $37,200.00 − $38,800.00 = −$1600.00. Commission charges and interest on margin would be the same as in Example A.[2] Net profits therefore would be

$$-\$1600 - \$100 - \$48 = -\$1748.$$

The speculator under these circumstances would have sustained a loss of $1748.

---

[1]Other types of orders are: (1) "Limit order," can be executed only at a specified (or better) price. (2) "Open order," also specifies the price, but the order stands until it is executed or cancelled. (3) "Day order," must be executed on a specified day—or is automatically cancelled. Unless otherwise stated, all orders are considered to be day orders. (4) "Stop (loss) order," stipulates a price, then, if or when the market reaches that price, a market order becomes effective.

[2]Additional margin probably would be required, but this is ignored for simplification.

There are, of course, many other types of speculative ventures which can be undertaken in futures markets. The type illustrated here is one of the most elementary, but the two assumed outcomes will serve to illustrate the place of speculation as an adjunct to hedging. Without speculation, hedgers at times would have difficulties in placing orders. This will be apparent in the following discussion.

Hedging Price Risk

Hedging price risk in the futures market is the use of futures market transactions in such a way as to offset the effects of adverse price movements in the cash market. Whether a price change is "adverse" depends upon the position one holds in the actual commodity (in the cash market). To the cattle feeder who has a pen of cattle on feed, it is clear that a price decline in cattle would be adverse. To a packer who has an agreement to deliver dressed beef carcasses at a stipulated price, but who has not yet purchased cattle for slaughter, a price rise in cattle would be adverse.

A hedge is accomplished by taking a position in the futures market opposite from the one held in the cash market. The cattle feeder mentioned above has "bought" the cash commodity (actual cattle). His position is said to be "long" the cash market. He would execute a hedge by taking an opposite position (selling) in the futures market. He then is said to be "short" the futures market and in trade terminology this would be called a "short hedge."

The position described for the packer above is said to be "short" the cash market. He would execute a hedge by taking a "long" position (buying) in the futures market. This is called a "long hedge."

How can the hedger be sure that he will be able to take the desired position in the futures market? Here is where speculation comes in. Any transaction requires the participation of two parties. A sale cannot be made in the futures market, or any other market, unless someone else takes the opposite side of the transaction (buys). A purchase cannot be made unless someone else sells. A hedger can never be sure that a speculator stands ready to take the opposite side of a transaction at a particular price the hedger may choose, but in an active, viable market he can be sure that a speculator will take the opposite side at some price. And, if the market has a relatively high volume of hedging and speculative activity, the hedger can be reasonably sure that a trade will be executed near the price level at which other trades are being made at that particular time. Speculators are needed to give markets the flexibility and breadth to absorb hedges without undue price reaction.

Theoretical Cash-Futures Price Relationships.—For the futures market to provide an adequate hedging mechanism, two fundamental cash-futures price relationships must hold: (1) over a period of time cash prices and futures prices for a given commodity must move in the same general direction, and (2) as a futures contract expires (during month of contract maturity) cash price (at par delivery points) and the price of the expiring futures must come reasonably close together. Figure 10.1 illustrates these relationships. The point of major concern with both of these relationships is the stability and predictability of the difference between cash and futures prices. That difference is known as the "basis."

In the first instance, a hedger expects that, over time, if one advances, the other will advance; if one declines the other will decline. Erratic

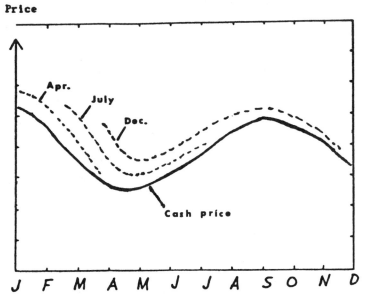

FIG. 10.1.   TYPICAL RELATIVE MOVEMENTS OF CASH AND FUTURES PRICES

day-to-day, or short period deviations from this proposition occur and are not particularly damaging, but sustained, unpredictable movements would render the futures market untenable for hedging purposes. For effective hedging, variations in the difference between the two prices (i.e., variations in the basis) must be less than variations in cash and futures prices. This has been shown to be the case (Leuthold 1977; Price et al. 1978).

Futures prices, basically, are determined by traders' (primarily speculators) expectations of supply and demand conditions as deducted from current and anticipated conditions.[3] Current cash prices are based upon appraisal of the current supply-demand situation, conditioned by

[3]It will be shown later that the futures market has not been an accurate forecaster of cash prices over an extended period, but it was not organized for price forecasting purposes.

future expectations. Thus, cash and futures prices usually react in the same direction to continuously developing supply and demand factors. Tomek and Gray (1970) state, "The element of expectations is imparted to the whole temporal constellation of price quotations, and futures prices reflect essentially no prophecy that is not reflected in the cash price and is in that sense already fulfilled." Records show that, over time, cash prices and futures prices usually move in the same general direction though not necessarily at the same rate (Blau 1944–1945).

The tendency for cash prices and futures prices at par delivery points[4] to come reasonably close together at expiration of a futures contract is based upon the fact that any point of time in the future inexorably becomes the present with the passage of time. As mentioned earlier, futures contracts can be liquidated by delivery of the actual commodity (though this seldom is done). The possibility of delivery sets up powerful speculative forces which prevent a wide divergence of prices at termination of futures contracts. To use an exaggerated example, assume that as a June cattle contract approached expiration (during the month of June) the cash price of contract grade cattle in Omaha, Nebr., or Sioux City, Iowa (designated par delivery points for both live slaughter cattle and live feeder cattle CME contracts) was $5.00 per cwt above the futures price. A strong incentive would exist for speculators to buy futures, hold the futures until actual cattle were delivered, then immediately sell these cattle on the higher cash market. On the other hand, if cash prices were $5.00 below futures prices as expiration of the futures approached, a strong incentive would exist for speculators to sell futures and when delivery was possible, to buy some actual cattle in the relatively low-priced cash market, then deliver them in satisfaction of the futures obligation. Thus, forces are set in motion which bring cash prices (of the contract grade) and futures prices of an expiring future fairly close together as trading in the contract month terminates. A study of Ehrich (1967) showed that futures prices tended to converge with the average price of 1100–1300-lb choice steers at Chicago. Of course, there are uncertainties in exact time of delivery, exact quality that might be delivered,[5] and in elapsed time between delivery and resale of the actual commodity so that cash and futures prices are not necessarily identical at contract determination, hence use of the term "reasonably" close.[6]

---

[4]Par delivery points are locations specified by futures markets as points where commodities may be delivered to satisfy futures contracts at the contract price. Prior to closing of the Chicago Stock Yards in 1971, Chicago was the par delivery point for live cattle futures contracts.

[5]Presumably a trader who chooses to deliver actual cattle in fulfillment of a futures contract would deliver the lowest quality which meets contract specifications. However, grading live animals is somewhat subjective.

[6]Occasionally a "squeeze" develops as a contract is being closed out. Traders become anxious to get out before delivery and futures deviate temporarily from cash prices. This, however, is a short-lived phenomenon.

While these two fundamental price relationships hold within reasonable limits there is a substantial amount of variability and uncertainty associated with the live animal basis. This will be discussed in a later section.

Since the cash price and futures price tend to come together at futures contract maturity, this implies that the basis relative to the "near" futures narrows as the near futures terminates. But which futures market and which cash prices are we concerned with? Any reference to basis should be specific with respect to both points. The several markets which handle livestock and livestock products were listed earlier. Since the bulk of futures trading in these commodities at this time is done in the Chicago Mercantile Exchange, that market will be used in the following discussion as a source of reference for futures price quotations. There are, however, many cash markets for livestock and meat throughout the United States. The earlier comment about cash and futures prices coming reasonably close together at contract maturity specified the cash price at par delivery points. A cattle feeder at Dodge City, Kan., who sells finished cattle to a local packer needs to know the Dodge City basis, i.e., the amount by which Dodge City cash price is below (or above) the CME cattle futures price for a specified contract month. To effectively appraise a potential hedging situation, every feeder needs an understanding of the concept of "basis" and a knowledge of the actual basis of the market in which he sells relative to the futures in which he contemplates placing a hedge. The concept is simple and futures prices are relatively easy to obtain. However, it is difficult to obtain adequate local cash prices at some local points for comparison with the futures prices. Not only are local price quotations by sex, grade, and weight often difficult to obtain, but studies have shown that the basis varies over time. Figure 10.2 shows the Kansas City live steer basis. Figure 10.3 shows the St. Joseph, Mo., live hog basis. A substantial variation has existed in the basis over the years with an apparent increase in variability in recent years.

To this point we have been concerned with the locational basis, i.e., we have assumed that local cash price being compared to futures price was for cattle of identical sex, weight, and grade as specified in the futures contract. On occasion, livestock producers may be interested in a basis which also includes other factors, such as steers of lower or higher grade, heifers (instead of steers), or weights and dressing yield different from those specified in the futures contract. Hog producers may desire the basis for hogs which do not meet contract specifications.

Very little empirical work is available on the applicable basis for any of these factors at local markets. However, producers often can obtain reasonable estimates from experienced producers and market personnel.

Mechanics of Hedging.—Four steps can be identified in every hedging

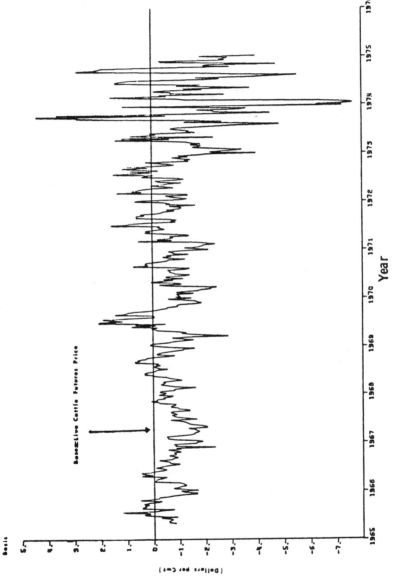

FIG. 10.2.   KANSAS CITY LIVE CATTLE BASIS, MAY, 1965—DEC. 1974

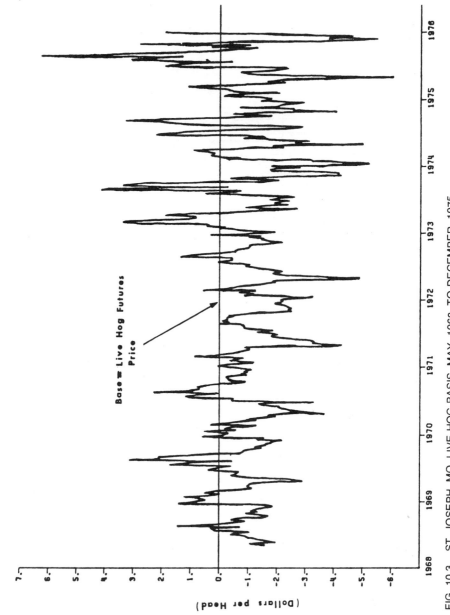

FIG. 10.3.   ST. JOSEPH, MO. LIVE HOG BASIS. MAY, 1968, TO DECEMBER, 1975

operation—2 in the cash market and 2 in the futures market. They are: (1) Assumption of original position in cash market. (2) Assumption of original position in futures market. (3) Liquidation of position in cash market. (4) Liquidation of position in futures market.

Steps 1 and 2 usually are carried out at, or about, the same time in a *traditional* hedge. At a later date Steps 3 and 4 are carried out at, or about, the same time in the traditional hedge. In the more modern (selective) hedging programs, exceptions are made to these generalizations (and also to most other generalizations), but for purposes of this exposition it will be assumed that transactions in Steps 3 and 4 also are executed simultaneously, but at a later date than Steps 1 and 2.

Steps 2 and 4 require the services of a broker. The livestock producer who contemplates hedging should "open an account" with a brokerage firm prior to the time he wishes to place the hedge. The producer also should acquaint himself with specifications of the futures contracts. The various exchanges publish and distribute free brochures with this information. Most brokerage firms have them for distribution.

In the examples immediately following, none of the transactions are dated. This is to simplify the presentation. It does, however, oversimplify one problem. In executing a hedge in the futures market it is necessary to specify the contract month in which the hedge is to be placed. In general, a hedge is placed in the contract month which coincides with anticipated sale month of the actual commodity, or the contract which matures next after the anticipated sale date of the actual commodity. For example, if feeder cattle are bought on Nov. 1, to be fed 140 days, the sale date would be March 18. Futures trading in slaughter cattle is not carried on in a March contract, so a hedge on such cattle usually would be placed in the April contract.

Examples of Hedges.—Assume in the following hypothetical examples that an account has been opened and initial margin requirements deposited with a brokerage firm at the time hedge transactions are initiated.

Example I: "Short" Hedge of Cattle on Feed.—Consider that 600-lb choice feeder steers are purchased, fed 140 days, and sold as 1000-lb choice slaughter steers. At the time feeder cattle are purchased, cash prices are $58 per cwt for choice feeders and $56 for choice slaughter steers at the market used by the cattle feeder.

The conditions set up are elementary. It is assumed that: (1) a short hedge is executed at the same time the feeder cattle are purchased (i.e., today) and the futures position is liquidated by an offsetting transaction at the same time the cattle are sold (i.e., 140 days after today); (2) cash and futures price movements are exactly parallel; (3) the operation is fully hedged during the entire feeding period; (4) there is no death loss;

(5) there are no marketing costs in the cash market; (6) margin requirements are $1200 per contract unit (margin requirements sometimes are less on hedge transactions than on speculative transactions) with interest on margin funds at 9%; and (7) commission is $50 per contract unit.

Three alternative price movements are considered: downward 1(a), upward 1(b), and no change 1(c).

Tables 10.1, 10.2, and 10.3 summarize hedging results under the three assumed price movements. It will be noted that the indicated net returns are identical ($1140) under each of the three alternatives. This follows from the assumption of identically parallel movements in the cash and futures markets—admittedly an unrealistic assumption, but deliberately chosen to illustrate the point that a hedge tends to offset the effects of price change, whether the change be a decline or an advance.

Results in Table 10.1 show that the hedge provided protection against a cash price decline. This adverse price movement was offset in the futures transaction. The feeding operation would have sustained a loss of $10,400 if carried out unhedged. Under hedged conditions (as assumed), the futures market gain ($12,000) offset (in fact more than offset) the cash market loss and resulted in a net return of $1140.

It is apparent in Table 10.2 that a hedged feeder will forego windfall profits that otherwise would accrue from a price advance. If the operation were unhedged, a gain of $13,600 would be realized from cash market transactions. But with advancing prices, the futures transactions would sustain a loss of $12,000. A hedge "protects" from a favorable price movement just as it does from an adverse price movement.

If prices remain steady (unchanged), nothing is gained from a hedging operation (see Table 10.3). In fact, the commission and interest on funds tied up in margin would reduce net return by those amounts.

In real world hedging operations, the cash and futures markets seldom move in exactly parallel fashion. In a short hedge, if the futures price drops more than the cash price, a hedger will gain more on the futures transaction than he loses on the cash transaction. If the futures price drops less than the cash price, he will gain less on the futures transaction than is lost on the cash transaction.

In the previous examples, the futures price at initiation of the hedge (i.e., $57 per cwt "today") was indicated to be higher than the average total cost per hundredweight of the finished cattle ($110,400 ÷ 2000 cwt = $55.20 per cwt). The latter is known as "break-even" price. This situation was deliberately built into examples 1(a), 1(b), and 1(c). Real world market conditions are not always that accommodating. At times the futures price is less than "break-even" price. When this situation exists, the execution of a hedge usually would result in a net loss. This is illustrated in Table 10.4 where it is assumed that cash prices decline during the feeding period—and the result was a net loss of $4860.

TABLE 10.1. HEDGING EXAMPLE 1(a): "SHORT" HEDGE OF CATTLE ON FEED—ASSUMING PRICE DECLINES DURING FEEDING PERIOD

Cash Market ($)	Futures Market ($)	Results ($)	
Today:	Today:		
(1) Buy 200 feeder steers @ $58/cwt	(2) Sell futures contracts, 5 units @ $57/cwt		
Cost of feeders 69,600			
Cost of grain @ $51/cwt 40,800			
Total cost $110,400	Total value of sale $114,000		
140 days from today:	140 days from today:	Loss in cash market	$10,400
(3) Sell 200 choice slaughter steers @ $50/cwt	(4) Buy futures contracts, 5 units @ $51/cwt	Gain in futures market	12,000
		Less commission	250
		Less interest on margin	210
Total returns $100,000	Total value of purchase $102,000	Net profit	$1140

TABLE 10.2. HEDGING EXAMPLE 1(b): "SHORT" HEDGE OF CATTLE ON FEED—ASSUMING PRICE INCREASES DURING FEEDING PERIOD

Cash Market ($)	Futures Market ($)	Results ($)
Today:	Today:	
(1) Buy 200 feeder steers @ $58/cwt	(2) Sell futures contracts, 5 units @ $57/cwt	
Cost of feeders 69,600		
Cost of gain @ $51/cwt 40,800		
Total cost $110,400	Total value of sale $114,000	
140 days from today:	140 days from today:	
(3) Sell 200 choice slaughter steers @ $62/cwt	(4) Buy futures contracts, 5 units @ $63	
Total returns $124,000	Total value of purchase $126,000	
	Gain in cash market	$13,600
	Loss in futures market	12,000
	Less commission	250
	Less interest on margin	210
	Net return	$1,140

TABLE 10.3. HEDGING EXAMPLE 1(c): "SHORT" HEDGE OF CATTLE ON FEED—ASSUMING PRICE REMAINS UNCHANGED DURING FEEDING PERIOD

Cash Market ($)		Futures Market ($)		Results ($)	
Today:		Today:			
(1) Buy 200 feeder steers @ $58/cwt		(2) Sell futures contracts, 5 units @ $57/cwt			
Cost of feeders	69,600				
Cost of gain	40,800				
Total cost	$110,400	Total value of sale	$114,000	Gain in cash market	$1,600
				Gain in futures market	0
140 days from today:		140 days from today:		Less commission	250
				Less interest on margin	210
(3) Sell 200 choice slaughter steers @ $56/cwt		(4) Buy futures contracts, 5 units @ $57/cwt		Net returns	$1,140
Total returns	$112,000	Total value of purchase	$114,000		

TABLE 10.4. HEDGING EXAMPLE 1(d): "SHORT" HEDGE OF CATTLE ON FEED—ASSUMING FUTURES PRICE AT INITIATION OF HEDGE IS BELOW BREAK-EVEN PRICE

Cash Market ($)	Futures Market ($)	Results ($)
Today:	Today:	
(1) Buy 200 feeder steers @ $58/cwt	(2) Sell futures contracts, 5 units @ $54/cwt	
Cost of feeders 69,600		
Cost of grain @ $51/cwt 40,800		
Total cost $110,400	Total value of sale $108,000	Gain in cash market $3,600
		Loss in futures market 8,000
		Less commission 250
		Less interest on margin 210
Break-even price		
$\dfrac{\$110,400}{2,000\ \text{cwt}} = \$55.20/\text{cwt}$		
		Net loss $4,860
140 days from today:	140 days from today:	
(3) Sell 200 choice slaughter steers @ $57/cwt	(4) Buy futures contracts, 5 units @ $58/cwt	
Total returns $114,000	Total value of purchase $116,000	

It needs to be emphasized that hedging in itself does not guarantee a profit. A hedge only sets the selling price. In this sense, it can be used as a forward pricing device. If that price is above the break-even price, a profit is indicated, but if it is below the break-even price, a loss is to be expected. It will be noticed that in this last sentence the reference to profit and loss is qualified. This is necessary because if cash and futures prices do not move in a parallel manner, i.e., if the basis does not remain constant, then profits (or losses) will vary as mentioned in the paragraph immediately above.

In each of the four examples (Tables 10.1, 10.2, 10.3, and 10.4), it can be seen that the assumed cash sale price of the cattle was $1 below the liquidation price of the futures. That is, the basis was minus $1 at the time the transactions were closed out. This conforms to the earlier statement that cash prices and futures prices "tend to come reasonably close together at par delivery points at contract maturity." The basis tends toward an amount equal to delivery costs. At non-par delivery points freight costs add to the basis. Furthermore it usually is not possible (and in fact it is not desirable) to attempt to liquidate the hedge during the final days of a terminating contract, at which time cash and futures prices are expected to approach each other. A hedge usually is liquidated as the cattle are sold which may come prior to maturity of the contract month in which the hedge is placed. The basis is less predictable at such times than near contract termination.

The principles and mechanics used by a producer in hedging feeder cattle and hogs on feed are similar to those explained for hedging slaughter cattle. No explicit examples are given.

Example II: "Short" Hedge of Hogs Bought on Contract.—Packers at times buy hogs from producers on forward cash contract before the hogs are finished out. Terms of the contract call for delivery when the hogs are finished, but the price may be set upon signing the contract.

A primary reason packers enter such contracts is to line up supplies in advance. Producers may be willing to contract to reduce price uncertainty. The producer may prefer such a cash contract in lieu of hedging in the futures market. The packer in this situation may want protection against a price decline. Why? If hog prices decline, this packer's competitors will be able to obtain their kill at prices lower than the contract price to which this packer is committed. In sale of the meat, then, this packer would be at a disadvantage.

By hedging in the futures market the packer can shift this price risk. Upon signing the cash contract the packer assumes a *"long"* position in the *cash market*. He can hedge this position by taking an opposite (*"short"*) position in the *futures market*. Table 10.5 illustrates the results of a hedge in which a contract price was set at $44 per cwt on July 1 but

TABLE 10.5. HEDGING EXAMPLE II: "SHORT" HEDGE OF HOGS BOUGHT BY PACKER ON CONTRACT-ASSUMING PRICE DECLINES $4 PER CWT IN CASH HOG MARKET

Cash Market ($)	Futures Market ($)	Results ($)
On July 1:	On July 1:	
(1) Packer contracts 300 hogs for Sept. 20 delivery @ 44/cwt	(2) Packer sells Oct. futures, 2 units @ 45.50/cwt	
Total value of contract $26,400	Total value of sale $27,300	
On Sept. 20:	On Sept. 20:	
(3) Hogs are delivered to packer who pays contract price of 44/cwt, but cash price has dropped to 40/cwt	(4) Packer buys Oct. futures, 2 units @ 41.50/cwt	Contract cost of hogs $26,400
		Less gain in futures market 2,400
		Plus commission 80
		Plus interest on margin 30
Total market value $24,000	Total value of purchase $24,900	Net cost to packer $24,110

cash price had dropped to $40 per cwt upon delivery Sept. 20. Market sale weight is assumed to be 200 lb per head. A contract unit is 30,000 lb liveweight. Commission is calculated at $40 per unit (round-turn) and margin requirements (for a hedged transaction) at $750 per unit with interest at 9%.

The results in Table 10.5 show that gains in the futures transaction offset declines in the cash market price. The packer's net cost would be $24,100. The cost unhedged would have been $26,400. In this example, the gain in the futures market would have offset the $2400 decline in cash value. In a sense the $80 commission and $30 interest on margin could be classed as insurance premium against the price decline in the cash market.

Example III: "Long" Hedge of Forward Beef Sales.—Packers, at times, enter contracts with large retail and wholesaler buyers for the sale of meat of certain specifications—the meat to be delivered at some specified later date—and the price is established at negotiation of the contract. If the packer has livestock already bought, he knows the cost and can price the meat accordingly. On the other hand, if he has not yet bought livestock, he could, if he chose, enter the contract, price the meat, and utilize the futures market to protect himself from an advance in livestock prices.

Unofficial reports indicate this type of hedge is seldom used by packers, but the possibility merits a brief presentation. Flour millers have long used such hedges in contracting for delivery and pricing flour before actually purchasing the wheat from which flour will be milled.

In this example, assume (1) the meat sold under contract is 72,000 lb of USDA choice steer beef carcasses weighing 600 lb each at a contract price of $90 per cwt with delivery in 60 days; (2) the packer has not yet bought cattle to fill the contract; and (3) cattle prices advance $3 per cwt by the time cattle are purchased. The price risk of such meat sale conceivably could be hedged in either beef carcass futures or live cattle futures. Futures markets have been established for beef carcasses but have never attracted sufficient volume for effective hedging. However, it is relatively easy to translate carcass requirements into live cattle requirements and a high correlation exists between wholesale carcass prices and live cattle prices. The sale of beef carcasses, therefore, can be hedged in the live cattle futures. Commission and margin costs for live cattle futures hedging transactions are calculated at $50 and $1200, respectively, per contract unit.

Table 10.6 illustrates how this contract sale could be hedged by a purchase in the live cattle futures. The $3 advance in cash live cattle prices is assumed to be matched by an equivalent advance in the futures market, so the additional cost of cattle would be offset by a gain in the futures market. Unhedged, the $3 per cwt advance in the cash market

TABLE 10.6.   HEDGING EXAMPLE III: "LONG" HEDGE OF FORWARD SALE OF BEEF—ASSUMING PRICE ADVANCE OF $3 PER CWT IN CASH CATTLE MARKET

Cash Market ($)		Futures Market ($)		Results ($)	
Today:		Today:			
(1) Sell 72,000 lb choice beef on contract @ $90/cwt (equivalent to 120 head of steers [1000 lb each] @ $56/cwt[1])		(2) Buy futures contracts, 3 units @ $57/cwt			
Total value of sale	$67,200	Total value of purchase	$68,400	Loss on cash market	$3,600
				Gain on futures market	3,600
				Less commission	150
				Less interest on margin	50
56 days from today:		56 days from today:			
(3) Buy 120 choice steers @ $59/cwt		(4) Sell futures contracts, 3 units @ $60/cwt		Net cost	$200
Total cost	$70,800		$72,000		

[1]$56 per cwt is considered to be live weight equivalent of dressed cost.

would have resulted in an additional cost of $3600 to the packer. This would be offset by the hedge so the only additional cost would be $200, the cost of commission and interest on margin. As in the previous example, this can be looked upon as a premium for price risk insurance.

This same type hedge could be used by large retailers in (1) advertising programs where retail prices are announced before wholesale meat purchase contracts have been made, or (2) fixing the purchase price of wholesale meat at a time when their market analysis shows the likelihood of advancing prices. In both instances the price risk is one of advancing prices and the initial step in the futures market would be a long position. This position would be liquidated as actual supplies were purchased in the cash wholesale meat market. There is in actuality little evidence that retailers use such hedges. In most cases retailers are able to obtain a firm agreement on meat prices prior to advertising and as a matter of operating policy retail prices (with the exception of specials) are held relatively steady compared to wholesale meat and live animal prices.

Example IV: Hedging Feed Costs.—Feed cost is a principal component of the total cost of finishing livestock, and the major share is grain cost. A livestock feeder who purchases grain for feeding can, if he chooses, use the futures market to protect himself against a rise in grain prices. Attention is called to an earlier assertion that hedging livestock prices does not in itself guarantee a profit. Likewise, the hedging of feed prices does not in itself guarantee a profit. Hedging grain prices, in effect, sets the price of grain and this can be done at the start of the feeding period (or earlier). Whether or not a feeder chooses to hedge will depend on (a) whether the current grain futures prices permit him to set (lock in) prices which are favorable and (b) the feeder's analysis of whether grain prices are likely to rise. Corn and grain sorghum (milo) futures are available. Corn is actively traded, but little interest has been shown in the grain sorghum futures. Since cash prices of corn and grain sorghum are closely correlated, a feeder who desires price protection in grain sorghum can hedge in the corn futures.

Explicit step-by-step procedures will not be shown in this case. It is clear that the risk to be avoided is a rise in grain prices. This calls for a long hedge (a purchase of futures) against grain requirements. A feeder would (a) calculate his grain requirements, (b) place a hedge by purchasing futures, then (c) as grain is purchased on the cash grain market he would liquidate an equivalent quantity of futures, hold the remainder until additional grain is purchased, then liquidate more futures.

Soybean meal and some other feed ingredient prices may be hedged in futures markets in a like manner.

Use of a Hedge Worksheet.—It is emphasized, again, that hedging in itself does not guarantee a profit. Each hedging situation should be evaluated by an analysis of estimated costs and estimated returns to labor, management, and profit before a hedge is executed. Profit is defined here as an amount over and above an acceptable return to labor and management. In other words, it is considered that an operator will classify return to his labor and management as a cost to be covered before profits are calculated. Some operators may prefer to place a zero value on labor and management and consider everything over operating costs as profit. This is up to each individual.

Table 10.7 is the author's version of a worksheet designed to evaluate the feasibility of a hedge in terms of profit and returns to labor and management. Even a cursory review of this worksheet reveals two critical components—costs, and basis. Costs are listed as: cost of feeder cattle, cost of gain, marketing costs (in cash market), hedging costs, and if an operator chooses to enter it, an "acceptable operator's return to labor and management."

Basis, it will be recalled, is the difference between the applicable cash market price (i.e., the cash market in which the livestock will actually be sold) and the futures price. As the worksheet shows, this difference can be affected by market location, grade, sex, weight, and dressing yield. If a feeder were located near a par delivery point, feeding choice steers that would weigh 1050–1150 lb with a carcass yield 61%, each of the indicated items of basis adjustment would be minimal, for these characteristics meet par specifications of the CME futures contract. However, there are costs associated with delivery, e.g., grading, yardage, fee for handling by firm on the yard, etc., and there will be some trucking cost regardless of how close the feedyard may be to the delivery yards. If the feeder operated near Greeley, Colo., or Lubbock, Texas, for example, he would need to enter a larger locational differential. If he were feeding cattle of lower grade than choice, he would need to enter a grade differential, and so on for sex, and liveweight and dressing yield.

A question might be raised whether heifers can be hedged in a futures which specifies steers as the base commodity. Obviously, heifers cannot be delivered to satisfy the contract, but delivery is not anticipated. The steer futures could be used, and satisfactorily—providing cash heifer price movements generally parallel the steer futures price movements, even though they are at different absolute levels. The futures still could be used for hedging on a selective basis (even though cash and futures price movements are not parallel) if they move in a predictable pattern so a determination could be made as to when it is advantageous to hedge and when not to hedge.[7] Records show that slaughter steer and heifer prices for given grade and comparable weights do move in a generally parallel fashion. Deviations from parallel movements occur with variations in relative supplies, but such change occurs at a slow rate.

TABLE 10.7. HEDGE WORKSHEET FOR CATTLE

Buy _____ lb _____ feeder _____; feed _____ days; sell _____ lb _____ slaughter cattle
       (grade)   (steers or   (grade)
                heifers)

A. Estimated Costs:
  (1) Cost of feeder per head (_____lb @ $_____)......................... $_____
  (2) Cost of gain:
  (3) (a) Feed costs per head...................................... $_____
  (4) (b) Nonfeed costs per head................................. $_____
  (5) (c) Total cost of gain per head [line (3) + line (4)] .... $_____
  (6) Marketing costs per head (in cash market)............................. $_____
  (7) Hedging costs per head... $_____
  (8) Total all costs per head [line (1) + line (5) + line (6) + line (7)]................ $_____
  (9) Acceptable operators return per head for labor and management............. $_____
  (10) Break-even amount, total per head [line (8) + line (9)]........................... $_____
  (11) Break-even price, per cwt [line (10) ÷ sale wt]   .... $_____

B. Estimated Hedge Selling Price:
  (12) Current Chicago futures price, _____ future, per cwt........... $_____
                          (month)
  (13) Basis—adj for market location differential per cwt. .... $_____
  (14) Basis—adj for grade differential per cwt................................. $_____
  (15) Basis—adj for avg. livewt and dressing yield............................. $_____
  (16) Basis—adj for heifer differential per cwt................................. $_____
  (17) Effective futures hedge price
     [line (12) ± line (13) ± line (14) ± line (15) − line (16)]......................... $_____

C. Estimated Profit:
  (18) Est hedged profit per cwt [line (17) − line (11)] .... $_____
  (19) Est hedged profit per head [line (18) × sale wt]   .... $_____

D. Estimated Hedged Return to Labor, Management, and Profit per Head:
  (20) Line (9) + line (19).. $_____

Worksheet Evaluation of a Cattle Feeding Hedge.—In Table 10.8, an example is shown of a completed cattle hedge work sheet. It is suggested that such an analysis be completed prior to purchasing feeder cattle. Every feeder should have a solid estimate of his cost of gain, including nonfeed costs. It is recognized that cost of gain varies with weather

---

[7]This is precisely what is involved in hedging steers on a local basis. For example, a feeder at Guymon, Okla., does not expect his local price level to equal that at Chicago, but if the Chicago futures is to work satisfactorily as a hedging mechanism, his local cash price movements must generally parallel futures prices or, if not parallel, then move in a predictable relationship.

conditions, inherent gainability of cattle, etc., but many factors are controllable. Over a period of time, with the aid of a good set of records and efficient management, cost of gain can be estimated with an acceptable degree of accuracy. Without such an estimate the evaluation of a hedging potential is seriously handicapped.

In calculating basis adjustments, the first prerequisite is a familiarity with futures market contract specifications. Any deviation from those specifications gives rise to variations in the basis. Items listed in lines 13, 14, 15, and 16 of the worksheet (Tables 10.7 and 10.8) indicate the major factors, but not all will apply in every case. Information needed for basis adjustment usually is not available in published form. However, experienced producers and market personnel have satisfactory knowledge of the price effect of such factors.

In the example of Table 10.8 costs (line 8) are shown to be $591.80. Line 9 provides a space for entering an amount for operator's labor and management. As mentioned earlier, this is a matter for each operator to

TABLE 10.8.   HYPOTHETICAL EXAMPLE: HEDGE WORKSHEET FOR CATTLE

Buy 700 lb choice feeder steers; feed 135 days; sell 1060 lb choice slaughter cattle
     (Grade)   (steers or   (grade)
                heifers)

A. Estimated Costs:
   (1) Cost of feeder per head (700 lb @ $54.00) $378.00
   (2) Cost of gain:
   (3) (a) Feed cost per head $156.00
   (4) (b) Nonfeed costs per head................................$ 52.00
   (5) (c) Total costs of gain per head [line (3) + line (4)].............. $208.00
   (6) Marketing costs per head (in cash market) .... $ 4.00
   (7) Hedging costs per head ....$ 1.80
   (8) Total all costs per head [line (1) + line (5) + line (6) + line (7)].............. $591.80
   (9) Acceptable operators return per head for labor and management[1]......... $ 12.00
  (10) Break-even amount, total per head [line (8) + line (9)] $603.80
  (11) Break-even price, per cwt [line (10) ÷ Sale wt] .... $ 56.96

B. Estimated hedge selling price:
  (12) Current Chicago futures price, April future, per cwt......... $ 59.75
                             (month)
  (13) Base-adj for market location differential per cwt .... $ −1.50
  (14) Basis-adj for grade differential per cwt........................... $ −0.50
  (15) Basis-adj for avg. livewt and dressing yield .... $ —
  (16) Basis-adj for heifer differential per cwt........................... $ —
  (17) Effective futures hedge price
     [line (12) ± line (13) ± line (14) ± line (15) − line (16)]........................ $ 57.75

C. Estimated profit:
  (18) Est hedged profit per cwt [line (17) − line (11)].................................... $ 0.79
  (19) Est hedged profit per head [line (18) × Sales wt] .... $ 8.37

D. Estimated Hedged Return to Labor, Management, and Profit per Head:
  (20) Line (9) + line (19)... $ 20.37

[1]Entry for operators labor and management is optional—each individual must decide whether to enter such a charge. If entry is made the amount will vary with individual's objective.

decide. Most operators hope to obtain some reimbursement for their labor and management. It makes little difference in some respects whether this is set out as a separate item or lumped in with profits. From a business standpoint, however, there is merit in setting an explicit goal for labor and management. In this example, $12 per head was entered as an acceptable return for labor and management (line 9). The total amount needed to return all costs ($591.80 plus $12 for labor and management), then, is $603.80 (line 10). This amounts to $56.96 per cwt (line 11), referred to as the break-even price. Note that break-even price is calculated from "sale weight," that is "pay weight." Steers handled in a feeding program as described in this example would have to be fed to a weight of around 1100 lb to realize a sale weight of 1060 lb. This is due to weight shrinkage in marketing—either actual or "pencil" shrink.

The entry in line 12 should be the futures price for the contract month in which the hedge would be placed. The current futures price assumes a hedge could be placed at about the most recently quoted price. This ordinarily would be the case unless the market were actively moving either up or down.

Lines 13, 14, 15, and 16 are for basis adjustments as previously described. In the example (Table 10.8), an adjustment of − $1.50 was indicated for location and −0.50 for grade differential. Nothing was entered in line 15 because the anticipated average weight and dressing yield were assumed to be equivalent to contract specifications. Line 16 does not apply in this case since the example deals with steers. The *effective* futures hedge price (line 17) is line 12 adjusted for basis factors. In this example it is $57.75. This immediately can be compared with the break-even price of $56.96 (line 11). When the effective futures hedge price equals or exceeds the calculated break-even price, a potentially favorable situation is indicated. In this case it is seen in line 19 that the estimated hedged profit is $8.37 per head, and in line 20 this $8.37 is added to line 9 (acceptable operator's return to labor and management), making a total return of $20.37 per head.

Again, this is not a guaranteed return. However, if expected cash-futures price relationships hold and if costs and bases have been accurately estimated, the estimate of line 20 will approximate feeding returns on a hedged operation.

Feeder cattle programs could be hedged in a manner similar to that described above by use of the feeder cattle futures. Feeder cattle growing operations, feedlot finishing, and feed procurement could be tied together in a virtually completely hedged operation by use of the respective futures.

Worksheet Evaluation of a Hog Feeding Hedge.—Table 10.9 illustrates worksheet evaluation of hedging a hog feeding operation. The format used is similar to the previous example for cattle. In this case, however, the relationships among costs, bases, and futures prices are intentionally set up so the effective futures hedge price ($51.80 on line 16) is less than break-even price ($56.61 on line 11). That situation would prevail even if nothing were entered for operator's labor and management (line 9).

In general, an evaluation which shows a negative return to labor, management, and profit (line 19) would be considered a nonhedging situation. Under hedged conditions, the operator would expect to take a loss. However, hedging still might be feasible if, for some reason, an operator (or his banker) preferred to take the indicated loss rather than assume risks of an even greater loss if the feeding operation were carried out unhedged.

The data used in Table 10.8 and 10.9 are hypothetical. Each feeder should use cost and basis data pertinent to his individual situation.

Worksheet Analysis of Maximum Price to Pay for Feeder Cattle.— The futures market can be used to approximate the maximum price an operator can pay for feeder cattle. This applies to the situation where the operator would follow through with a hedge on the feeding operation. Table 10.10 illustrates an approach to the calculations.

This analysis, like those preceding, necessitates a knowledge of costs

TABLE 10.9.   HYPOTHETICAL EXAMPLE: HEDGE WORKSHEET FOR HOGS

Buy 40 lb #2 feeder pig; feed 120 days; sell 220 lb #2 slaughter hog
          (grade)   (grade)

A. Estimated costs:
   (1) Cost of feeder per head (40 lb @ $1.13) $45.00
   (2) Cost of gain:
   (3) (a) Feedcosts per head $55.60
   (4) (b) Nonfeed costs per head $20.00
   (5) (c) Total cost of gain per head [line (3) + line (4)] $75.60
   (6) Marketing costs per head .... $ 1.50
   (7) Hedging costs per head .... $ 0.45
   (8) Total all costs per head [line (1) + line (5) + line (6) + line (7)]............. $122.55
   (9) Acceptable operators return per head for labor and management........... $ 2.00
   (10) Break-even amount, total per head [line (8) + line (9)] $124.55
   (11) Break-even price, per cwt [line (10) ÷ Sale wt]................................. $ 56.61

B. Estimated hedge selling price:
   (12) Current Chicago futures price, June future, per cwt........... $ 52.80
                                      (month)
   (13) Basis-adj for market location differential per cwt................. $−1.00
   (14) Basis-adj for grade differential per cwt....................... $ —
   (15) Basis-adj for weight differential per cwt .... $ —
   (16) Effective futures hedge price
        [line (12) ± line (13) ± line (14) ± line (15)]......................... $51.80

C. Estimated Profit:
   (17) Est hedged profit per cwt [line (16) − line (11)]................................ $ −4.81
   (18) Est hedged profit per head [line (17) × Sales wt] .... $−10.58

D. Estimated Hedged Return to Labor, Management, and Profit per Head:
   (19) Line (9) + line (18) .... $−8.58

and bases. A number of the same items estimated in Table 10.8 are identical for this analysis, but they are used in a different order. The approach used here takes the "effective futures hedge price" (line 6 in Table 10.10) as the estimated selling price of finished cattle, as would be the case if the cattle were hedged. This is in line with earlier discussion where it was shown that selling price can be established by a short hedge in the futures market, providing the appropriate basis is considered.

With selling price established, market value of finished steers would be the product of selling price and sale weight (see line 8, Table 10.10). By deducting all costs except the feeder steer itself (line 17) from market value (line 8) an estimate can be obtained of the maximum amount per head that could be paid for a feeder steer (see line 18). Division of this amount by the weight of a feeder steer gives maximum price per hundredweight (line 19).

The format of Table 10.10 also is readily adaptable for estimating the maximum amount a hog feeder could pay for feeder pigs, providing the futures market is used for a short hedge.

It should be noted at this point that the above use of this worksheet is not equivalent to accepting the futures quotation as a forecast of prices.

TABLE 10.10.   HYPOTHETICAL EXAMPLE: HEDGE WORKSHEET—MAXIMUM TO PAY FOR 650-LB FEEDER STEERS

A. Estimated Hedge Selling Price:
  (1) Current futures prices, <u>Oct.</u> futures, per cwt................. $ <u>56.45</u>
                            (month)
  (2) Basis-adj for market location differential per cwt................. $ <u>−1.50</u>
  (3) Basis-adj for grade differential per cwt............................ $ <u>−0.75</u>
  (4) Basis-adj for avg livewt and dressing yield .... $ <u>—</u>
  (5) Basis-adj for heifer differential per cwt............................ $ <u>—</u>
  (6) Effective futures hedge price
      [line (1) ± line (2) ± line (3) ± line (4) − line (5)] .... $ <u>54.20</u>

B. Estimated Sale Weight per Head:
  (7) From production records............................ <u>1060</u> lbs

C. Estimated Market Value Finished Cattle per Head
  (8) [line (6) × line (7)].. $574.52

D. Estimated Costs (Exclusive of Feeder Animal):
  (9) Cost of gain:
  (10) (a) Feed costs per head................................. $156.00
  (11) (b) Nonfeed costs per head .... $ 52.00
  (12) (c) Total cost of gain per head [line (10) + line (11)]............. $208.00
  (13) Marketing costs per head.. 4.00
  (14) Hedging costs per head .... $ 1.80
  (15) Total costs per head [line (12) + line (13) + line (14)] $213.80
  (16) Acceptable operators return per head for
      labor and management .... $ 12.00
  (17) Total of [line (15) + line (16)] $225.80

E. Estimated Maximum Amount for Feeders:
  (18) Est amount per head [line (8) − line (17)].. $348.72
  (19) Est price per cwt [line (18) ÷ 6.50][1] .... $ 53.65

[1]In general form this term is [line (18 ÷ wt of feeder in hundredweights].

However, by making a sale (a short hedge) in the futures market, an operator can "lock-in" that price—with qualifications as noted earlier regarding possible variation in the basis.

Research Findings on Hedging Strategies

A critical question in evaluating the feasibility of hedging is, what effect does it have upon (1) the level of profits and (2) the stability of profits? Other things being equal, most operators would prefer higher profits. Some may prefer more stable profits at somewhat lower average levels to higher average profits if the latter also are highly variable. The optimum hedging program, however, would be one which increases profit level and simultaneously reduces variability of hedged operation. A considerable body of research literature has been developed on hedging strategies in recent years. The studies generally agree that routine hedging of every lot of cattle or hogs placed on feed will reduce variability of profits over a period of years, but will not enhance average profits. Price (1976) in a simulated analysis of a continuously operated cattle feedlot found weekly average profits of routinely hedged operations to be $2.45 per head, while profits on unhedged operations was $13.82 per head. Even though variability of profits was reduced significantly, the reduction in profit level would be unacceptable by most standards. While there are notable short-run exceptions, cattle prices have been on a general uptrend since initiation of live cattle futures in late 1964. That largely accounts for the failure of routine hedging to enhance average profits. The costly and disruptive effect of numerous short-run downturns in cattle and hog prices, especially the 1973–1976 period, has prompted the development of strategies which provide for selective hedging.

Two basic approaches to selective hedging are: (1) develop criteria which specify *whether* to place a hedge, then *if* a hedge is placed, leave it on until the livestock are sold, and (2) develop criteria which specify *whether* to place a hedge and *if* a hedge is placed, further criteria to provide for possible lifting of the hedge (and possible subsequent replacing/relifting, etc.) before the livestock are sold. Several possibilities have been examined within each of those classifications.

Selective Hedge: Futures > Break-even.—This strategy specifies that a hedge will be placed only when the futures price (adjusted for basis) is greater than the calculated break-even price (calculated at the beginning of the feeding period). For those cases where the criteria for placing a hedge are not met, a feedlot operator has two alternatives (a) feed unhedged, or (b) place no livestock on feed that period. McCoy and Price (1975) in a simulated commercial cattle feedlot operation covering the period May 1965–December 1974, found that this strategy with alternative (b) resulted in higher profits than unhedged operation, but the difference was statistically non-significant. Alternative (b) resulted in lower profits than alternative (a) if both fixed and variable costs were considered, but alternative (b) with only variable costs considered resulted in profits greater than alternative (a).

Selective Hedge: Futures > Cash.—This strategy specifies that a hedge will be placed only when the futures price (adjusted for basis) at the beginning of a feeding period is greater than current cash price for contract grade cattle. Profits under this program were greater than the futures > breakeven program and significantly greater than unhedged operations under alternative (a) as described above. Alternative (b) produced significantly lower profits.

Selective Hedge: Futures > Break-even and Cash.—Under this strategy a hedge is placed only when the futures price (adjusted for basis) at the beginning of a feeding period is greater than the calculated break-even price *and* at the same time is greater than the current cash price. This program produced the highest profits of the three selective hedging strategies discussed up to this point. The variance of profits under each of the three strategies was lower than variance in profits of unhedged operations, but only in the futures > cash program was the variance significantly less than that of unhedged operations.

Selective Hedge: Flexible Application.—The selective hedges discussed previously followed the pattern that *if or when* the criteria being used specified that a hedge be placed, that a hedge was not lifted until the livestock were sold. That is the traditional pattern. It sometimes is argued that if the profit margin locked in by the hedge is satisfactory at the initiation of the hedge, it ought to be good enough to stay with till the end of the feeding period. However, the livestock feeder with a short hedge, for example, who follows that approach will forego additional profits if livestock prices turn upward after the hedge is placed. A substantial amount of research has been done on development of strategies designed to identify turning points in price trends so that a short hedge may be lifted once an uptrend is established, and replaced if or when the trend subsequently turns downward.[8]

Among the most commonly used approaches for identifying price trends and turning points are: (a) quantitative price prediction models, (b) point and figure charts, (c) bar charts, and (d) moving averages.

---

[8] In the case of a long hedge criteria are developed for lifting the hedge if the price trend turns downward and replacing it if prices subsequently turn upward.

Brown and Purcell (1978) used approach (a) and approach (d), both separately and in combination in evaluating hedging programs for feeder cattle. Price and McCoy (1978) used approach (d) in evaluating hedging strategies for slaughter steers, feeder steers, and hogs. In both of these studies flexible selective hedging produced greater profits than unhedged or traditional selective hedging. Further improvement can be expected in the development of these techniques.

A word of caution is in order. At this stage, it is not known whether the U.S. Internal Revenue Service will classify flexible selective hedging as a bona fide hedge for income tax reporting. In a bona fide hedge gains and/or losses on the futures transactions are treated as ordinary gains/losses. Gains and/or losses on futures transactions that do not qualify as bona fide hedges are treated as capital gains/losses and the length of time involved in most livestock hedges would put them in the classification of short-term capital gains/losses. Income tax consequences of short-term capital gains/losses are considerably different from ordinary gains/losses. In due time this question will be clarified.

Problems and Characteristics of the Basis.—Basis usually is defined as the difference between futures price and cash price at a specified cash market and specified time. That definition is somewhat ambiguous as it is not apparent whether, or when, the basis is positive or negative. It is more precise to define basis as the amount by which cash price is below the futures price (i.e., negative) or above the futures price (i.e., positive). As shown in Fig. 10.2 and Fig. 10.3 the basis for cattle and hogs may be positive or negative.

In futures markets for storable commodities which are produced at one time of the year and stored for use throughout the subsequent period, a logical reason exists for hypothesizing that the futures price for a given contract month will be above the cash price at par delivery points at any particular time by the amount of storage charges that would carry the commodity to termination of the futures contract.[9] The basis at non-par markets would include, in addition to storage, a transportation charge. In that sense, the basis for storable commodities can be construed to be a market determined charge. Live animals in condition for slaughter are not storable. Some efforts have been made at theoretical explanation of cash-futures price relationship for live cattle. Skadberg and Futrell (1966) pointed out the unique differences between livestock and storable commodities, i.e., livestock change form over time, holding time is limited as

---

[9]A considerable body of literature exists on this aspect of cash-futures price relationships. The subject is complicated including consideration of risk premium, and convenience yield, as well as out-of-pocket storage costs. Periods of time when cash grain prices exceeded futures grain prices prompted the development of the theory of "inverse" carrying charges. No effort is made in this text to develop these aspects to livestock and meat futures.

livestock approach market condition, and there is no necessary tie between today's cash price and the futures price for deferred delivery. Paul and Wesson (1967), in a theoretical treatment, concluded that the difference between a futures price and the sum of feeder cattle and feed prices on the cash market is the price of feedlot services. A refinement of this approach was pursued by Ehrich (1969). Leuthold (1977) makes a good case for the proposition that the basis for nonstorable commodities is not market determined, but rather that cash prices and futures prices of nonstorables are determined in independent markets and the basis (the difference between them) is a residual.

Suffice it to say that at this stage there is not a universally accepted basis theory for livestock futures. However, that does not mean the futures cannot be used for hedging. A number of studies have shown practical application of selective hedging and flexible selective hedging as described earlier, e.g., Brown and Purcell (1978), Erickson (1978), Heifner (1972), McCoy and Price (1975), McCoy and Price (1976), and Price (1976).

Do Futures Market Prices Forecast Cash Prices?

It might be argued that traders base their evaluation of futures prices on expected supply-demand conditions and that the price level at any given time represents a consensus of expected prices in the future. An almost equal case could be made that the futures price at a given time is a consensus of what the actual price will *not* be at termination of the contract. Speculative buyers trade on the expectation that prices will move higher and speculative sellers expect prices to move lower. Tomek and Gray (1970) state that ". . . prices of futures 6 months from maturity, 1 month from maturity, and cash prices are all 'forecasts' or 'nonforecasts' in approximately the same degree . . .

"Of course, unforeseen developments occur between futures expiration dates. The new but unpredictable information may indeed change expectations, but this is reflected in the entire constellation of prices. This in no sense implies that the expired future was incorrectly priced before its expiration, nor therefore that the pre-existing price relationship was in error. The range of prices at a point in time is based on the information then available, but new events and information may make the forecast of the previous period incorrect."

If futures prices did not move, there would be little incentive for either speculation or hedging. It is apparent that traders take a position because they believe the price will move. Regardless of whether one accepts the theory that futures prices represent traders' best estimate of what prices actually will be at a future date, or the opposite—that it represents their judgment of what it will not be—the record clearly shows it has not been an accurate forecaster of prices. This, however, does not negate its usefulness as a hedging mechanism. As long as a market attracts sufficient speculation to absorb hedges without undue price reaction, it can perform the hedging function, which in actuality is a price-setting function. Price forecasting is a responsibility of the personnel who trade on the futures market, and not vice versa.

Does Futures Trading Stabilize Livestock Prices?—Observation of the record would immediately prompt the conclusion that livestock prices have not stabilized since initiation of futures trading. Price changes are a reflection of changes in supply and demand. If it could be shown that the futures market tends to stabilize supply or demand (or both) then a reason would exist for suggesting that futures trading would stabilize prices. However, there is nothing inherent in the market that tends to stabilize demand. Factors which affect demand were discussed in Chap. 3, and are beyond the scope of the futures market.

If futures prices were relatively stable and if producers in large volume used the market as a price-setting mechanism (by hedging) then the futures market would impart a degree of stability in prices. Tomek and Gray (1970) state that "for commodities such as potatoes, which lack continuous inventories, prices of distant futures are less variable than cash prices and hence producer-hedging in such markets tends to stabilize revenue." This property is yet to be shown for livestock futures. Livestock futures might continue to exist with somewhat less fluctuation than at present. However, if a futures market became highly stable it probably would disappear. Without price movement there would be no incentive for either speculation or hedging.

While the futures may not be expected to stabilize prices, hedging can contribute to stability of profits (McCoy and Price 1975; McCoy and Price 1976).

Hedging and Livestock Credit

Large-scale feeding operations require large-scale financing. Summers (1967) commented that "not long ago, most cattle loans were to farmers who had feed in their cribs and who seldom asked to borrow more than 20% of their net worth for the purchase of feeders. Today, cattle feeders are more likely to have a modern feedlot, and may have part of the facilities paid for but have a small net worth in relation to the requested loan, yet they ask us to furnish money for the cattle and the feed as well." Grain merchandisers and storage interests long have found bankers more prone to grant credit on hedged than unhedged commodities. With the inauguration of live cattle futures it was anticipated that cattle feeders who hedged would be in better position to obtain credit than formerly. Theoretical considerations link credit availability to the degree of risk, and since hedging is presumed to lessen price risk, it follows that credit would be more readily available to a hedged operator.

Little work has been done to test this hypothesis. In a 1965 survey, Waldner (1965) reported that of the banks he contacted regarding loan policy on hedged cattle, ". . . no bank had a defined loan policy nor was any bank prepared to alter its customary loan policy to accommodate 'hedged' cattle." This was relatively soon after initiation of futures contracts in live cattle. In a 1968 study of South Dakota lending agencies, Powers (1968) found that hedging and forward contracting (referred to in this text as cash contracting) had little influence on the ability of a producer to obtain a loan. However, if a producer qualified for a loan otherwise, hedging permitted an increase in the size of the loan. He found, also, that hedging had no effect on interest rates. Powers' study is summarized as follows: "The data . . . indicate that the average increases in loans on hedged livestock ranged from 12.2 to 17.5% of the value of assets. On contracted livestock the average increases ranged from 11.9 to 18.3%. All of these increases are significantly greater than zero, thus indicating that hedging and forward contracting of livestock assets do aid farmers in obtaining capital by increasing the amount loaned on given livestock assets." With respect to interest rate, Powers commented, ". . . not a single agency which had made loans on hedged and contracted collateral reduced the interest rates on such loans . . .

"There are probably three major reasons for these results. First, some of the lenders probably believe that hedging and forward contracting do not reduce their risk . . . Second, a number of the respondents indicated that they based the interest rate on their cost of money, not on the different amounts of risk presented by farmers or firms. Third, it is quite likely that the risk reduction has been fully accounted for by the increase in the size of the loan."

Lending institutions have given substantially greater emphasis to livestock hedging in the years since the price break of 1973. Lenders are more knowledgeable of the market than formerly and many insist that low equity customers hedge their livestock operations.

BIBLIOGRAPHY

BAILEY, F., Jr. 1968. The cattle feeder and the futures market. Banking Jan., 57–59.

BAKKEN, H. H. (Editor). 1970. Futures Trading in Livestock—Origins and Concepts. Mimir Publishers, Madison, Wisc.

BLAU, G. 1944–1945. Some aspects of the theory of futures trading. Rev. Econ. Studies *12*, 7.

BROWN, R. A. and PURCELL, W. D. 1978. Price prediction models and related hedging programs for feeder cattle. Oklahoma Agr. Expt. Sta. Bull. B-734.

EHRICH, R. L. 1969. Cash-futures price relationships for live beef cattle. Am. J. Agr. Econ. Febr., 26–40.

ERICKSON, S. P. 1978. Selective hedging strategies for cattle feeders. Univ. of Illinois, Dept. Agr. Econ., Illinois Agr. Econ. Vol. 18, No. 1, 15–20.

FUTRELL, G. A., and SKADBERG, J. M. 1966. The futures market in live beef cattle. Iowa Coop. Agr. Ext. Serv. Circ. *M-1021.*

GRAY, R. W. 1961. The search for a risk premium. J. Political Econ. *LXIX*, No. 3, 250–260.

JOHNSON, D. A. 1970. The use of live steer futures contracts and their effect on cattle feeding profits. M.S. Thesis, Kansas State Univ.

KIMPLE, K. C. *et al.* 1978. The live cattle basis at selected locations in Kansas. Kansas Agr. Expt. Sta. Bull. 621.

LEUTHOLD, R. M. 1977. An analysis of the basis for live beef cattle. Univ. of Illinois, Dept. Agr. Econ. Staff Paper, No. 77 E-25.

McCOY, J. H. and PRICE, R. V. 1975. Cattle hedging strategies. Kansas Agr. Expt. Sta., Bull. 591.

McCOY, J. H. and PRICE, R. V. 1976. Hog hedging strategies. Kansas Agr. Expt. Sta., Bull. 604.

McCOY, J. H. and PRICE, R. V. 1978. Flexible selective hedging of livestock. Dept. Econ. Kansas State Univ. Unpublished data.

PAUL, A. B., and WESSON, W. T. 1967. Pricing feedlot services through cattle futures. USDA Agr. Econ. Res. *29*, No. 2, 33–45.

POWERS, M. 1968. Hedging forward contracting and agricultural credit. S. Dakota Agr. Expt. Sta. Bull. *545.*

POWERS, M., and JOHNSON, A. C. 1968. The frozen pork belly futures market: An analysis of contract specifications and contract viability. Dept. Agr. Econ., Univ. Wisconsin Agr. Econ. Bull. *51.*

PRICE, R. V. 1976. The effects of traditional and managed hedging strategies for cattle feeders. M.S. Thesis, Kansas State Univ.

SCHNEIDAU, R. E. 1967. Hedging on the live hog futures market. Indiana Coop. Agr. Ext. Serv. Circ. *EC-312.*

SKADBERG, J. M. and FUTRELL, G. A. 1966. An economic appraisal of futures trading in livestock. Amer. J. Agr. Econ. 48, 1485–1489.

SUMMERS, M. M. 1967. Hedging and how it helps reduce loan risk. Agr. Banking Finance May/June 32.

TOMEK, W. G., and GRAY, R. W., 1970. Temporal relationships among prices on commodity futures markets: Their allocative and stabilizing roles. Am. J. Agr. Econ. *52*, No. 3, 372–380.

WALDNER, S. C. 1965. Will hedging offer financing advantages? Feedlot Sept. 54.

WORKING, H. 1960. Speculation on hedging markets. Food Res. Inst. Studies *1*, No. 2, 185–220.

WORKING, H. 1962. New concepts concerning futures markets and prices. Am. Econ. Review *LII*, No. 3, 434–436.

WORKING, H. 1963. Futures markets under renewed attack. Food Res. Inst. Studies *4*, No. 1, 13–24.

11

Grades and Grading

Producers, slaughterers, processors and distributors are cognizant of live animal and meat grades. Market news reports utilize federal grade nomenclature in reporting prices, and within the meat trade, both federal and private (brands) grades are used in wholesale transactions. Consumers are exposed to federal grades—though most know little about them (Hutchinson 1970)—and private brands through advertising media and in grocery store shopping. In short, grades and grading are an integral part of livestock and meat marketing.

This chapter is concerned primarily with federal government grades, but private brands which are based on consistent quality specifications are tantamount to private grades.

There is evidence that some consumers confuse inspection with grading (Hutchinson 1970). This point should be clearly understood. Both the USDA inspection mark and USDA grade mark are applied in color on meat, but that is the only similarity. Inspection is concerned with the wholesomeness of meat and meat products—such things as diseased animals, sanitation in slaughtering and processing plants, prevention of adulteration, correct labeling, etc. This is a separate and distinct activity from grading. The inspection mark stamped on a carcass is a circle, and if the carcass passed inspection the circle contains the abbreviations "U.S.

GRADE MARKS

INSPECTION MARK

← Yield grade

← Quality grade

INSP'D & P'S'D." The grade mark is a shield in which is inscribed the abbreviation USDA and also the grade name or number as the case may be. When viewed side by side the difference is obvious as here shown.

DEFINITION AND PURPOSE

Grading is the segregation of units of a commodity into lots, or group-ings, which have a relatively high degree of uniformity in certain specified attributes associated with market preferences and valuation.

The relevant attributes differ widely among commodities in number, in range of variability, and in susceptibility to objective measurement. Eight grades are used to cover the range in quality of steer and heifer carcasses, but only three for ready-to-cook poultry. In addition to quality grades, five separate yield, or cutability, grades (i.e., yield of boneless, closely trimmed retail cuts) are used in conjunction with quality grades in beef and lamb. Pork grades incorporate yield and quality considerations without separate quality and yield designations. Poultry grades are based entirely on quality.

Quality grades in beef are determined basically on subjective consider-ations. Quality is based on the "eatability" or palatability indicating characteristics of the lean meat.[1] Cutability or yield grades for beef are determined primarily by objective measurements. Pork carcass grades rely heavily on objective measurements to determine expected combined yield of lean cuts, used in conjunction with a subjective determination of "acceptable" or "unacceptable" quality of the lean.

Thus, it can be seen that even though the definition of grading is relatively straightforward, the drawing up of standards or specifications for a system of grades is a complex problem. Some of the attributes upon which grades are based can be evaluated directly, others indirectly, through so-called indicators. For example, the yield of lean cuts of pork is related to (indicated by) backfat thickness, carcass weight, and carcass length. A great amount of research and consultation is involved in de-veloping grade standards.

The *purpose* of grading is related to whether the grading is a private or public function. Grading is not the sole prerogative of government. Most national packers and many regional packers have private grades and grade standards. The federal government has an extensive system of live animal and meat grades.

The objectives of a firm in setting up private grades are only partially compatible with objectives of a public grading system. One primary con-

---

[1]Prior to 1976, conformation of the carcass also was considered in beef quality grades, but that factor was eliminated with grade changes of that date.

cern of a private firm is a grading system which adequately describes the quality of the product so that trade can be conducted with a minimum of time, effort, and expense. Grade designations which consistently and fully describe a product do away with the necessity of personal inspection by buyers. Transactions can be made by telephone which lessens the expense of both buyers and sellers. In this respect, private and public grading have a common purpose. Private firms also use grades to differentiate their products from those of competitors. To the extent the products are different, this is an informational service which assists buyers and, if the difference commands a higher market price, this is a matter of buyer discretion. On the other hand, if the differentiation is spurious, buyers may be misled and the result may be individual and social loss.

From the standpoint of public interest, a profusion of grades and grade standards is confusing. Without some official supervision of the application of specified standards, doubts and suspicions sometimes arise as to whether products meet the standards. The general public also has an interest in the effect grades have on the efficiency with which the market system allocates resources among alternative production possibilities.

Producers' interests are associated with the degree to which use of grades assists them in obtaining equitable prices. Consumers primarily are interested in obtaining products commensurate with their preferences and with the prices they pay.

In view of the multiplicity of considerations, it was deemed to be in the public interest to inaugurate a system of federal grades and grading. At the same time individual firms have the privilege of using private (or house) grades, either alone or in combination with U.S. grades.

The principal purposes of government grades can be placed under the general headings of improving (1) pricing efficiency and (2) operational efficiency.

Grades and Pricing Efficiency

As indicated in Chap. 1, pricing efficiency is concerned with such things as the accuracy, effectiveness, and speed with which a price system measures product values to buyers and reflects these values back through marketing channels to producers. Involved in this concept are issues of efficiency in allocation of resources among alternative uses, and distribution of income among producers and others involved in various economic activities of an industry. The uniform application of an appropriate set of grades can contribute to pricing efficiency. If consumers indicate a willingness and ability to pay more (or less) for a certain quality of meat, that information must be transmitted back to producers before they can react.

A market economy depends upon price differentials to relay the message, and grades provide a meaningful, common language whereby the price differential can be associated with the desired (or undesired) quality characteristic.[2] It is recognized that, at a given time, price differentials among quality levels of a particular product can arise due to variations in relative supplies as well as to variations in demand. Over a period of time, however, the pricing system is presumed to allocate resources in such a way as to optimize supplies of various qualities of the product. This is implied in pricing efficiency.

The use of grades also has a place in improving efficiency in distribution of products of varying quality among alternative uses. Differential pricing among grades enables wholesalers and retailers to move given supplies into their highest order use at rates which clear the market. Thus Canner and Cutter grades are diverted into processed meats. Prime grade steaks and roasts go to the highest class night club, hotel, restaurant, and steamship trade. Choice and Good grades form the primary retail cuts. As a result, total revenue from the entire supply of meat is enhanced; and increased competition tends to distribute the benefits among all sectors of the livestock and meat industry.

The use of grades can increase the degree of competition in a market by increasing the level of knowledge. Buyers and sellers in a perfectly competitive market are assumed to possess perfect knowledge of market conditions. Grades cannot be expected to impart perfect knowledge, but their use can "enhance" the knowledge of traders. This applies not only to quality aspects but also to yield (cutability) characteristics of meat animal carcasses. The difference in retail sales value between adjacent yield grades, though it varies with relative supply and demand, can easily amount to $5.00 per cwt of choice carcass beef. Therefore, a knowledge of yield grades and associated value differentials can strengthen competition between wholesalers and retailers. Uniformity in application of grades which adequately describe a product allows trade by description (as constrasted to inspection) via telephone or wire service. This enhances competition as wholesalers and retailers can shop over wide areas. The use of uniform federal grades permits small packers to compete with large packers in terms of quality of product—if not in volume. This increases the degree of competition and pricing efficiency in the industry.

Grades and Operational Efficiency

Operational efficiency is concerned with costs of performing marketing functions. The concern is not necessarily with reduction in total market-

---

[2]Grades which transmit quality differentials to consumers would enhance total utility even though no price differentials existed.

ing costs but with reduction in per unit marketing costs. This may be accomplished by marketing a given quantity of goods at lower total costs, an increased quantity at the same total cost, or an increased quantity at a less than proportionate increase in total cost.

In some respects, grading can improve operational efficiency, while in others it may have the opposite effect. Grades permit trading by description, which can result in a reduction in transaction expenses (travel, labor, and associated expense) for both buyer and seller, compared to trading by inspection. Generally, it is assumed this means a reduction in total marketing expense, which probably is correct; however, the statement must be qualified. Grading does not come free. Total marketing costs will be reduced providing the reduction in transactions costs are not exceeded by costs of grading, additional telephone expenses, costs associated with possible misunderstanding, and rejection of shipments which do not meet grade specifications.

Government grading of fresh beef and lamb largely has forestalled the use of private grade advertising of these products. As a result, the total advertising bill undoubtedly is less than it would be otherwise. This reduction in marketing cost can be construed to be a gain in operational efficiency.[3]

Federal grading also has had an impact on decentralization and specialization in the packing and processing industries. Use of federal grades has permitted small packers to compete well with large operations, and smaller independent packers led the way in decentralization. The opening of plants near livestock production centers contributes to cost reduction in a number of ways. Under the prevailing freight rate structure, transportation costs are less for dressed meat than for live animals. Often, operating costs are less at interior points than at terminal locations due to lower wage rates, taxes, utility rates, etc.

Specialization also is enhanced as grading permits more accurate description and identification of products. Specialization in products is complemented by specialization in labor, equipment, plant layout, and work techniques which, in turn, contribute to cost reduction and operational efficiency.

DEVELOPMENT OF LIVESTOCK AND MEAT GRADES

During earlier periods a variety of market terms evolved in particular localities which were more or less descriptive of certain quality charac-

---

[3]This is not to be construed as an indictment of all advertising, but brand name advertising designed to alter consumers' preferences to match the characteristics of a product or to develop fictitious differences among products can result in social costs from misallocation of resources (see Farris 1960).

teristics of livestock. Associated with expansion of the large terminal markets of that era was the private publication of market prices using local class and quality terminology. This, of course, was a significant advancement over no published quotations at all; and the reports were useful to livestock interests. However, weaknesses were apparent. There was a lack of uniformity in terminology and standards among markets. A price quotation on "native cattle," for example, has limited meaning when the specifications for "native cattle" vary from one market to the next, or when some markets do not have a classification of "native cattle."

In many cases an accepted term was not used uniformly from person to person, or throughout the year at a given market. These conditions are symptomatic of growing pains in any developing economy. Trade is handicapped to the extent that buyers must physically inspect all lots of animals. Market price quotations cannot serve their full potential of usefulness.

The handicaps of this situation were apparent for many years and some pioneering efforts were made to develop uniform grades and standards. The Agricultural Experiment Station of Illinois was a leader in this area. Between the years 1901 and 1908, a series of investigations were made of actual conditions and ways and means of alleviating the situation. Five Illinois bulletins were issued with following titles: "The Market Classes of Horses," 1901; "Market Classes and Grades of Cattle, with Suggestions for Interpreting Market Quotations," 1902; "Market Classes and Grades of Swine," 1904; "Market Classes and Grades of Horses and Mules," 1908; and "Market Classes and Grades of Sheep," 1908. These studies did not precipitate immediate governmental action, but soon a movement developed for official market news reporting. The ensuing debates over implementation of market reporting could not be separated from consideration of official grades, because the lack of uniformity in local grade names and laxity in local grade specifications rendered them unsatisfactory for market reporting purposes. In response to this need (and for other reasons) the USDA established an Office of Markets in 1913.[4] That office began work in the development of official grades for live animals and meat about 1915 (Dowell and Bjorka 1941). The work done at Illinois provided a starting point, but vested interests in local grades and lack of agreement on grade standards were serious obstacles in obtaining a consensus on uniform grades. Nevertheless, unofficial and unpublished grade specifications were drawn up and used for market reporting of meat prices beginning in 1916. Responsibility for developing grade standards and implementing their use has shifted from one office to another within USDA, usually following administrative reorganization. Currently

---

[4]Over the years the federal grading service has been under a succession of different offices resulting from renaming, merging, and reorganizing.

responsibility for live animal grades is in the Agricultural Marketing Service, while responsibility for carcass grades is in the Food Safety and Quality Service.

Derivation of an acceptable and adequate set of grade standards requires an enormous effort. Information must be obtained from all interested parties to determine attributes which impart value to the product. There are differences of opinion on this point among producers, packers, processors, wholesalers, retailers, and consumers. Even when a consensus can be obtained, problems arise in selecting indicators by which the attributes can be measured in live animals and meat. This is especially difficult where only subjective measures are usable. The USDA held many conferences and hearings with all interested parties, including agricultural college personnel. Research was conducted and encouraged at colleges, and a considerable amount of experimentation was carried out using unofficial and tentative standards before grades were officially promulgated. And this work does not stop with establishment of official grade names and standards. Over a period of time, changes occur in quality of the product produced, in consumer preferences, in trade practices, etc. Continuous change requires continuous review of grade standards and revision or amendment as the occasion demands. Figure 11.1 shows the steps and precautionary measures involved in setting up standards.

The application of uniform grades requires some prior classification and subclassification of livestock. An obvious first classification is by specie—cattle, hogs, and sheep. Beyond that, for official grading purposes, livestock are segregated according to use, i.e., slaughter and feeder. Within use categories a further breakdown, referred to officially as *class*, is made on the basis of sex condition, then grades are applied to classes.

Designated classes[5] of the various specie and use categories of livestock are: (A) Slaughter cattle—steers, heifers, cows, bullocks, and bulls. (B) Feeder cattle—same as slaughter cattle (except bulls are not graded). (C) Slaughter swine—barrows, gilts, sows, boars, and stags (boars and stags are not graded). (D) Feeder pigs—same as slaughter swine (except sows, boars and stags are not graded). (E) Slaughter lambs, yearlings, and sheep—ram, ewe, and wether (these class names are used in conjunction with age groupings, i.e., lambs, yearlings, and sheep). (F) Feeder lambs, yearlings, and sheep—ewe and wether (in the case of feeder sheep, only the ewe class applies).

The first grades to be formulated were for dressed beef in 1916. "They provided the basis for uniformly reporting the dressed beef markets according to grades, which work was inaugurated as a national service early

---

[5]Some classes are designated and defined, but are not graded.

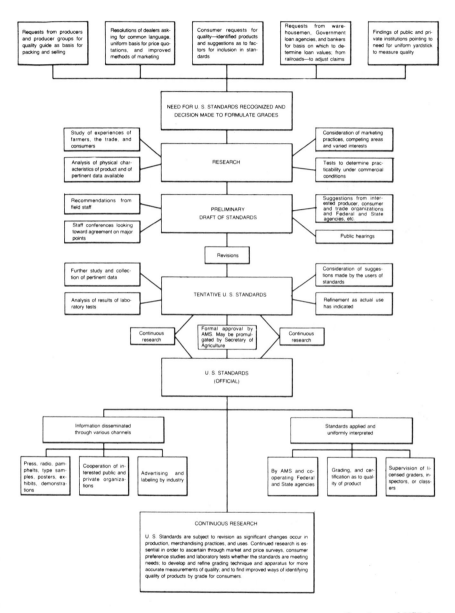

Courtesy of USDA

FIG. 11.1.   STEPS IN SETTING UP QUALITY STANDARDS FOR FARM PRODUCTS

in 1917" (USDA 1965A). These beef grades were used in unofficial, tentative status until 1926, but in the meantime a number of revisions were made as experience and usage dictated. Their use also spread beyond market reporting. "The tentative standards, although designed primarily for meat market reporting purposes, were put to further practical test in numerous ways. During World War I they were used in the selection of beef for the Army, Navy, and Allies. Later they were included in the specifications of the Emergency Fleet Corporation for the purchase of its beef supplies. Soon thereafter they were incorporated in the specifications of many commercial concerns, including steamship lines, restaurants, hotels, dining car services, and hospitals" (USDA 1965A).

Official grades were promulgated for carcass beef on June 3, 1926. The use of federal grades has been on a voluntary basis from the beginning, except during World War II and the Korean Conflict when it was mandatory for all FI plants handling beef, veal, calf, and lamb. The voluntary stamping of graded beef was made available in 1927.[6] For a brief period at the beginning, grading costs were paid by the federal government, but since then costs are paid by the parties who request the service.

Eight changes (amendments or revisions) have been made in the official beef grades—1939, 1941, 1949, 1950, 1956, 1965, 1973, and 1976. The 1965 revision not only changed quality standards, but standards for "cutability" were added. Cutability is the consideration of percentage of trimmed, boneless, major retail cuts which can be derived from a carcass. Cutability has been recognized as an important value-determining characteristic for a number of years, and standards for cutability had been proposed and tested in a 1-yr trial period beginning July 1, 1962. With some modifications it was made official in 1965. A new class designated as "Bullock" was added in the 1973 revision. Bullocks are young bulls. This class was added in recognition of research which showed that young bulls were superior to steers in rate and efficiency of gain. Prior to 1973 meat from these animals was marked "Bull" beef. Consumer bias against that designation was considered a deterrent to its acceptance. A conviction remains among some that the term bullock also is a deterrent, but at the time of the revision the official position was that enough difference existed in the palatability of young bull and steer beef that the grading system should differentiate them and "bullock" was selected as an appropriate term. A revision was announced to become effective in April 1975, but legal proceedings delayed its implementation until February 1976. Four major changes of that revision were: (1) a reduction in marbling requirements for the Prime and Choice grades for all but the youngest cattle, (2) marbling requirements for the Good grade were in-

---

[6]For a comprehensive analysis of issues involved in early days of grade development and marking of beef see Rhodes (1960). Some opposing views are presented by Williams (1960).

creased for young cattle and decreased for older cattle, (3) conformation was removed as a factor in quality grades, and (4) yield grading was made mandatory on all carcasses that are quality graded. Previously carcasses could be quality graded, or yield graded, or given both designations.

Changes in slaughter cattle grades generally have paralleled those of carcass beef, but the timing varied slightly as shown in Table 11.1.

The development of other dressed meat grades (veal and calf carcasses; pork carcasses, and lamb, yearling mutton, and mutton carcasses) as well as grades for live slaughter and feeder animals followed a comparable course of formulation, tentative adoption, official promulgation, and subsequent amendment. Comments here are limited to a few major changes. Details may be found elsewhere.[7]

During the period of tentative feeder cattle grades, the grouping was referred to as "Feeder and Stocker Cattle." Although "stocker" is a commonly used term in the cattle trade, it is no longer used in official grade standards. The term "feeder" is intended to include those formerly classed as stockers and feeders. Table 11.2 illustrates changes in feeder cattle grade names which now correspond to slaughter cattle except for the bottom two grades. Grade terminology and changes for vealers and slaughter calves are shown in Table 11.3.

Several major changes have been made in slaughter barrow and gilt grades. In 1952 the former tentative Choice (meat type) was divided into two grades and renamed Choice No. 1 and Choice No. 2; the former Choice (fat type) was renamed Choice No. 3 (Table 11.4). These changes represented a significant recognition of preference for leaner, meat-type pork and the importance of yield of lean cuts. Changes in 1968 were designed to reflect improvements in quality which could not be properly evaluated under previous standards. Minimum requirements for backfat thickness were eliminated for U.S. No. 1 grade. A new U.S. No. 1 grade was specified for superior carcasses, which were graded either U.S. No. 1 or Medium under previous standards. Under the revision, most of the former U.S. No. 1, 2, and 3 were renamed U.S. No. 2, 3, and 4, respectively, while the former Medium and Cull grades were renamed Utility. The term Utility now has a generally uniform meaning in the grade pattern for all species. Feeder pig grades are of relatively recent origin, and, as with other species, their development has been an attempt to keep feeder grades comparable to grades for slaughter animals (Table 11.5).

The development of lamb grades has had a particularly stormy history. A major source of difficulty arose from using a single set of standards to evaluate diverse types of lamb ". . . western lambs, bred for their ability

---

[7]Details of grade specifications and changes are published in USDA Service and Regulatory Announcements which are issued with each amendment and revision.

TABLE 11.1.   CHANGES IN USDA GRADES FOR SLAUGHTER CATTLE[1]

1918	1928	1939	1950	Official 1956	1966[2]	1973[4]	1976[5]
Adopted for market reporting	Tentative 1928 Published						
	Prime	Prime	Prime[3]	Prime[3]	Prime[3]	Prime	Prime
	Choice	Choice	Choice	Choice	Choice	Choice	Choice
	Good	Good	Good	Good	Good	Good	Good
	Medium	Medium		Standard	Standard	Standard	Standard
			Commercial	Commercial	Commercial	Commercial	Commercial
	Common	Common	Utility	Utility	Utility	Utility	Utility
	Cutter	Cutter	Cutter	Cutter	Cutter	Cutter	Cutter
	Low Cutter	Canner	Canner	Canner	Canner	Canner	Canner

Source: USDA.

[1] Area of spaces occupied by grade names is not intended to be proportional to the relative importance of grades.

[2] Reduced marbling requirements in quality grades and added five yield grades (No. 1, 2, 3, 4, and 5) to identify differences in cutability.

[3] Only steers and heifers eligible for Prime grade.

[4] "Bullock" added as classification for young bulls; but provided for only five grades: Prime, Choice, Good, Standard, and Utility.

[5] Reduced marbling requirements for Prime and Choice grades for all but the youngest cattle, increased marbling requirements for Good grade for young cattle and decreased them for older cattle, removed conformation as a factor in quality grades, and yield grading was made mandatory if quality grade used.

TABLE 11.2.   CHANGES IN USDA GRADES FOR
FEEDER CATTLE[1]

1925	1934	Tentative 1938	1942	Official 1964
		Fancy	Fancy	Prime[2]
		Choice	Choice	Choice
		Good	Good	Good
		Medium	Medium	Standard
				Commercial
		Plain	Common	Utility
		Inferior	Inferior	Inferior

Source: USDA (1965B).
[1] Area of spaces occupied by grade names is not intended to be proportional to the relative importance of grades.
[2] Only steers and heifers are eligible for Prime grade.

TABLE 11.3.   CHANGES IN USDA GRADES FOR
VEALERS AND SLAUGHTER CALVES[1]

Tentative 1926	1928	Official 1951	1956
	Prime	Prime	Prime
	Choice		
	Good	Choice	Choice
	Medium	Good	Good
		Commercial	Standard
	Common	Utility	Utility
	Cull	Cull	Cull

Source: USDA (1957).
[1] Area of spaces occupied by grade names is not intended to be proportional to the relative importance of grades.

to use sparse ranges or mountain pastures and for their wool, are different from native lambs" (Fienup *et al.* 1963). While federal grade standards recognized differences due to maturity, they did not differentiate between western and native lambs. Western lamb producers felt that early official standards discriminated against their lambs, and in 1959 several sheep industry groups requested a suspension of federal grading. "Much of the pressure on federal grades for lamb in 1959 arose out of a disagreement over the quality levels that were included in U.S. Choice.

TABLE 11.4.   CHANGES IN USDA GRADES FOR SLAUGHTER BARROWS AND GILTS[1]

Formulated 1918	1930	Tentative 1940	1952[2]	Official 1955	1968[3]
					U.S. No. 1
Formulated for use in market reporting		Choice (fat type)	Choice No. 1	U.S. No. 1	U.S. No. 2
			Choice No. 2	U.S. No. 2	U.S. No. 3
		Choice (meat type)			
			Choice No. 3	U.S. No. 3	U.S. No. 4
		Good			
			Medium	Medium	
		Medium			
					U.S. Utility
		Cull	Cull	Cull	

Source: USDA (1968B).
[1] Area of spaces occupied by grade names is not intended to be proportional to the relative importance of grades.
[2] Former tentative Choice (meat type) subdivided and renamed Choice No. 1 and Choice No. 2; former tentative Choice (fat type) renamed Choice No. 3.
[3] U.S. No. 1 in 1968 was a new classification for superior carcasses. Former U.S. No. 1, 2, and 3 renamed U.S. No. 2, 3, and 4, respectively. Former Medium and Cull grade names dropped. Utility added for barrows and gilts with characteristics which indicate carcass lean of unacceptable quality. Four numbered grades based on expected combined carcass yield of the four lean cuts.

TABLE 11.5.   CHANGES IN USDA GRADES FOR FEEDER PIGS[1]

Tentative 1940	Official 1966	1969
		U.S. No. 1
	U.S. No. 1	
		U.S. No. 2
	U.S. No. 2	
		U.S. No. 3
	U.S. No. 3	U.S. No. 4
	Medium	U.S. Utility
	Cull	U.S. Cull

Source: USDA (1969A).
[1] Area of spaces occupied by grade names is not intended to be proportional to the relative importance of grades.

Federal standards for lamb were designed on the basis of meat-type characteristics more prominent in native lambs. A smaller proportion of western than native lambs met conformation standards for U.S. Choice prior to the 1960 revision. In attempt to qualify for U.S. Choice by improving quality to offset a lack of conformation, many lambs were fed to heavier weights and discounted in price . . . The lower conformation re-

TABLE 11.6. CHANGES IN USDA GRADES FOR SLAUGHTER LAMBS, YEARLINGS, AND SHEEP[1]

Formulated 1917	1936	Tentative 1940	1951	Official 1957[2]	1960[3]	1969[4]
Formulated for use in market reporting	Choice or No. 1	Prime		Prime[5]	Prime[5]	Prime[5]
		Choice	Prime[5]			
	Good or No. 2					
		Good	Choice	Choice	Choice	Choice
	Medium or No. 3					
		Medium	Good	Good	Good	Good
	Plain or No. 4					
		Common	Utility	Utility	Utility	Utility
	Cull or No. 5					
		Cull	Cull	Cull	Cull	Cull

Source: (1969B).
[1]Area of spaces occupied by grade names is not intended to be proportional to the relative importance of grades.
[2]No grade name changes. Quality requirements for Prime and Choice reduced for more mature lambs. Quality requirements for Good lambs increased slightly.
[3]No grade name changes. Quality and conformation requirements lowered for Prime and Choice. Limit placed on the extent superior quality may compensate for deficient conformation.
[4]Yield grades (No. 1, 2, 3, 4 and 5) added to pre-existing quality grades.
[5]Slaughter sheep older than yearlings not eligible for Prime grade.

quirements in the new standards (1960) increased the proportion of western lambs grading U.S. Choice" (Fienup et al. 1963).

A major change in 1969 was the establishment of yield grades for lamb. A chronological picture of changes in slaughter lamb, yearling, and sheep grades is shown in Table 11.6. Thus, cutability (yield of retail cuts) now is an integral factor in the grade standards of cattle, hogs, and lamb.

For centuries, wool has been an important product of the sheep industry. For much of history wool was more important than the meat produced by sheep. That generally was the situation in early America and, to a limited extent, still exists in some parts of the world. Meat is the primary product of the U.S. sheep industry, but wool can make the difference between profit and loss—constituting approximately 10% of sales from a ewe flock.

The fiber market distinguishes wool from wool top, and separate grades have been developed for each. Wool is defined as the fiber from the fleece of sheep. Wool top is a continuous untwisted strand of scoured wool fibers from which the shorter fibers, or noils, have been removed by combing.

Grade standards for wool were first promulgated in 1926. The basic grade factor was fiber diameter (i.e., fineness) as determined by visual inspection. Twelve grades were established. Revisions were proposed in 1955 and in 1963 but were not adopted. Standards were revised in 1966. Major changes were: (1) the addition of four grades—two intermediate degrees of fineness, and an additional grade at each end of the range of specified diameters. The latter two grades are open ended in that they encompass all degrees of fineness outside those specified in the 14 intermediate grades. (This is shown in the following section entitled *Annotation of Grade Standards*.) (2) Provision for objective measurement of fiber diameter (as well as visual determination), and (3) Recognition of the distribution of various fiber diameters in a fleece, or group of fleeces, (as represented by the standard deviation from average diameter of a sample of fibers).

Tentative grade standards for wool top were drawn up in 1893— providing for seven grades. Official USDA standards for wool top were established in 1926. At that time the number of grades was increased to 12, corresponding to the range in recognized fiber diameters in official USDA wool grades. "Since 1926, the wool top standards have been revised three times. In 1940, a grade 62s was added and average fiber diameter and fiber diameter distribution specifications were issued for eight of the 13 grades. These revised standards also provided that grade could be determined either by inspection—visual comparison of samples with practical forms of the official grade standards—or by measurement of samples. In 1955, a grade 54s was added, bringing the total to 14 grades. During that year, all grades were assigned average fiber diameter and fiber diameter distribution specifications. Until 1955, grades were determined by either the inspection or measurement method. At this time, however, an addition to the standards provided that in cases where these methods resulted in different grades, the grade determined by measurement would prevail.

"The current wool top standards became effective January 11, 1969. On that date two new grades, 'Finer than Grade 80s' and 'Coarser than Grade 36s,' were added. A dual grade designation was also provided for wool top in which the average fiber diameter and fiber diameter distribution do not meet the requirements of the same grade" (USDA 1971).

ANNOTATION OF GRADE STANDARDS[8]

In the previous section a brief description was given of historical developments of the U.S. grading system. In this section emphasis will be placed on standards currently in use. Both aspects are important. A knowledge of the background of grade development affords a better understanding of how and why the system evolved to its current state. This type of information is helpful, if not essential, in further amendments

---

[8]Material in this section draws heavily upon USDA publications giving official U.S. grade standards and a series of USDA market bulletins which present a more popularized version of grade information.

which inevitably will come in the future. Readers who are interested in greater detail on development are referred to the references cited at the end of this chapter.

A general knowledge of current grade standards is essential to everyone engaged in livestock and meat marketing. As mentioned in earlier chapters, extensive use is made of official USDA grades in wholesale meat marketing, even though grading is strictly on a voluntary basis. Many large retail chain buyers insist on USDA graded meat in their wholesale purchasing program. Many retailers carry through by using grade labels on retail cuts.

Live animal grades are not officially applied in the marketing of livestock, but the USDA Market News Service publishes price quotations by grade. In direct selling by carcass grade and weight, USDA standards may be used in grading the carcasses (some packers insist on using their own grading system; others use USDA grades). Even an intelligent interpretation of price quotations calls for some understanding of official grades. Selling (or buying) by carcass grade and weight, or involvement in the wholesale meat trade, makes a knowledge of grade standards a necessity. But a knowledge of grades and grading also is important in buying and selling by liveweight methods. Packer buyers are experienced in estimating the potential carcass grade of live animals. Producers also need to be able to estimate grade to know whether they are getting a price commensurate with the quality and cutability attributes of their livestock. In developing a breeding herd it is imperative to be able to select stock for herd improvement, and this includes improvement in grade of animals produced.

No attempt is made here to present methods and techniques of actual grading. This is a complex problem requiring extensive study and training. An attempt is made, however, to point up the attributes, or characteristics, of live animals and meat upon which grades are based—and indicators used in evaluating these attributes.

In this section both carcass and live animal grades will be discussed. Separate standards are established and published for carcass and live animal grades, but a high correspondence exists between them. Historically, carcass grades have been established first, then slaughter animal grades made to conform to carcass grades. Feeder animal grades also are made to conform to slaughter animal grades. For example, a U.S. No. 1 feeder pig is evaluated on its potential for feeding out to a U.S. No. 1 slaughter hog which has the potential of producing a U.S. No. 1 carcass.

It is apparent that some of the indicators used to evaluate a carcass may be different, and presumably more precise in application, from those used on the live animal, but the basic attributes which give the carcass value and utility are the same as those of the live animal.

Carcass Beef

Grades of carcass beef are based on separate evaluation of two general considerations: (1) palatability characteristics of the lean, which is referred to as "quality," and (2) the indicated percentage of trimmed, boneless, major retail cuts, which is referred to as "yield grade" or "cutability." Prior to 1976 quality grades included a consideration of conformation. Dropping of that factor was a reversion to previous standards. "In previous grade standards for beef and in the standards for grades of other kinds of meat, the Department uses the term 'quality' to refer only to the palatability—indicating characteristics of the lean, without reference to conformation" (USDA 1965A).

Standards are established for eight quality grades: Prime, Choice, Good, Standard, Commercial, Utility, Cutter, and Canner. Five cutability (or yield) grades have been defined: No. 1, 2, 3, 4, and 5, with No. 1 representing the highest cutability. Not all classes are eligible for every quality grade. Cows are not eligible for Prime grade and in addition bull beef is not quality graded. Bullocks are excluded for Commercial, Cutter, and Canner grades. All classes are eligible for all yield grades.

Steer, heifer, and cow carcasses are graded and stamped with no reference to sex condition. Separate grade standards are provided for bull and bullock carcasses, and when the carcasses are grade stamped they also are stamped as "BULL" or "BULLOCK" as the case may be. As mentioned earlier, the bullock classification was provided in 1973 to designate young bull beef which previously had to be marked as bull beef. At the same time the term "stag" was eliminated and beef formerly in that class was redesignated as bullock or bull depending on its evidence of maturity.

Prior to the revision of 1976, quality and yield grades were separate considerations. Grading is not mandatory, but the 1976 revision stipulated that if a carcass is quality graded it must also be yield graded, and the carcass must carry both stamp marks. While the standards for both quality and cutability are defined primarily for carcass beef, the quality standards apply also to certain wholesale and primal cuts—forequarters, hindquarters, rounds, loins, short loins, loin ends, ribs, and chucks. Portions of primal cuts also may be graded if still attached to the primal cut. Although not designated as retail grades, retailers do retain the original grade in cuts fabricated from wholesale and primal cuts. In addition to carcasses, cutability grades may be applied to hindquarters, forequarters, ribs, loins, and short ribs, but not to rounds and chucks.

Carcass Quality Factors.—The quality grade of a beef carcass is based on the palatability—indicating characteristics of the lean. Determination of quality of the lean is accomplished by considering the degree of marbling and firmness in conjunction with maturity. Maturity is determined by evaluating the size, shape, and ossification of bones and cartilages, and by color of the lean. Marbling refers to flecks of fat interspersed among muscle fibers in the lean. The degree of marbling is considered to be positively associated with flavor, tenderness, juiciness, and palatability

Quality Grade Marks

in general. Marbling requirements for a given grade increase with maturity. Seven different degrees of marbling and five different maturity groupings are recognized in the application of grade standards. Figure 11.2 illustrates the relationship between marbling, maturity, and quality. Marbling requirements for bottom USDA Choice vary from a minimum of the range set up for "small" degree of marbling for the youngest animals, to a maximum of the range set up for "modest" degree for the second stage of maturity. Prime, Choice, Good, and Standard grades are limited to beef from relatively young animals.

Carcass Yield Grade or "Cutability" Factors.—Official grade standards provide for five cutability groupings in recognition of the fact that carcasses of identical quality grade and weight can have substantial difference in value due to differences in the percentage of closely-trimmed, boneless, retail cuts obtainable from the carcasses. These groupings, called "yield" grades, are numbered 1 through 5, with No. 1 having the highest percentage of retail cuts and No. 5 the lowest. Value, of course, varies with the level of market prices, but $5.00 per cwt of carcass is not unusual.

The chief factors which account for variation in yield of retail cuts are (1) the amount of fat that must be trimmed, and (2) the thickness and fullness of the muscling. Extensive research has shown that four indicators highly correlated with yield are as follows: (1) amount of external fat, (2) amount of kidney, pelvic, and heart fat, (3) area of the rib eye muscle, and (4) carcass weight. The following mathematical equation for calculating yield grade shows relationships among the four factors:

Yield Grade Marks

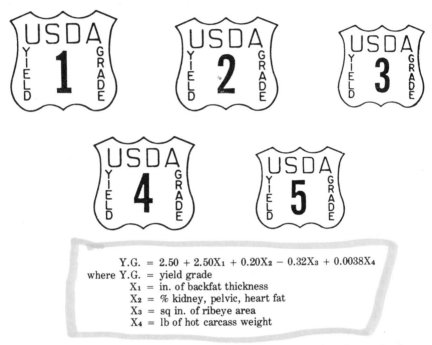

$$Y.G. = 2.50 + 2.50X_1 + 0.20X_2 - 0.32X_3 + 0.0038X_4$$

where Y.G. = yield grade
X_1 = in. of backfat thickness
X_2 = % kidney, pelvic, heart fat
X_3 = sq in. of ribeye area
X_4 = lb of hot carcass weight

The result is expressed as a whole number (any fractional part is dropped). A simpler means of evaluating objective measurements is by use of a slide-rule type device, designed by the Livestock Division, Consumer and Marketing Service, USDA, called the Beef Carcass Yield Grade Finder.

In contrast to quality grade indicators, all of these yield factors can be measured objectively. This was of great help in developing yield grade standards although in application actual measurement is unnecessary. Descriptions given in the standards (USDA 1965A) ". . . facilitate the subjective determination of the cutability group without making detailed measurements and computations. The cutability group for most beef carcasses can be determined accurately on the basis of a visual appraisal."

The establishment of yield grades culminated a long period of discussion and research. It had been known for a number of years that differences existed in cutability. Large wholesale buyers, including buyers for large chains, personally selected carcasses when buying from packers. Many still do this to meet more exact specifications within both yield and

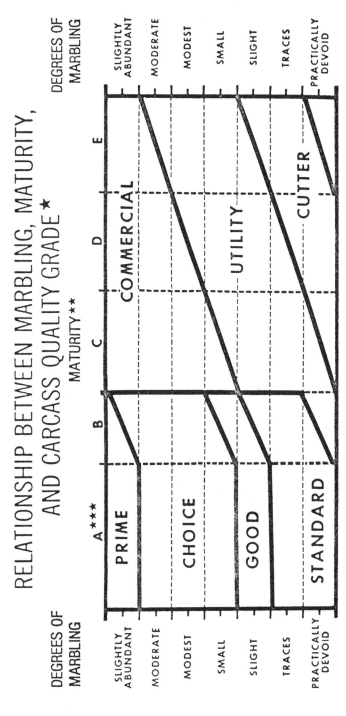

RELATIONSHIP BETWEEN MARBLING, MATURITY, AND CARCASS QUALITY GRADE ★

DEGREES OF MARBLING

MATURITY ★★

DEGREES OF MARBLING

SLIGHTLY ABUNDANT
MODERATE
MODEST
SMALL
SLIGHT
TRACES
PRACTICALLY DEVOID

PRIME
CHOICE
GOOD
STANDARD
COMMERCIAL
UTILITY
CUTTER

A ★★★   B   C   D   E

★ Assumes that firmness of lean is comparably developed with the degree of marbling and that the carcass is not a "dark cutter."

★★ Maturity increases from left to right (A through E).

★★★ The A maturity portion of the Figure is the only portion applicable to bullock carcasses.

FIG. 11.2.   MARBLING AND MATURITY RELATIONSHIP FROM THE OFFICIAL GRADE STANDARDS REVISED IN 1976

quality grades, but as will be pointed out in a following section, a rapid growth has occurred in use of yield grades since their inauguration in 1965.

The economic significance of variation in yield grades is shown by USDA calculations in Table 11.7. The retail value per cwt of carcass varied from $111.70 for Yield Grade 5 to $139.07 for Yield Grade 1 based on 1978 prices. That would amount to $164.16 on a 600 lb carcass. An efficient marketing system will reflect such differences back to producers. The use of yield grades tends to impart pricing efficiency in the system. The USDA publishes current retail value of Choice beef carcasses of various yields in their *Livestock, Meat, Wool Market News*.

Table 11.7 points up differences in percentage yield and retail value by cuts of beef from the several yield grades. Retail price per pound is not affected by yield grade, but rather by quality grade. In this example all carcasses are identical in quality. Retail value per cut varies with the yield of retail meat in the cut. The only "cut" which increases with lower yield grade is fat—a dubious distinction.

Slaughter Cattle

The term "slaughter cattle" does not include vealers and slaughter calves. Special efforts have been made to make grades for slaughter animals conform directly to the grades of carcasses produced from the live animals. Eight quality grades and five yield grades, identical in name with carcass grades, have been established for slaughter cattle. Eligibility of the various classes for particular grades are identical for both live animals and carcasses.

Quality Grade Factors.—Slaughter cattle quality grades are based on factors related to the palatability of the lean. In evaluating quality, attention is directed primarily to finish (its amount and distribution), muscling (its firmness and fullness), and physical characteristics associated with maturity. By necessity, appraisal of finish is achieved primarily by observable characteristics of external finish.

From the standpoint of maturity, approximate maximum age limits for steers and heifers, are 42 months for Prime and Choice, 48 months for Good and Standard. If these classes are over 48 months of age, Commercial grade applies. No age limits are recognized for any class in the Utility, Cutter, and Canner grades.

Figure 11.3 illustrates differences in slaughter steers of a partial range in quality grades.

SLAUGHTER STEERS
U.S. GRADES
(QUALITY)

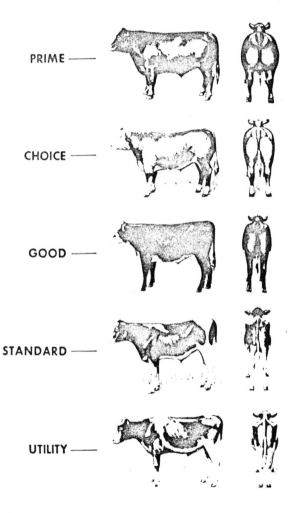

PRIME

CHOICE

GOOD

STANDARD

UTILITY

COMMERCIAL, CUTTER, AND
CANNER GRADES ARE OMITTED

Courtesy of USDA

FIG. 11.3.   U.S. GRADES OF SLAUGHTER STEERS (QUALITY)

TABLE 11.7. COMPARISON OF YIELDS OF RETAIL CUTS AND RETAIL SALES VALUES CHOICE BEEF AND LAMB CARCASSES, BY YIELD GRADE*

Retail Cut	Price per pound	1		2		3		4		5	
Yield Grades		% of carcass	Value/ Cwt. carcass	% of carcass	Value/ Cwt. carcass	% of carcass	Value/ Cwt. carcass	% of carcass	Value/ Cwt. carcass	% of carcass	Value/ Cwt. carcass
Rump boneless	2.00	3.7	7.40	3.5	7.00	3.3	6.60	3.1	6.20	2.9	5.80
Inside round, boneless	2.17	4.9	10.63	4.5	9.76	4.1	8.90	3.7	8.03	3.	7.16
Outside round, boneless	2.03	4.8	9.74	4.6	9.34	4.4	8.93	4.2	8.53	4.0	8.12
Round tip, boneless	2.02	2.7	5.45	2.6	5.25	2.5	5.05	2.4	4.85	2.3	4.65
Sirloin, bone-in	2.18	9.1	19.84	8.7	18.96	8.3	18.09	7.9	17.22	7.5	16.35
Short loin, bone-in	2.72	5.3	14.42	5.2	14.14	5.1	13.87	5.0	13.60	4.9	13.33
Blade chuck, bone-in	1.09	9.9	10.79	9.4	10.25	8.9	9.70	8.4	9.16	7.9	8.61
Rib, short cut (7"), bone-in	2.15	6.3	13.54	6.2	13.33	6.1	13.12	6.0	12.89	5.9	12.68
Chuck, arm boneless	1.49	6.4	9.54	6.1	9.09	5.8	8.63	5.5	8.19	5.2	7.75
Brisket, boneless	1.72	2.5	4.30	2.3	3.96	2.1	3.61	1.9	3.27	1.7	2.92
Flank steak	2.51	.5	1.26	.5	1.26	.5	1.26	.5	1.26	.5	1.26
Lean trim	1.38	12.3	16.97	11.3	15.59	10.3	14.21	9.3	12.83	8.3	11.45
Ground beef	1.08	13.3	14.36	12.2	13.18	11.1	11.99	10.0	10.80	8.9	9.61
Kidney	.52	.3	.16	.3	.16	.3	.16	.3	.16	.3	.16

	1		2		3		4		5	
	% of car-cass	Value/Cwt.	% of car-cass	Value/Cwt.	% of car-cass	Value/Cwt.	% of car-cass	Value/Cwt.	% of car-cass	Value/Cwt.
Fat	7.6	.46	12.7	.76	17.8	1.07	22.9	1.37	28.0	1.68
Bone	10.4	.21	9.9	.20	9.4	.19	8.9	.18	8.4	.17
Total	100.00	139.07	100.0	132.23	100.0	125.38	100.0	118.54	100.0	111.70

Difference in retail value between yield grades—$6.84 per cwt of carcass.

CHOICE LAMB
Yield Grades

Retail Cut	Price per pound	1		2		3		4		5	
		% of car-cass	Value/Cwt.	% of car-cass	Value/Cwt.	% of car-cass	Value/Cwt.	% of car-cass	Value/Cwt.	% of car-cass	Value/Cwt.
Leg, short cut	2.14	23.6	50.50	22.2	47.51	20.8	44.51	19.4	41.52	18.0	38.53
Sirloin	3.00	6.7	20.10	6.4	19.20	6.1	18.30	5.8	17.40	5.5	16.50
Short loin	3.61	10.4	37.54	10.1	36.45	9.8	35.38	9.5	34.31	9.2	33.21
Rack	3.38	8.1	27.38	7.9	26.70	7.7	26.03	7.5	25.35	7.3	24.67
Shoulder	2.02	24.9	50.30	23.8	48.07	22.7	45.85	21.6	43.63	20.5	41.41
Neck	1.11	2.2	2.44	2.1	2.33	2.0	2.22	1.9	2.11	1.8	2.00
Breast	1.00	9.8	9.80	9.8	9.80	9.8	9.80	9.8	9.80	9.8	9.80
Foreshank	1.58	3.5	5.53	3.4	5.37	3.3	5.22	3.2	5.05	3.1	4.90
Flank	1.48	2.3	3.40	2.3	3.40	2.3	3.40	2.3	3.40	2.3	3.40
Kidney	1.05	0.5	.52	0.5	.52	0.5	.52	0.5	.52	0.5	.52
Fat	.06	4.6	.29	8.2	.49	11.8	.71	15.4	.92	19.0	1.14
Bone	.02	3.4	.07	3.3	.07	3.2	.06	3.1	.06	3.0	.06
Total		100.0	207.87	100.0	199.94	100.0	192.00	100.0	184.07	100.0	176.14

Difference in retail value between yield grades—$7.93 per cwt. of carcass.

*The comparisons reflect average yields of retail cuts from beef and lamb carcasses typical of the midpoint of each of the USDA yield grades and average prices (including salepriced items) for USDA Choice beef and lamb during April 1978.
Source: (USDA 1978) Livestock, Meat and Wool, Vol. 46, No. 22, June 6, 1978.

Yield Grade Factors.—The factors used in establishing yield grades for slaughter cattle are identical with those used for carcass beef, i.e., thickness of fat over ribeye; percentage of kidney, pelvic, and heart fat; carcass weight; and area of the ribeye muscle. These characteristics cannot be measured directly. They can be estimated and yield calculated by equation, but ". . . a more practical method of appraising slaughter cattle for yield grade is to use only two factors normally considered in evaluating live cattle—muscling and fatness.

"In the latter approach, . . . evaluation of the thickness and fullness of muscling in relation to skeletal size largely accounts for the effects of two of the factors—area of ribeye, and carcass weight. By the same token, an appraisal of the degree of external fatness largely accounts for the effects of thickness of fat over the ribeye and the percentage of kidney, pelvic, and heart fat" (USDA 1966B).

Figure 11.4 illustrates differences in steers of the full range in yield grades.

Veal and Calf Carcasses

Veal and calf production has been decreasing for a number of years. It amounted to only 834 million pounds of a total of 25,279 million pounds of beef, veal, and calf slaughter (slightly over 3%) under commercial slaughter in 1977. However, veal and calf production are important in some areas.

The basic factor which differentiates veal, calf, and beef is maturity, and the three respective sets of standards designed for grading are intended to represent the entire range in bovine maturity. For grading purposes, veal, calf and beef carcasses are distinguished (USDA 1956) "primarily on the basis of the color of the lean, although such factors as texture of the lean; character of the fat; color, shape, size and ossification of the bones and cartilages; and the general contour of the carcass are also given consideration. Typical veal carcasses have a grayish pink color of lean that is very smooth and velvety. . . . By contrast, typical calf carcasses have a grayish red color of lean . . ." Color of the lean in beef carcasses becomes progressively darker red with increasing maturity.

Grade and Class Mark

CALF

SLAUGHTER STEERS
U.S. GRADES
(YIELD)

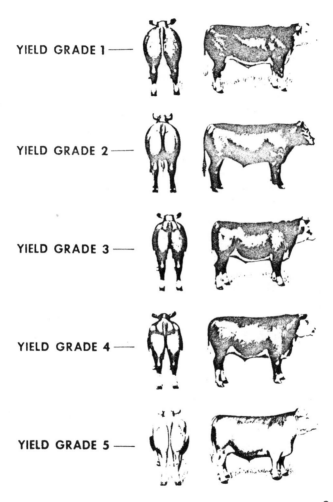

YIELD GRADE 1

YIELD GRADE 2

YIELD GRADE 3

YIELD GRADE 4

YIELD GRADE 5

Courtesy of USDA

FIG. 11.4.   U.S. GRADES OF SLAUGHTER STEERS (YIELD)

Official grade standards recognize three classes of veal and calf carcasses on the basis of sex condition—steers, heifers, and bulls. Standards are established for six grades: Prime, Choice, Good, Standard, Utility, and Cull. Each class is eligible for the full range in grades. In addition to the grade stamp, veal and calf also are stamped "VEAL" or "CALF" as the case may be.

Yield grades have not been established for veal and calf carcasses.

"Veal and calf carcasses are graded on a composite evaluation of three general grade factors—conformation, finish, and quality. These factors are concerned with the proportions of the various wholesale cuts and the proportion of fat, lean, and bone in the carcass and the quality of the lean" (USDA 1956). Without explicitly using the term "cutability," this factor is recognized—primarily in the consideration of conformation. Quality of the lean is specified and is identified by consideration of color, texture, firmness, and marbling of the lean. Palatability is recognized, not only in the above-mentioned characteristics of the lean, but also in the degree of finish. Finish refers primarily to the degree of fatness, which is presumed to be associated with marbling and, hence, with palatability.

Vealers and Slaughter Calves

According to USDA (1957), "The basis for differentiation between vealers and calves is made primarily on age and certain evidence of type of feeding. Typical vealers are less than three months of age and have subsisted largely on milk . . . they have the characteristic trimness of middle. . . . Calves are usually between 3 and 8 months of age, have subsisted partially or entirely on feeds other than milk for a substantial period of time, and have developed the heavier middles and physical characteristics associated with maturity beyond the vealer stage."

The three classes of vealers and calves are steers, heifers, and bulls. Eligible grades for each class are Prime, Choice, Good, Standard, Utility, and Cull. Grade is determined by a composite evaluation of conformation, finish, and quality. "Conformation refers to the general body proportions . . . and to the ratio of meat to bone . . . Finish refers to the fatness of the animal . . . Quality in the slaughter animal refers to the refinement of hair, hide and bone and to the smoothness and symmetry of the body. Quality is also associated with carcass yield and the proportion of meat to bone" (USDA 1957). Thus, it is apparent that yield of lean meat (as well as carcass yield) is inherent in the grade standards by considerations of conformation, finish, and quality. The intent of evaluating quality of the meat is explicit in the statement that "the quality, quantity, and distribution of finish are all closely associated with the palatability and quality of the meat" (USDA 1957).

Feeder Cattle

The official distinction between feeder cattle and slaughter cattle is in the intended use. Feeder cattle are intended for further feeding (growing or finishing) before slaughter. It is recognized that some so-called two-way cattle do not fit exclusively into the feeder or slaughter classification, but this is no obstacle to grading. Any cattle may be graded for slaughter purposes—based on slaughter grade standards. However, if used as feeder animals, they will be graded on feeder cattle grades.

Grades of feeder cattle are related to slaughter cattle grades in that the primary characteristics considered are the feeder animals' logical slaughter potential as beef. Thriftiness is another factor. The final grade is a composite evaluation of the two with greater weight given to slaughter potential. Conformation is a key to slaughter potential. According to USDA (1965B), ". . . conformation or inherent muscular development is the most important single factor affecting the grade of a feeder animal . . . conformation is determined by appraising the development of the muscular system in relation to development of the skeletal system." While degree of fatness is not a grade factor, it affects appearance and is recognized in evaluating conformation.

"Thriftiness (USDA 1965B) refers to the ability of a feeder animal to gain weight and fatten rapidly and efficiently." Shape-for-age and alertness are indicators of thriftiness. This factor, however, is not used in grading except when the animal is less thrifty than normal.

Grade names used for feeder cattle are Prime, Choice, Good, Standard, Commercial, Utility, and Inferior. Figures 11.5 and 11.6 illustrate differences in a partial range of grades for feeder steers and feeder steer calves.

The introduction of so-called "exotic breeds" into the U.S. and wide adoption of cross-breeding programs has rendered the feeder cattle grade standards of 1964 somewhat obsolete. Little use is made of official grades in commercial trade.

A revision of feeder cattle grade standards is under consideration which would base grades on frame size and muscling, with thriftiness as a possible third factor. Three categories of muscling—numbered 1, 2, and 3—are suggested as closely related to yield grade. Three frame sizes—large, medium, and small—also are considered. Frame size is related to the length of feeding period required to make Choice quality grade. Thriftiness is considered as a possible modifying factor.

FEEDER STEERS
U.S. GRADES

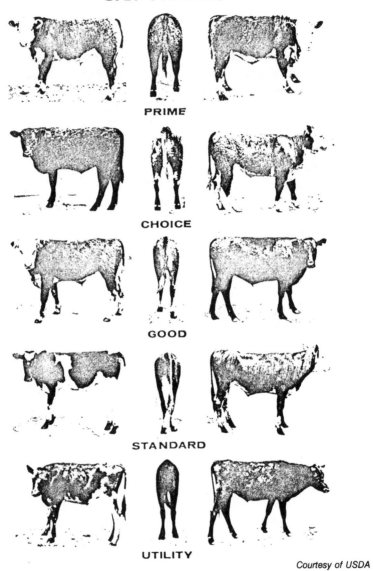

Courtesy of USDA

FIG. 11.5.   GRADES OF FEEDER STEERS

FEEDER STEERS (CALVES)

U.S. GRADES

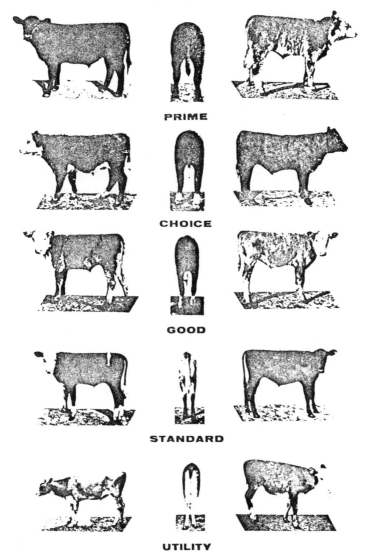

Courtesy of USDA

FIG. 11.6.   U.S. GRADES OF FEEDER STEERS (CALVES)

Pork Carcasses—Barrows and Gilts

Market classes of pork carcasses are barrows, gilts, sows, boars, and stags; however, the following comments are limited to barrows and gilts. In grade standards (and in market prices) no distinction is made between barrows and gilts. Specified grades are U.S. No. 1, 2, 3, 4, and Utility. According to the revised (USDA 1968A) standards of April 1, 1968,

"grades for barrow and gilt carcasses are based on two general considerations: (1) quality-indicating characteristics of the lean, and (2) expected combined yields of the four lean cuts (ham, loin, picnic shoulder, and Boston butt)." From the standpoint of quality, two general levels are specified: "acceptable" and "unacceptable." By direct observation of a cut surface, acceptability is based on considerations of firmness, marbling, and color. Indirect indicators are firmness of fat and lean, feathering between the ribs, and color. The degree of external fatness, as such, is not considered in evaluating the quality of the lean (USDA 1968A). Suitability of the belly for bacon (in terms of thickness) also is considered in quality evaluation as is "softness" and "oiliness" of the carcass. Carcasses which have unacceptable quality of lean, and/or bellies that are too thin, and/or carcasses which are soft and oily are graded U.S. Utility.

If a carcass qualifies as acceptable in quality of lean, and belly thickness, and is not soft and oily, then it is graded U.S. No. 1, 2, 3, or 4 ". . . entirely on the expected carcass yields of the four lean cuts . . ." (USDA 1968A). Carcasses grading U.S. No. 1 are expected to yield 53% or more of the four lean cuts based on chilled carcass weight; U.S. No. 2 carcasses, 50–52.9%; U.S. No. 3 carcasses, 47–49.9%; and U.S. No. 4 carcasses, less than 47%. Variations in yield of the four lean cuts are associated with "variations in their degree of fatness and in their degree of muscling . . . (USDA 1968A). Since many carcasses have a normal distribution of fat and a normal distribution of muscling for their degree of fatness, in determining their grade the actual average thickness of backfat and the carcass length or weight are the only factors considered" (USDA 1968A). Figure 11.7 illustrates these relationships in objective measurements of the variables. Note that carcass weight *or* carcass length may be used but the standards (USDA 1968A) state further that "in cases where length and backfat thickness indicate a different grade than weight and backfat, the grade shall be determined using length." The grade standards also specify the extent to which superior muscling can compensate for greater than average backfat thickness.[9]

Basically, grades for pork carcasses, like those for beef carcasses, are designed to evaluate both quality and yield. The significance of this to the livestock and meat industry follows the same pattern as discussed under

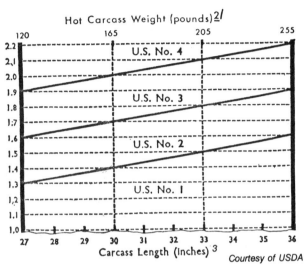

FIG. 11.7. RELATIONSHIP BETWEEN AVERAGE THICKNESS OF BACKFAT, CARCASS LENGTH OR WEIGHT, AND GRADE FOR CARCASSES WITH MUSCLING TYPICAL OF THEIR DEGREE OF FATNESS

[1] An average of three measurements including the skin made opposite the first and last ribs and the last lumbar vertebra. It also reflects adjustments, as appropriate, to compensate for variations from normal fat distribution.
[2] Carcass weight is based on a hot packer style carcass.
[3] Carcass length is measured from the anterior point of the aitch bone to the anterior edge of the first rib.

the section on beef carcasses. A USDA (1968A) calculation showed that during July, 1969, the value differences between adjacent grades of 150-lb carcasses averaged nearly $1.25 cwt.

Slaughter Hogs

The classes of slaughter hogs and associated grade names are identical with classes and grades of swine carcasses discussed in the previous section.

---

[9]For descriptive discussion see USDA (1970A).

Standards used in grading slaughter animals are consistent with those used in grading carcasses. (A U.S. No. 1 slaughter barrow or gilt is expected to produce a U.S. No. 1 carcass, and so on through the other grades.) The same attributes are evaluated in the live animal as in the carcass: (1) quality of the lean and (2) yield of the four lean cuts. As was the case with carcasses, quality of the slaughter animal includes, in addition to quality per se of the lean meat, a consideration of belly thickness as an indicator of its adequacy for bacon production and a determination of whether the carcass will be soft and oily. "Since carcass indices of lean quality are not directly evident in barrows and gilts, some other factors in which differences can be noted must be used to evaluate quality. Therefore the amount and distribution of external finish, firmness of fat, and firmness of lean are used as quality-indicating factors" (USDA 1968B).

Slaughter barrows and gilts which do not meet standards of acceptable quality are graded Utility. Those which meet the standards for acceptable quality are then assigned grade U.S. No. 1, 2, 3, or 4 on the basis of expected combined carcass yield of the four lean cuts, with U.S. No. 1 representing the highest yield. The reproductions in Fig. 11.8 illustrate the appearance of hogs of the various grades.

Factors used to evaluate yield are average backfat thickness in relation to carcass length *or* live weight. In the live animal only weight can be readily determined. However, with experience these factors can be evaluated visually. Studies have shown objective relationships between these variables, as illustrated in Fig. 11.9, for slaughter barrows and gilts having a normal development of muscling for the degree of fatness. The standards recognize that the degree of muscling varies, and provision is made for superior muscling to compensate for greater fatness (and vice versa) with specified limitations.

Feeder Pigs

Following the revision in grade standards for pork carcasses and slaughter swine in 1968, the grade standards for feeder pigs were revised in 1969. This change was made to keep feeder pig grades consistent with slaughter swine and pork carcass grades. Official standards do not differentiate between barrow, gilt, or boar pigs, but it is assumed boar pigs will be castrated prior to development of secondary physical characteristics of a boar. Sows, stags, and mature boars are seldom used as feeders and are not included in the following. The criteria upon which feeder pigs are evaluated are comparable to those for feeder cattle, namely logical slaughter potential and thriftiness. As pointed out by USDA (1969A), "the logical slaughter potential of a thrifty feeder pig is expected slaugh-

SLAUGHTER SWINE

U.S. GRADES

U.S. NO.1   **U.S. NO.2**

U.S. NO.3   **U.S. NO.4**

U.S. UTILITY   *Courtesy of USDA*

FIG. 11.8.   U.S. GRADES OF SLAUGHTER SWINE

ter grade at a market weight of 220 lb after a normal feeding period . . . Thriftiness in a feeder pig is its apparent ability to gain weight rapidly and efficiently."

Designated grades for feeder pigs are U.S. No. 1, 2, 3, 4, Utility, and Cull. Unthrifty pigs are graded Utility and Cull, depending on the degree of unthriftiness, Cull being the lower end. Pigs classed as thrifty are graded U.S. No. 1, 2, 3, or 4 on the basis of logical slaughter potential.

An indication of slaughter potential is derived by a composite evaluation of development of the muscular and skeletal systems. Indicators of thriftiness are size for age, health, and other general characteristics. Figure 11.10 illustrates the appearance of feeder pigs of all grades except Cull.

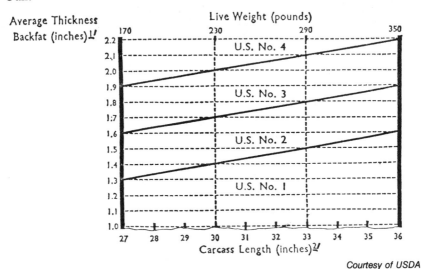

Courtesy of USDA

FIG. 11.9. RELATIONSHIP BETWEEN AVERAGE THICKNESS OF BACKFAT, WEIGHT OR CARCASS LENGTH, AND GRADE FOR BARROWS AND GILTS WITH MUSCLING TYPICAL OF THEIR DEGREE OF FATNESS

[1]An average of three measurements including the skin made opposite the first and last ribs and the last lumbar vertebra. It also reflects adjustment, as appropriate, to compensate for variations from normal fat distribution.
[2]Carcass length is measured from the anterior point of the aitch bone to the anterior edge of the first rib.

Lamb, Yearling Mutton, and Mutton Carcasses

Revision of the grade standards for lamb, yearling mutton, and mutton carcasses in 1969 included establishment of explicit yield grades. This change followed a lengthy period of discussion, conferences, and research. And, as was the case with other meats, not all interests agreed to the change, but USDA investigations indicated a consensus in favor of it.

The groupings of lambs, yearlings, and sheep (which produce the above indicated carcasses) are based on age, or maturity differences—characterized by differences in development of the muscular and skeletal systems. As enunciated by USDA (1969B), "the grade of an ovine carcass

FEEDER PIGS

U.S. GRADES

U.S. NO.1   **U.S. NO.2**

U.S. NO.3   **U.S. NO.4**

CULL GRADE IS
OMITTED   **U.S. UTILITY**

Courtesy of USDA

FIG. 11.10. U.S. GRADES OF FEEDER PIGS

is based on separate evaluations of two general considerations: (1) palatability-indicating characteristics of the lean and conformation, herein referred to as quality; and (2) the estimated percentage of closely trimmed, boneless, major retail cuts to be derived from the carcass, herein referred to as yield." The relevant retail cuts are from the leg, loin, hotel rack, and shoulder.

Quality Grade Marks

The definitions of such terms as "quality of the lean" and "conformation," as well as the indicators used to evaluate them, are consistent with those used in grading beef carcasses. Quality grades specified are Prime, Choice, Good, Utility, and Cull. Mutton carcasses, however, are not eligible for Prime grade. In addition to the grade stamp, yearling mutton and mutton are stamped "YEARLING MUTTON" or "MUTTON" as the case may be.

Five yield grades, numbered 1 through 5, cover the range in cutability, with Yield Grade 1 indicating the highest yield. "A carcass which is typical of its yield grade would be expected to yield 3.5% more in total retail cuts than the next lower yield grade, when USDA cutting and trimming methods are followed" (USDA 1970C). Variations in yield are attributed primarily to the same two considerations referred to in the case of beef carcasses—namely, the amount of fat trimmed off in making retail cuts, and the thickness and fullness of muscling.

"The yield grade of an ovine carcass is determined by considering three characteristics: the amount of external fat, the amount of kidney and pelvic fat, and the conformation grade of the legs" (USDA 1970C).

The grade standards include an equation by which yield grade can be calculated. The equation may be written as:

$$Y.G. = 1.66 - 0.05X_1 + 0.25X_2 + 6.66X_3$$

Where $Y.G.$ = yield grade
X_1 = leg conformation grade code
X_2 = % kidney and pelvic fat
X_3 = adjusted fat thickness over the ribeye in inches

Any fractional part of a yield grade is dropped in this calculation and the result is reported as a whole number. Leg conformation used in the equation is the evaluation used in quality grading, but expressed as a coded number ranging from 1 to 15, with 15 being the most desirable conformation (i.e., conformation applicable to high Prime). The amount of external fat, an indicator of retail trim, can be measured but it also can be accurately estimated. Kidney and pelvic fat cannot be measured without removal from the carcass; however, this too can be evaluated subjectively with an acceptable degree of accuracy.

Yield grades are applicable to wholesale and primal cuts as well as to carcasses.

The economic importance of yield is indicated in Table 11.7 where quality grade and carcass weight are assumed to be constant and the only difference between the two carcasses is two yield grades. The calculated difference in retail sales value between yield grades is $7.93 per cwt of carcass.

Wool and Wool Top

Grades of wool are indicated by ". . . numerical designations of fineness based on average fiber diameter and variation of fiber diameter" (USDA 1966). Grades of wool top are similarly defined as "A numerical designation of wool top fineness based on average fiber diameter and fiber diameter dispersion" (USDA 1971). Official specifications for each provide for two methods of determining grade: (a) by measurement, and (b) by visual inspection. Both methods are official. However, if the grade determined by inspection differs from the grade determined by measurement, for a given lot, then the grade determined by measurement prevails.

Grading by visual inspection could appropriately be classified as an art. Skilled and experienced commercial graders can, almost instantaneously, grade fleeces which to an amateur may appear identical. Specifications for official sampling, measuring, and grading are spelled out in detail in USDA (1966C) and USDA (1971). Official U.S. wool grades and associated standards are given below, (from USDA 1966C).

OFFICIAL STANDARDS OF THE UNITED STATES FOR GRADES OF WOOL

Sec. 31.0 *Official grades.* The official grades of wool shall be those established in Secs. 31.1 through 31.16: *Provided, however,* That the wool which qualifies for any of the grades in Secs. 31.1 through 31.15 on the basis of its average fiber diameter shall be reduced in grade to the next coarser grade if its standard deviation in fiber diameter exceeds the maximum specified for the grade to which the average fiber diameter corresponds.

Sec. 31.1 *Finer than grade 80's.* Wool with an average fiber diameter of 17.69 microns or less and a standard deviation in fiber diameter of 3.59 microns or less.

Sec. 31.2 *Grade 80's.* Wool with an average fiber diameter of 17.70 to 19.14 microns, inclusive, and a standard deviation in fiber diameter of 4.09 microns or less.

Sec. 31.3 *Grade 70's.* Wool with an average fiber diameter of 19.15 to 20.59 microns, inclusive, and a standard deviation in fiber diameter of 4.59 microns or less.

Sec. 31.4 *Grade 64's.* Wool with an average fiber diameter of 20.60 to 22.04 microns, inclusive, and a standard deviation in fiber diameter of 5.19 microns or less.

Sec. 31.5 *Grade 62's.* Wool with an average fiber diameter of 22.05 to 23.49 microns, inclusive, and a standard deviation in fiber diameter of 5.89 microns or less.

Sec. 31.6 *Grade 60's.* Wool with an average fiber diameter of 23.50 to 24.94 microns, inclusive, and a standard deviation in fiber diameter of 6.49 microns or less.

Sec. 31.7 *Grade 58's.* Wool with an average fiber diameter of 24.95 to 26.39 microns, inclusive, and a standard deviation in fiber diameter of 7.09 microns or less.

Sec. 31.8 *Grade 56's.* Wool with an average fiber diameter of 26.40 to 27.84 microns, inclusive, and a standard deviation in fiber diameter of 7.59 microns or less.

Sec. 31.9 *Grade 54's.* Wool with an average fiber diameter of 27.85 to 29.29 microns, inclusive, and a standard deviation in fiber diameter of 8.19 microns or less.

Sec. 31.10 *Grade 50's.* Wool with an average fiber diameter of 29.30 to 30.99 microns, inclusive, and a standard deviation in fiber diameter of 8.69 microns or less.

Sec. 31.11 *Grade 48's.* Wool with an average fiber diameter of 31.00 to 32.69 microns, inclusive, and a standard deviation in fiber diameter of 9.09 microns or less.

Sec. 31.12 *Grade 46's.* Wool with an average fiber diameter of 32.70 to 34.39 microns, inclusive, and a standard deviation in fiber diameter of 9.59 microns or less.

Sec. 31.13 *Grade 44's.* Wool with an average fiber diameter of 34.40 to 36.19 microns, inclusive, and a standard deviation in fiber diameter of 10.09 microns or less.

Sec. 31.14 *Grade 40's.* Wool with an average fiber diameter of 36.20 to 38.09 microns, inclusive, and a standard deviation in fiber diameter of 10.69 microns or less.

Sec. 31.15 *Grade 36's.* Wool with an average fiber diameter of 38.10 to

40.20 microns, inclusive, and a standard deviation in fiber diameter of 11.19 microns or less.

Sec. 31.16 *Coarser than grade 36's*. Wool with an average fiber diameter of 40.21 microns or more.

Official U.S. wool top grades and associated standards are shown in Table 11.8.

Slaughter Lambs, Yearlings, and Sheep

As with other species, USDA (1970C) specifies "grades of slaughter ovines are intended to be directly related to the grades of the carcasses they produce. To accomplish this, these slaughter ovine grade standards are based on factors which are directly related to the quality grades and the yield grades of ovine carcasses." Quality grade names and yield grade designations for the live animals are identical with those for carcasses. The factors upon which live quality and yield grades are based also are consistent with those of carcass grades. However, the presence of the fleece makes evaluation more difficult and necessitates handling as well as visual observation.

TRENDS IN USE OF FEDERAL GRADES

Federal grading has been on a voluntary basis except during World War II and the Korean War. The service was not readily accepted at the beginning, but a slight annual increase was evident. There was a normal reluctance in the industry to accept something new, and further, this involved an expense since packers pay for the cost of grading. In addition, some packers had vested interests in their own private grading systems and viewed federal grades as detrimental to their interests. Following World War II the proportion of graded meat dropped significantly; but it did not drop back to its former level. The same thing happened following the Korean War. Since the Korean War, the trend in grading of beef and lamb was steadily upward until the 1976 beef grade revision which specified that any beef carcasses quality graded also must be yield graded. That resulted in a decrease in quality grading, but the proportion graded still is approximately 60%. Yield grading of beef increased rapidly following initiation of the service in 1965, and the grade revision of 1976 gave it a sharp boost. Quality grading and yield grading both facilitate the marketing function and will continue in general usage.

TABLE 11.8. WOOL TOP GRADES AND SPECIFICATIONS

Grade	Average Fiber Diameter Range, Microns	Fiber Diameter Distribution, percent:[1]								Number of fibers required per test[2]
		25 Microns and under, min.	30 Microns and under, min.	40 Microns and under, min.	25.1 Microns and over, max.	30.1 Microns and over, max.	40.1 Microns and over, max.	50.1 Microns and over, max.	60.1 Microns and over, max.	
Finer than 80s	Under 18.10	95	—	—	5	1	—	—	—	400
80s	18.10–19.59	91	—	—	9	1	—	—	—	400
70s	19.60–21.09	83	—	—	17	3	—	—	—	400
64s	21.10–22.59	—	92	—	—	8	—	—	—	600
62s	22.60–24.09	—	86	—	—	14	1	—	—	800
60s	24.10–25.59	—	80	—	—	20	1.5	—	—	800
58s	25.60–27.09	—	72	—	—	28	2	—	—	1,000
56s	27.10–28.59	—	62	—	—	38	—	—	—	1,200
54s	28.60–30.09	—	54	—	—	46	—	1	—	1,400
50s	30.10–31.79	—	44	—	—	56	—	1	—	1,600
48s	31.80–33.49	—	—	75	—	—	25	2	—	1,800
46s	33.50–35.19	—	—	68	—	—	32	2	1	2,000
44s	35.20–37.09	—	—	62	—	—	38	—	1	2,200
40s	37.10–38.99	—	—	54	—	—	46	—	2	2,400
36s	39.00–41.29	—	—	44	—	—	56	—	3	2,600
Coarser than 36s	over 41.29	—	—	—	—	—	—	—	4	2,600

Source: USDA (1971).
[1]The second maximum percent shown for any grade is a part of, and not in addition to, the first maximum percent. In each grade, the maximum percent and the first maximum percent total 100 percent.
[2]Research has shown that when wools of average uniformity in fiber diameter are measured, the prescribed number of fibers to measure per test will result in confidence limits of the mean ranging from approximately ±0.4 to ±0.5 micron at a probability of 95 percent.

IMPROVEMENTS IN GRADES OF LIVESTOCK PRODUCED

An efficiently operating market system is presumed to price products over a period of time in a manner that reflects changing demand and supply characteristics. Theoretically, prices on less preferred products will tend to decline, and production of those products also will tend to decline. The reverse is expected on the more preferred products. Grades and grading provide a means of imparting knowledge which presumably tends to make the system more efficient. The pork market provides evidence that this has occurred in spite of much criticism and impatience with the rate of change.

Long-run wholesale pork prices show a clear-cut increasing trend for the four major lean cuts as a percentage of live hog values, while the price of lard has declined. That is the consumers' way of saying, through the marketing system, they prefer lean pork. Data in Table 11.9 indicates that producers received the message. Through breeding and feeding programs, average lard production has declined. The trend has been underway for decades, but in just the 17 yr period of Table 11.9 lard production per cwt of hogs slaughtered dropped from 14.1 lb to 6 lb—a 57% decline. The pounds of pork produced per animal increased from 149.8 to 165.8, with average liveweight remaining essentially unchanged. This type of information indicates general improvement in production of meat-type hogs, and indirectly suggests that improvements have been made in the grade of hogs, but it does not show changes within the grouping of generally acceptable hogs.

Data are not regularly collected on hog slaughter by grade, since the wholesale trade makes little use of pork grades. However, the USDA has made special studies in which determinations were made of the percentage of barrow and gilt carcasses in each of the various grades. These studies showed a substantial increase in percentage of U.S. No. 1 carcasses and a substantial decrease in percentage of U.S. No. 3 carcasses.

Data on the distribution of beef by grade has been collected for a number of years. Changes in grade standards have affected the historical grade distribution—several changes have tended to increase the proportion of beef grading Choice. However, even when grade standard changes are accounted for there is an apparent trend of increase in the proportion of beef grading Choice and a decrease in beef grading Standard. The proportion grading Good held fairly stable until the revision of 1976, then dropped as a result of that change. An example of how a change in standards can affect distribution by grade is found in the case of lamb. A revision in 1960 relaxed the requirements for Prime grade. As a result (Fienup et al. 1963), "prime increased from 1% in the 1959 marketing year

TABLE 11.9.   CHANGES IN SELECTED CHARACTERISTICS OF HOGS SLAUGHTERED
UNDER FEDERAL INSPECTION 1959, 1969, AND 1976

Characteristic	1959 (Lb)	1969 (Lb)	1976 (Lb)
Lard per cwt	14.1	9.7	6.0
Lard per animal	33.7	23.2	14.4
Pork per animal[1]	149.8	161.9	165.6
Avg dressed weight	183.5	185.1	186.8
Liveweight	239.5	239.5	238.0
Dressing yield (%)	76.6	77.3	78.4

Source: USDA (1977).
[1]Pounds of pork per animal calculated.

(March through the following February) to 16% in 1960 and 12% in
1961 . . . Good dropped from 11% in 1959 to 3% in 1960 and 1961. Choice
dropped from 87% in 1959 to 81% in 1960 and 82% in 1961." However, in
the case of lamb, other factors such as cut-out of retail cuts, tend to
confirm a genuine improvement in quality of lamb produced.

Factors other than use of grades probably had an effect on grade improvement, but grades and grading provide a means for improvement of
market efficiency and an efficient market tends to induce the production
of preferred products. The evidence clearly indicates this has happened
and use of grades undoubtedly was a beneficial contribution.

CONSUMER PREFERENCES AND KNOWLEDGE
OF MEAT GRADES

In most discussions concerned with grades and grading it usually is
assumed (explicitly or implicitly) that (1) consumers prefer one grade to
another, (2) they are able to distinguish between and among grades, and
(3) they act accordingly. Numerous studies have been made to test
hypotheses along these lines. This section is not intended to be a comprehensive review of results, but rather to point up some generalizations
derived from such studies. The area of consumer preference involves
many subjective considerations which are exceedingly complex, and results have not been entirely consistent among studies or with some preconceived notions on the subject.

Consumer preference studies ordinarily are not designed to determine
whether consumers can grade meat (place meat of various characteristics
in the correct grade). There is clear evidence that consumers are not
well-informed in grade names, let alone grade standards. Consumer preference studies are more concerned with a determination of the attributes

in meat which consumers prefer, and in consumers' ability to differentiate between meats possessing varying degrees of those attributes.

Consumer evaluation of meat is largely based on (1) visual preferences and (2) sensory preferences—but these are not mutually exclusive. Other possible considerations are situational factors and imaginary differences. Most research has centered on visual and sensory preferences. In studies of visual preferences, consumers usually are shown photographs of cuts of meat (sometimes actual cuts of meat are used) selected in a manner intended to show differences between grades and/or within grades. In most of these studies consumers generally have indicated a preference for leaner cuts whether the meat be beef, pork, or lamb (Branson 1957; Perdue *et al.* 1958; Rhodes *et al.* 1955; Seltzer 1955; Stevens *et al.* 1956).

With beef, the leaner-appearing cuts are the so-called "lower" grades—Commercial, Standard, and Good. A first reaction might be that consumers were irrational, but a more considered judgment is that studies asking consumers' preferences within a range from Commercial to Prime places the respondent in a position not experienced in actual buying.

Most stores carry only one grade of beef—usually Choice, or its equivalent, if not officially graded.[10] The only alternative in this situation (and the only experience) consumers have is selection within the given grade. Many consumers select the leaner-appearing cuts to avoid trimming waste, with the expectation that eating quality will be relatively uniform. When confronted with Prime grade which has the appearance of more fat, and Good, Standard, and Commercial which appear to be leaner in that order (even though trim specifications may be comparable), many consumers seem to follow the pattern of selecting the leaner, which in these circumstances are cuts of different quality. In this situation their decisions based on visual appearance are rational.

When it comes to sensory tests (eating tests), consumers tend to show a preference for Prime and Choice grades of beef. Kiehl and Rhodes (1956) reported that "the eating preference patterns contradict the visual preferences found by many researchers, including ourselves. Many visual preferences have been made for the leaner grades at equal prices or even some with price differentials against the leaner cuts. Eating preferences were very rarely for the leaner grades."

---

[10]Some stores carry other USDA grades, but few advertise the fact. During parts of the 1973–1976 period, when grain prices were high and fed beef supplies relatively scarce, many stores experimented with the equivalent of "good" grade beef sold under their own brand name. That practice dwindled subsequently when grain prices dropped and the volume of choice beef increased.

Ikerd *et al.* (1971) in a semantic differential factor analysis of the consumer image of pork found, "The overall image of pork indicated by the study was that pork was wholesome and healthy, low in purchase cost, practical, tasty, and generally acceptable. However, it was considered lacking in stimulative quality and pork was considered to be fatter than consumers desired."

Consumer studies of pork have shown a consistent preference for lean cuts. However, a substantial variation was apparent among the studies on the degree of preference (Birmingham *et al.* 1954; Nauman *et al.* 1959; Stevenson and Schneider 1959; Trotter and Engelman 1959; Vrooman 1952).

Several studies have shown that consumers, by and large, are not familiar with grade names and often confuse inspection with grading. In a study including beefsteak and bacon (along with other nonmeat items) Hutchinson (1970) found that 90% of the persons interviewed said "yes" to the question, "Is there a government grade for beefsteak?" and a like percentage said they usually bought government-graded beefsteak. However, some skepticism arises when 70 and 81%, respectively, of the respondents answered "yes" to the same questions for bacon. Bacon is not sold by government grade. The author (Hutchinson 1970) commented that "possibly a 'halo effect' exists; the consumer reasons that since some beef is graded, all meat is graded. When asked to identify the shape of the USDA grade mark, 21.5% correctly identified the shield-shaped mark, but 29.5% said it was a circle, indicating a high probability that they were confusing the inspection mark with the grade mark.

In summary remarks, Hutchinson (1970) stated, "Most U.S. consumers know little about federal grades for the foods they buy, but those who are aware of grades say they find them helpful in buying decisions . . . the use of both adjective and letter grades for different food items apparently was not confusing."

Weidenhamer *et al.* (1968), with reference to data collected in a study of opinions about meats reported, "homemakers are indeed confused about inspection and grading. Although virtually all respondents reported that the meat they buy is inspected and that beef is graded, half of the respondents stated that pork is graded; and a considerable number ascribed functions of grading to inspection, and vice versa. A question, therefore, arises about which statements regarding these functions were made on the basis of knowledge, which were based on assumptions, and which stemmed from misrepresentations—such as may be fostered by the way some stores or packers indicate that the meats they sell are inspected or graded." These authors also found homemakers were not well informed on grade names. When confronted with a list of 10 grade names for beef—including 5 correct grade names and 5 spurious grade names—

respondents included 1 spurious grade name among the 3 most frequently mentioned.

Despite inconsistencies among some studies and lack of pronounced consensus among consumers on some attributes of meat, consumer preference studies have served, and will continue to serve, an extremely useful purpose. They have pointed up some identifiable preference characteristics, such as preference for leaner cuts and a positive relationship between marbling and palatability. Probably of importance equal to any positive results was the revelation that quality is an extremely complex subject to study. Meat itself is a complex product. Nevertheless, intelligent improvement in grade standards necessitates continued study of quality attributes which consumers can identify with reasonable accuracy and consistency. The ordering of preferences should be immaterial to researchers and public agencies involved in developing grade standards. If consumers were to prefer Standard grade beef over Good, and Good over Choice, this is the consumers' prerogative. From a grading standpoint, it is important that the grades are based on characteristics by which consumers can differentiate between grades regardless of the order of preference.

ADDITIONAL USDA MEAT SERVICES

The USDA provides two additional services closely related to its grading services: (1) Meat Acceptance Service, and (2) Carcass Data Service. Both are voluntary, provided upon request, and paid for by users of the services.

Meat Acceptance Service

This is a service provided for large institutional buyers of meat (hospitals, steamship lines, government institutions, restaurant chains, etc.). Under this program, which began about 1960, an official USDA grader will certify that meat and meat products are in compliance with the purchaser's specifications. As an assist in the program, USDA has prepared a series of Institutional Meat Purchase Specifications (IMPS) covering some 300 fresh, cured, and processed meat and meat products. Plans call for the inclusion also of frozen, canned, and portion control meats. The specifications were developed in conjunction with members of the meat trade; and detail such things as kind of cut, quality grade, size and weight of cut, and style of trim. If the predesigned "specs" do not meet a particular buyer's need, USDA will cooperate in drawing up specifications which meet a particular buyer's need.

With detailed specifications, a buyer is in position to ask suppliers for bids. This facilitates the "bid and acceptance" method of purchasing as mentioned in Chap. 9. After the purchaser has accepted a bid, a USDA grader examines, accepts, and certifies the shipment if it does in fact meet the specifications.

Purchasers and suppliers must make prior arrangements for the service and the USDA is reimbursed for its costs. Normally the supplier pays the cost which presumably is recovered in the price of the meat.

This service can benefit both purchaser and supplier. From the purchaser's standpoint he can obtain (1) assistance, if necessary, in determining specifications which meet his particular needs, and (2) professional assistance which relieves him of employing experts to do the buying (buying can be done by "bid and acceptance") and to examine it upon arrival. The service also can benefit a supplier in that official certification assures him that the shipment will not be rejected upon arrival.

Meat which is accepted is stamped, either directly on the meat if it can be individually stamped, or on the carton containing products that cannot be stamped individually. Stamps used are shown here. Meat which can be

Acceptance Stamps

individually stamped bears the insignia shown on the left. Cartons which are stamped bear both stamps. The letters AC in this example identify the grader who accepted the product.

Beef Carcass Data Service

For a number of years, cow-calf producers and feedlot operators have indicated an interest in obtaining data on carcass characteristics of their cattle after slaughter. This is vital information to a cow-herd operator in his herd improvement program. It is equally vital to a feedlot operator's feeding program. In a very direct way this enters into their marketing programs. Both classes of producers experienced difficulties in obtaining this type of information prior to initiation of the service. Maintaining identity of animals during growing, finishing, and slaughtering is difficult. Without some agreement with packers and graders, producers (and feeders) had problems in lining up all the necessary arrangements. During the early 1960's the USDA made a move designed to alleviate the latter difficulty.

This was a program designated Beef Carcass Evaluation Service. Packers were informed of the program and graders were instructed to cooperate. It was the producer's responsibility to make arrangements with a packer and with a federal grader and to place an identification on each animal. After slaughter the grader filled out a form giving carcass characteristics and relayed the information to the producer.

The program worked, but from the producer's standpoint it was cumbersome and sometimes frustrating in making arrangements. Participation was limited in spite of the need and desire for carcass information.

In an effort to improve the service, USDA in early 1970 initiated a revised program on an experimental basis—renamed the Beef Carcass Data Service. Under the new setup, producers can purchase bright orange-colored USDA ear tags with identification marks—not directly from USDA, but from a cooperating organization which has established itself with USDA as a participating agency. That may be a cattle producer or feeder association, an agricultural organization, a state department of agriculture, etc. A cow-herd operator can tag his calves, or a feedlot operator can tag his feeders, and if the ear tag does not become lost or removed, that animal will be positively identified when it reaches a packing plant, even though it may have passed through several different ownerships and may be transported far from the home ranch or feedlot. Producers have no further responsibilities. Prearrangements have been made with USDA inspectors and graders whereby the inspector will remove the tag from the ear on the slaughter floor and attach it to the carcass. The grader at the plant will examine the carcass for quality grade and yield grade along with such relevant details as the evaluation of marbling; backfat thickness; ribeye area; kidney, pelvic, and heart fat; carcass weight; quality grade; and yield grade. This data, along with tag number, will be forwarded to a clearing center which forwards it to the cooperating organization which, in turn, delivers it to the tag owner.

BIBLIOGRAPHY

ABRAHAM, H. C. 1977. Grades of fed beef carcasses. USDA. Agr. Mktg. Serv. Mktg. Res. Rept. No. 1073.

AGNEW, D. B. 1969. Improvements in grades of hogs slaughtered from 1960–61 to 1967–68. USDA Econ. Res. Serv. Marketing Res. Rept. *849*.

BIRMINGHAM, E. B. *et al.* 1954. Fatness of pork in relation to consumer preference. Missouri Agr. Expt. Sta. Bull. *549.*

BRANSON, R. E. 1957. The consumer market for beef. Texas Agr. Expt. Sta. Bull. *856.*

CLEMEN, R. A. 1923. The American Livestock and Meat Industry. Ronald Press Co., New York.

DOWELL, A. A., and BJORKA, K. 1941. Livestock Marketing. McGraw-Hill Book Co., New York.

FARRIS, P. L. 1960. Uniform grades and standards, product differentiation and product development. J. Farm Econ. Feb. 854–863.

FIENUP, D. F. *et al.* 1963. Economic effects of U.S. grades for lamb. USDA Econ. Res. Serv. Agr. Econ. Rept. *25.*

HENDRIX, J. *et al.* 1963. Consumer acceptance of pork chops. Missouri Agr. Expt. Sta. Res. Bull. *834.*

HUTCHINSON, T. Q. 1970. Consumers' knowledge and use of government grades for selected food items. USDA Econ. Res. Serv. Marketing Res. Rept. *876.*

IKERD, J. E. *et al.* 1971. The consumer image of pork. Missouri Agr. Expt. Sta. Res. Bull. *978.*

KIEHL, E. R., and RHODES, V. J. 1956. Techniques in consumer preference research. J. Farm Econ. *38,* No. 5, 1335–1345.

NAUMAN, H. D. *et al.* 1959. A large merchandising experiment with selected pork cuts. Missouri Agr. Expt. Sta. Res. Bull. *711.*

PERDUE, E. J. *et. al.* 1958. Some results from a pilot investigation of consumer preferences for beef. Oklahoma Agr. Expt. Sta. Process. Ser. *P-304.*

RHODES, V. J. 1960. How the marking of beef grades was obtained. J. Farm Econ. *42,* No. 1, 133–151.

RHODES, V. J. *et. al.* 1955. Visual preference for grades of retail cuts. Missouri Agr. Expt. Sta. Res. Bull. *583.*

RHODES, V. J. *et al.* 1956. Consumer preferences and beef grades. Missouri Agr. Expt. Sta. Res. Bull. *612.*

SELTZER, R. E. 1955. Consumer preferences for beef. Ariz. Agr. Expt. Sta. Res. Bull. *267.*

STEVENS, I. M. *et al.* 1956. Beef consumer use and preference. Wyoming Agr. Expt. Sta. Res. Bull. *343.*

STEVENSON, J., and SCHNEIDER, V. 1959. Consumer reaction to price differentials for meat-type pork. Econ. Marketing Inform. Indiana Farmers Oct. 30.

TROTTER, C., and ENGELMAN, G. 1959. Consumer response to graded pork. Penn. Agr. Expt. Sta. Res. Bull. *650.*

USDA. 1956. Official United States standards for grades of vealers and slaughter calves. USDA Agr. Marketing Serv. Regulatory Announcements *114.*

USDA. 1957. Official United States standards for grades of vealers and slaughter calves. USDA Agr. Marketing Serv. Regulatory Announcements *113.*

USDA. 1965A. Official United States standards for grades of carcass beef. USDA Consumer Marketing Serv. Regulatory Announcements *99.*

USDA. 1965B. Official United States standards for grades of feeder cattle.

USDA Consumer Marketing Serv. Regulatory Announcements *183*.

USDA. 1966A. Official United States standards for grades of veal and calf carcasses. USDA Agr. Marketing Serv. Regulatory Announcements *113*.

USDA. 1966B. Official United States standards for grades of slaughter cattle. USDA Agr. Marketing Serv. Regulatory Announcements *112*.

USDA. 1966C. Official standards of the United States for grades of wool. USDA Consumer and Marketing Serv. SRA—C & MS, No. 135.

USDA. 1968A. Official United States standards for grades of barrow and gilt carcasses. USDA Consumer Marketing Serv. Reprinted in Federal Register.

USDA. 1968B. Official United States standards for grades of slaughter barrows and gilts. USDA Agr. Marketing Serv. Regulatory Announcements *172*.

USDA. 1968C. USDA yield grades for beef. USDA Consumer Marketing Serv. Marketing Bull. *45*.

USDA. 1969A. Official United States standards for grades of feeder pigs. USDA Consumer Marketing Serv. Regulatory Announcements *189*.

USDA. 1969B. Official United States standards for grades of lamb, yearling mutton, and mutton carcasses; slaughter lambs, yearlings and sheep. USDA Agr. Marketing Serv. Regulatory Announcements *168*.

USDA. 1970A. USDA grades for pork carcasses. USDA Consumer Marketing Serv. Marketing Bull. *49*.

USDA. 1970B. USDA grades for slaughter swine and feeder pigs USDA Consumer Marketing Serv. Marketing Bull. *51*.

USDA. 1970C. USDA yield grades for lamb. USDA Consumer Marketing Serv. Marketing Bull. *52*.

USDA. 1971. USDA grade standards for wool top. USDA Consumer and Marketing Serv. Mktg. Bull. No. 53.

USDA. 1975A. Official United States standards for grade of slaughter cattle. USDA Agr. Marketing Serv. (Unnumbered).

USDA. 1975B. Official United States standards for grades of carcass beef. USDA Food Safety and Quality Serv. (Unnumbered).

USDA. 1977. Livestock and meat statistics. USDA Stat. Rept. Serv., Supp. for 1976 to Stat. Bull. No. 333.

USDA. 1978. Livestock, meat, wool market news. USDA Agr. Mktg. Serv. Vol. 46. No. 22.

VROOMAN, C. W. 1952. Consumer report on pork production. Oregon Agr. Expt. Sta. Bull. *521*.

WEIDENHAMER, MARGARET *et al.* 1968. Homemakers' opinions about selected meats—a preliminary report. USDA Statist. Reporting Serv. *SRS-12*.

WILLIAMS, W. F. 1960. Note on how the marking of beef grades was obtained. J. Farm Econ. *42*, No. 4, 878–886.

Market Intelligence

It has become increasingly evident that success in farming and ranching requires more than technical production equipment and skill. The fruits of efficient production can be lost by inadequate market information. Just as production knowledge and practices have become more and more sophisticated, so has the marketing of the products. Market information is available, but one must be familiar with the sources, know how to obtain the information, and devote some time to its study and analysis. To meet this challenge many producers allocate part of their time to market analysis as rigorously as they pursue technological production developments.

In a competitive market economy the bargaining position of buyers and sellers is conditioned by the degree to which they are informed. The classical economic concept of a perfectly competitive market assumes perfect knowledge, which of course, is unattainable. Buyers and sellers utilize information which is available, and the relative position of buyers and sellers may be far from equal. Traditionally, agricultural producers have been labelled as "price takers." Part of this stems from lack of effective organization, but part also results from lack of market information, which in a systematic sense may be called market intelligence.

There was always a need for market information, but the urgency has increased with the trend toward more and more direct marketing. Livestock producers are dealing directly with professional buyers in the sale of finished stock and often with professional sellers in the purchase of feeder stock. The concept of competition implies a notion of rivalry, or struggle, and a large share of the struggle is in keeping at least as well informed as other parties. Other parties include competitors in the industry as well as participants in buying and selling transactions. In nationwide and worldwide markets, no single person or firm can possibly make enough observations to keep abreast of the ever-changing economic and political forces which shape the markets. The livestock and meat markets, like many others, are affected by supply-demand trends, import-export regulations, and public policy decisions not only in the United States but in Australia, New Zealand, Argentina, Canada, Ireland, Common Market Countries and many others. Private sources gather and disseminate much information, but it was recognized at an early stage

332

that government services were justified not only from the standpoint of practicality, but also as a matter of public interest.

MARKET DIMENSIONS

Market intelligence needs may be considered in two general dimensions: (1) the short run and (2) the long run. It is impossible to specify short and long run in weeks, months, or years, but in general the short run is concerned with informational needs in marketing to best advantage a production output which virtually is complete in quantity and quality (essentially no time remains for altering the product). The long run envisions enough time to adjust production programs which coincidentally are related to marketing programs. The time required for such changes is greater in cow-calf operations than for sow herds; and is still different in a calf wintering program.

A feeder with a pen of livestock about finished out and ready for sale needs to know current price levels for his class, grade, and weight of livestock (implied here is the ability to evaluate grade). He needs to know variations in relevant prices among alternative market outlets. Information on current conditions of supply and demand and market movements help him size up the current market situation, although individually he has no control over them. The market intelligence need in this short-run situation sometimes is referred to as "price and sales information," or as "market news."

The position a producer finds himself in at a given time, however, is, in part, determined by decisions he and others made some months earlier— i.e., when he bought feeder animals, outlined his feeding program or initiated a breeding program. It was then that the wheels were set in motion which largely determined the quantity and quality of livestock which would be available at later dates. Information is needed which assists producers in making wise planning decisions in this longer run context. This type of information also is needed by packers, processors, cold storage interests, wholesalers, and retailers. Consumers, too, have an interest, in that better informed market participants make for a more efficient market system which, in turn, allows consumer demands to guide production. The informational needs in this situation call for data and analyses which project probable supply and demand conditions; information which attempts to look ahead rather than at the present. This type of market intelligence often is referred to as "outlook and situation information," and is construed to include analyses of changing consumer preferences.

DEVELOPMENT OF INFORMATION SERVICES

The history of development of meat and livestock market reporting in the United States is almost as old as the livestock industry itself.[1] The earliest source of market information were newspaper quotations of provisions (meat, lard, and tallow) prices. The press, then as now, recognized the news worthiness of market reports. The reporting of live animal prices followed that of meat by a considerable number of years. The exportation of provisions gave that market a semblance of organized arrangements long before enough consistency was generated in live markets to warrant price reports.

Private market reporting grew with the industry largely in the form of newspapers, magazines, trade publications, and publications sponsored by organized markets. Some of the problems inherent in these reports were mentioned in the last chapter. The heterogenity of terms used among markets, and the inconsistency in use of terms, both within given markets and between markets, seriously handicapped the interpretation of reports.

In addition to the reporting of current market information, the need for basic industry data also was recognized at an early date. The merits of turning this job over to the federal government were evident from the beginning. Although the beginning was on an elementary scale, the first Census of Agriculture was taken in 1840 by the U.S. Office of Patents. In later years, responsibility for census enumeration was placed in the Bureau of Census, U.S. Department of Commerce. Early census data did not provide information of immediate application to the marketing of livestock, but it established the federal government as the data collecting and disseminating agency. Subsequent organization of the U.S. Department of Agriculture in 1862 placed additional agricultural data collection, interpretation, and dissemination in that branch of the federal government. However, its early efforts were directed at increasing production rather than improving marketing.

Producer dissatisfaction with market conditions was particularly evident during the latter half of the 1800's and the first decade of the 1900's. Producers, then as now, were troubled by the difference between prices received at the farm and prices charged at retail. Deficiencies in market price reporting were indicated to be one of the major problems. During this period many conferences were held and investigations were carried out. Among the spokesmen for producers, Henry C. Wallace, editor of *Wallace's Farmer*, stands out. Wallace also was secretary of a farmer organization, the Corn Belt Meat Producers Association. This organization, together with another, the American National Live Stock Associa-

---

[1]Three excellent accounts of early historical developments of market information are Dowell and Bjorka (1941), USDA (1969A), and Smeby (1961).

tion, carried on active campaigns in attempts to alleviate what were considered detrimental market reporting practices. The accusations ranged from inadequate and inaccurate reporting to deliberate misrepresentation. Henry Wallace participated in the conferences, carried on discussions in his magazine, and stimulated interest in producer organizations. During the fall of 1915 he published an editorial in which he presented, in some detail, suggestions for investigating the situation and for establishing a federal market reporting service.

A logical branch of government for marketing work was in existence at that time, although it was relatively new. The USDA Office of Markets had been established in 1913, and within it a Division of Livestock, Meats and Wool in early 1915. As a result of studies made by that office and guided extensively by Wallace's work, the federal government in 1915 set up arrangements for collecting market information. Congress appropriated $65,000 in 1916, for collection and dissemination of information on the number and grade of livestock being produced; prices, receipts, and shipments by class and grade at central markets; prices of meat and meat products; and the amount of meat in storage. However, at administrative discretion, the initial emphasis was on meat and lard. The first federal report on meat marketing was issued in December, 1916. Eighteen months later the first livestock marketing report was issued. These early market reports were limited to prices and market conditions at a few central markets. By this time tentative grades had been established on meat and livestock, so for the first time a uniform language was available for use in price quotations. Although the original emphasis was on meat market reporting, this gradually changed over the years. By the early 1940's, major attention was on the live animal market. As the trend toward direct marketing picked up momentum, particularly since World War II, interest again shifted to the meat trade. Direct selling, especially by the carcass grade and weight method, necessitates a good knowledge of conditions in the wholesale meat market. At the same time, interest is high in the live animal market so currently the Federal Market News Service attempts to maintain a balanced reporting of meat, livestock, and wool markets.

Market reporting, as carried on by the Federal Market News Service, primarily is aimed at reporting current prices, movements, and market conditions. This is intended to serve short run informational needs. Many states also carry out market reporting functions—some in cooperation with the federal service, others entirely as a state effort. Private firms and agencies have continued efforts in this area. Newspapers, radio and TV stations, farm magazines, farm organizations, commission firms, auction operators, and private market agencies expend an enormous effort in providing market intelligence to all sectors of the livestock and meat trade. Many of these firms and agencies are engaged only in dissemination of market news, using as their major source the data collected by government agencies. But there are exceptions. The *National Provisioner*'s "Yellow Sheet" and the *Meat Market Research and Reporting Service* "Pink Sheet" are compilations of privately-collected wholesale meat market information. The American National Cattleman's Association's "CATTLE-FAX" depends largely on privately-collected data and its interpretation. Several well established private research-consulting firms with specialization in livestock are primarily market analysts utilizing data which generally is available to the public.

Comprehensive work in the collection of situation type information followed shortly after that of market news. Following World War I, and continuing to the present, has been a deep interest in knowing the forces behind the market trends. The vast amount of data needed for this type of analysis requires the government services. Involved here, in addition to livestock and meat data, are such things as feed supplies, population (by age classifications), incomes (distributed by income levels), prices and supplies of products which compete with meats, trends in consumer preferences, imports, exports, etc. As mentioned earlier, a start on this type of information was made in 1840 in the first agricultural census. The agricultural census, however, does not collect nearly all the needed information, and furthermore, it is taken only at 5-yr intervals. In certain data series, the census serves an extremely useful purpose in providing a benchmark with which to check data collected from other sources, but by itself census data are inadequate for outlook and situation analyses. For many analyses, annual, quarterly, and monthly data are required. Although some work was done in this area earlier, the USDA has placed particular emphasis on it following World War II. The office with this responsibility has, over the years, undergone several name changes—currently major data collection is done by the Crop Reporting Board of the Economics, Statistics and Cooperatives Service (ESCS), and the Livestock Division of the Agricultural Marketing Service (AMS).

Analysis and interpretation of livestock and meat marketing data is performed by many governmental agencies. Of particular interest are the Livestock and Meat Situation, the Cotton and Wool Situation, and the Fats and Oils Situation, published regularly by the Commodity Economics Division of ESCS.

PUBLIC AGENCIES AND INSTITUTIONS

Federal-State Market News Service, AMS, USDA[2]

The Federal Market News Service is an arm of the Agricultural Marketing Service within the USDA hierarchy. Individual states may at their

---

[2]Much of the material in this section was furnished by Dr. Paul M. Fuller, Chief, Livestock Market News Branch, Agricultural Marketing Service, USDA.

option cooperate with the federal agency in activities within their respective states.

The major responsibility of the Federal-State Market News Service is collection and dissemination of current, unbiased, market information. The agency gathers information on many agricultural products, but this discussion is limited to activities in reporting prices, market conditions, and current supplies and demand for livestock, meat, and wool at selected markets throughout the United States.

Organizational Features.—The Federal Market News Service cooperates, where arrangements can be made, with state marketing in setting up Federal-State offices. In other instances, the federal service maintains offices not associated with state agencies. In 1978 there were 29 state marketing agencies cooperating with the federal agency. In total 61 livestock, meat, and wool field offices were maintained throughout the United States. This provided coverage of 19 terminal livestock markets and 309 auctions. Information was collected on direct sales of livestock from 26 major areas. Data on meat sales were collected from 10 trading areas in the Midwest, Colorado, Texas, the East Coast and West Coast. Wool and mohair marketing data were obtained from 14 areas with the primary office located in Denver. Arrangements also are operative for collection and dissemination of data on offal sales, daily slaughter estimates, weekly meat production estimates, daily pork cut-out values, and beef cut-out values.

The Federal Market News Service has attempted to keep pace with changing times, but this is a continuing problem. Funds and personnel are limited, and lags typically occur before changes can be made. The situation was relatively simple when terminals dominated the market scene. A single reporter could personally cover the highly concentrated activities. However, with expansion in auctions and direct marketing, the opposite situation prevails. To maintain the established standards of excellence within the fund restrictions, the reporting of auction sales has been limited to 309 major markets. The expansion in direct selling of livestock presented still another problem. Direct sales from farms are widely scattered, generally nonobservable, private transactions. Sales out of commercial feedlots are somewhat more concentrated, but adequate coverage still is a problem. Direct sales of slaughter livestock, of course, involve packing plants which, in effect, are points of concentration, but decentralization of the packing industry presents a widely scattered pattern—and there is no legal compulsion for buyers and sellers to divulge purchase and sale information. This points up an important aspect of the entire governmental marketing intelligence system. Federal Market News Service depends primarily upon voluntary cooperation of all segments of the industry in gathering information and also of the news media in disseminating it.[3] Data are obtained on direct sales—largely by telephone, with some visits by market reporters at points of sale.

At scheduled times the assembled market information is fed into a leased teletypewriter system for dissemination to the press, radio, television, and trade sources. There are about 100 automatic self-answering, taped, telephone reports. In addition, mimeographed and printed reports are sent by mail.

Operational Features.—The method of operation is conditioned by the type of market being covered. At most markets, trained, professional reporters personally observe sales, interview buyers and sellers, and check receipts and movements which have a bearing on supplies and demand. Where sales are sparce or distances too great, information is gathered by telephone and teletypewriter with some follow-up personal contacts and inspection of sales records.

Terminal Markets.—Experienced market reporters are stationed at 19 terminal markets. These men are capable livestock graders who tour the yards during trading hours interviewing buyers and sellers, and observing sales. This includes both private treaty and auction sales at terminals which have auction facilities. Reporters know the market personnel and have working relationships which permit discussions not afforded general market visitors. The actual transaction, ordinarily, is a private affair between buyer and seller, but generally this confidence is shared with market reporters. However, reporters realize that in some instances market personnel may attempt to bias the report for personal gain. An experienced reporter can recognize this and make appropriate adjustments.

Three reports are issued each day at terminals: (1) early morning, (2) mid-session, and (3) market closing. These reports are flashed over the wire network to other livestock markets and to news media. In some cases the reporter may also broadcast directly over a local radio station. The early morning report consists of an estimate of that day's receipts, an evaluation of opening price trends, and observations of early sales. After filing this report on the wire network, the reporter returns to the yards for observations which will comprise the mid-session report. By midsession a firm report can be made on receipts, price trends, and quotations for the bulk of sales by class, grade and weight. Further observations and checks give indications of possible late changes which go into the closing report.

---

[3]Some data collected by regulatory agencies, e.g., the Meat Inspection Division and the Packers and Stockyards, USDA, are mandatory and finds use in certain aspects of marketing, but these agencies are not a part of the Federal Market News Service.

Auction Markets.—Reporters visit auction markets and take a position which permits an unrestricted view of the sales ring. They use a prepared form for recording sale price and an evaluation of grade and condition of the animals. Weights often can be observed on the auction scales. As is the case at terminals, not all lots are uniform in grade and other value factors. Reporters are trained to recognize these situations and make appropriate adjustments. For the auctions covered, a local report is issued for each sale. In addition, a summary is forwarded to an appropriate central office for dissemination on a statewide or area basis.

Direct Livestock Sales.—Information is assembled on direct sales of slaughter cattle, feeder cattle, hogs, and sheep from 26 major market areas. Contract sales are covered as well as current transactions. Much information is gathered by telephone and teletypewriter, but interviews also are made at feedlots, packing plants, and ranches. Reports are issued daily, semiweekly or weekly, depending upon trading volume.

Wholesale Meat Markets.—Wholesale prices on carcass and primal cuts are reported daily from selected points, including the Midwest, Colorado, Texas, the East Coast, and the West Coast. A recent innovation is the collection and reporting of prices on fabricated cuts of beef (wholesale and oven-ready cuts prepared by packers) from the Midwest, the East Coast, the West Coast, and Texas. Chief reliance in this operation is placed on telephone interviews.

Wool and Mohair.—A weekly wool market report is issued from Denver. An experienced reporter collects information by telephone and by visiting with ranchers and local market agents.

Other Products.—The Federal Market News Service also reports a weekly hide and offal value which is released from Princeton. Daily estimates also are made on pork and beef cut-out values. The pork cut-out is the combined value of pork cuts, based on prices of 150 lb and 180 lb hog carcasses grading U.S. No. 1 through U.S. No. 4. The beef cut-out value is based on the combined value of beef cuts from choice, yield grade 3, 600–700 lb carcasses.

In addition to consistent use of official grade terminology, the Federal Market News Service adheres to uniform use of terms in describing market conditions. Following are the commonly-used terms and their meanings (USDA 1975).

"**MARKET**—A term with several meanings:

A geographic location where a commodity is traded.

The price, or price level, at which a commodity is traded.

To sell (verb).

MARKET ACTIVITY—The pace at which sales are being made.

ACTIVE—Available supplies (offerings) are readily clearing the market.

MODERATE—Available supplies (offerings) are clearing the market at a reasonable rate.

SLOW—Available supplies (offerings) are not readily clearing the market.

INACTIVE—Sales are intermittent with few buyers or sellers.

PRICE TREND—The direction in which prices are moving in relation to trading in the previous reporting period(s).

HIGHER—The majority of sales are at prices measurably higher than the previous trading session.

FIRM—Prices are tending higher, but not measurably so.

STEADY—Prices are unchanged from previous trading session.

WEAK—Prices are tending lower, but not measurably so.

LOWER—Prices for most sales are measurably lower than the previous trading session.

SUPPLY/OFFERING—The quantity of a particular item available for current trading.

HEAVY—When the volume of supplies is above average for the market being reported.

MODERATE—When volume of supplies is average for the market being reported.

LIGHT—When the volume of supplies is below average for the market being reported.

DEMAND—The desire to possess a commodity coupled with the willingness and ability to pay.

VERY GOOD—Offerings or supplies are rapidly absorbed.

GOOD—Firm confidence on the part of buyers that general market conditions are good. Trading is more active than normal.

MODERATE—Average buyer interest and trading.

LIGHT—Demand is below average.

VERY LIGHT—Few buyers are interested in trading.

MOSTLY—The majority of sales or volume.

UNDERTONE—The situation or sense of direction in an unsettled market situation."

Market News Reports.—Figure 12.1 shows the system of approximately 23,500 miles of leased wire facilities over which the Federal Market News Service makes its reports available to communications media. Not all of the office locations shown are concerned with livestock, meat, and wool information, but the network is available for transmission of reports. It will be noted that Kansas City is a critical relay point, connecting the network of the eastern half of the nation with Mountain, Southwest, and West Coast market points. While the widest and most rapid dissemination is over the wire network, Federal Market News Service also makes extensive use of mimeographed, mailed reports.

Each week, data from field offices are summarized and forwarded to Washington where they are compiled in a weekly publication entitled *Livestock Meat and Wool Market News*. This publication is one that by most standards should be on the "must" list for most livestock producers and market personnel. It is useful, not only for relatively up-to-date compilation of data for selected markets, but also for digests of current releases and reports, including outlook and situation reports issued by other government agencies.

Special mention should also be made of the automatic, taped, self-answering telephone service. Unquestionably, this is a valuable improvement in market news reporting. These devices accept long distance as well as local calls. Tapes are updated several times during each day and the service is available on a 24-hr basis.

The estimated daily slaughter under federal inspection is released at 2:30 p.m. Eastern time for cattle, calves, hogs, and sheep. The actual slaughter by class and specie is available two weeks later, and live and dressed weight information is available 3 weeks later. The weekly total meat production is released each Friday afternoon.

Livestock market news reporters are also designated as the officials to apply appropriate specifications for the settlement of futures and Commodity Credit Corporation contracts.

Crop Reporting Board, ESCS, USDA

Major responsibility of the Crop Reporting Board of ESCS is data gathering—as contrasted with analysis and interpretation. Most of the data are collected by state offices which are referred to as (state) Crop and Livestock Reporting Service. The information is forwarded to a Washington office of the Crop Reporting Board, where, in closed door sessions, the board reviews state and aggregated compilations prior to release.

Vast amounts of data are collected on all aspects of crops and livestock including such general categories as production, distribution, storage, inventories, average prices received by farmers, values, animals slaughtered, meat production, etc. Most of the data are related to marketing.

The type of data collected, and schedules of distribution make the market-related information particularly adapted for outlook and situation analyses. The research and analysis functions, however, are usually carried out in other agencies. The Crop Reporting Board makes the data

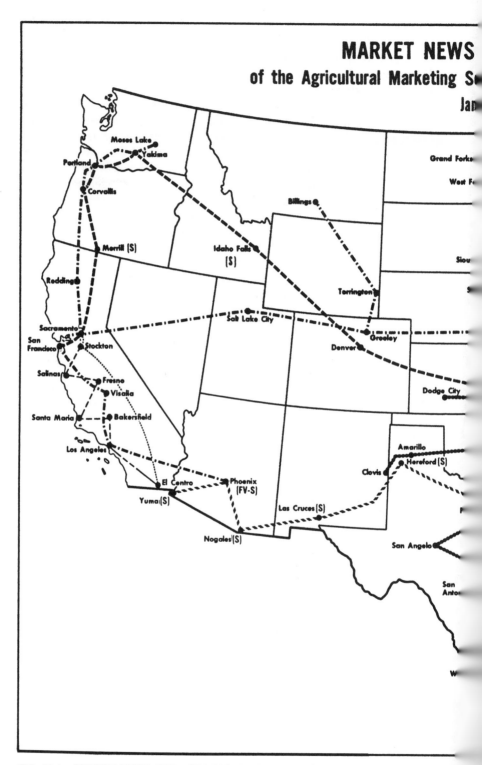

FIG. 12.1. MARKET NEWS SERVICE LEASED WIRE NETWORK

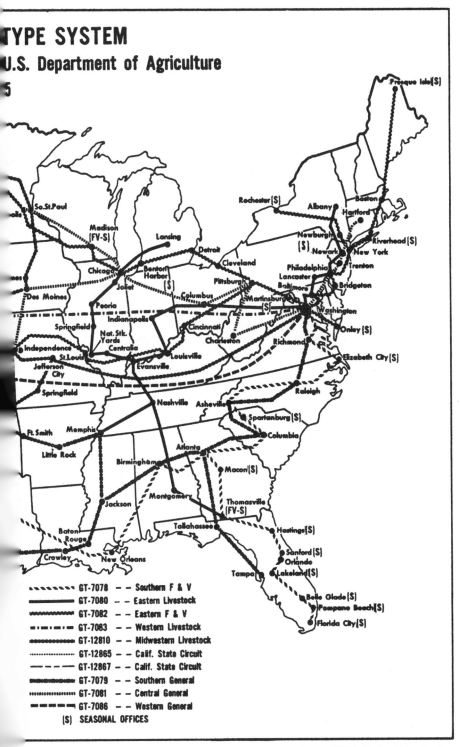

TYPE SYSTEM

U.S. Department of Agriculture

.... GT-7078 – – Southern F & V
───────── GT-7080 – – Eastern Livestock
~~~~~~~~~~ GT-7082 – – Eastern F & V
─·─·─·─·─ GT-7083 – – Western Livestock
●●●●●●●●● GT-12810 – – Midwestern Livestock
·········· GT-12865 – – Calif. State Circuit
─ ─ ─ ─ ─ GT-12867 – – Calif. State Circuit
●■●■●■●■● GT-7079 – – Southern General
········· GT-7081 – – Central General
─ ─ ─ ─ ─ GT-7086 – – Western General
(S) SEASONAL OFFICES

Source: USDA, Market News Service

available, not only to other governmental agencies but also to the public. Many of the findings have considerable newsworthiness (without further analysis) and are widely disseminated by communications media. However, it should be understood that this is not the same current, up-to-the-minute price and market condition information reported by the Federal Market News Service.

(1) **Cattle**, January 1, $\overline{\text{(yr)}}$ (Jan.), and July 1, $\overline{\text{(yr)}}$ (July). Gives number of cattle and calves on farms by class and by state, and annual calf crop.

(2) **Cattle and Calves on Feed**—(Jan., April, July, Oct.). Gives number on feed at beginning of each quarter, by states; number by class and by weight group for leading states; and number of grain-fed cattle marketed.

(3) **Cattle and Calves on Feed**—(Monthly). Gives number on feed beginning of each month for seven important feeding states and number of grain-fed cattle marketed.

(4) **Hogs and Pigs**—(June). Gives inventory of hogs and pigs on farms June 1, an estimate of spring pig crop farrowings that year, and an estimate of intended farrowings for the following fall pig crop, by states.

(5) **Hogs and Pigs**—(Dec.). Gives inventory of hogs and pigs on farms Dec. 1, an estimate of fall pig crop farrowings that year, and an estimate of intended farrowings for the following spring pig crop, by states.

(6) **Hogs and Pigs**—(March, June, Sept., Dec.). Provides quarterly estimates of farrowings and inventory for 14 important hog-producing states.

(7) **Sheep and Goats**, January 1, $\overline{\text{(yr)}}$ —(Feb.). Gives number of sheep and lambs on farms Jan. 1, by class and by state. Gives lamb crop and sheep and lambs on feed by state for 26 important sheep states. Also gives U.S. inventory of goats.

(8) **Sheep and Lambs on Feed**, March 1, $\overline{\text{(yr)}}$ —(March), and Nov. 1 $\overline{\text{(yr)}}$ (Nov.). Reports number of sheep on feed, by weight classification and by state for 6 important lamb feeding states.

(9) **Lamb Crop and Wool**, $\overline{\text{(yr)}}$ —(July). Gives number of breeding ewes, lamb crop, and wool production by state.

(10) **Wool and Mohair**, $\overline{\text{(yr)}}$ —(March). Gives number of sheep shorn, wool production, and price of wool by state.

(11) **Meat Animals—Farm Production, Disposition and Income** $\overline{\text{(yr)}}$ —(April). Gives data on indicated items, by states.

(12) **Livestock Slaughter**—(Monthly and Annual). Gives number of head and live weight of cattle, calves, hogs, sheep and lambs slaughtered

in commercial plants, by states; meat production by species, and lard production for the United States.

In addition to reports just listed, the Economics, Statistics, and Cooperatives Service issues about mid-year, an annual summary of livestock and meat data for the previous year, along with historical series for several prior years. Each year adjustments and corrections are made in data of prior years, if necessary. This publication is entitled *Livestock and Meat Statistics* and contains a wealth of basic market-related data. Prior to 1957 these data were published under the title *Livestock Market News Statistics and Related Data*.

Commodity Economics Division, ESCS, USDA

The Commodity Economics Division of ESCS basically is responsible for analysis and interpretation of data in a situation-outlook framework. One of its major activities is the analysis of livestock, meat, and wool data. This agency issues a publication entitled *Livestock and Meat Situation* about 6 times a year, and about 4 times each year it issues the *Cotton and Wool Situation*. Periodically these situation reports include special articles on analyses of currently relevant problems in the livestock and meat industry.

This agency goes beyond the presentation of data. Its activities include production, demand, and price projections—though somewhat guarded. Retail meat price compilations are given in each issue of the *Livestock and Meat Situation*, and data, with commentary, also are given on imports and exports of livestock and meat. Other relevant publications emanating from this agency are the *Feed Situation*, and the *Fats and Oils Situation*.

These "Situation" reports are recommended as another "must" among government publications for livestock and meat industry personnel who have an interest in the market situation.

While not directed specifically at livestock and meat, ESCS's *Agricultural Outlook* (monthly) contains much relevant statistical data along with descriptive features of current interest.

States

It was mentioned previously that some states cooperate with the federal government in operation of the Federal Market News Service. Most important agricultural states carry on additional work under a department, division, or an office of marketing. Among the major livestock-producing states, these agencies conduct a variety of programs associated with the marketing of livestock, meat, and wool. No attempt is made here to enumerate individual state activities, but some of the general areas emphasized are promotional programs designed to increase meat consumption, programs designed to improve both quantity and quality of products produced, collection and dissemination of market news not covered by the federal service, and the preparation and dissemination of outlook-type information.

Colleges and Universities

Most Land Grant Colleges and Universities in important livestock-producing states are actively engaged in work related to livestock and meat marketing. This activity is shared by personnel in the Agricultural Experiment Station and in the Cooperative Agricultural Extension Service. The major responsibility of the former is research, while that of the latter is dissemination of information. In most cases these activities are closely coordinated. In some, the personnel hold joint appointments and often are under joint administrative arrangements. Research activities cover every facet of marketing. Virtually no effort is expended in the *collecting* of marketing news, but substantial effort is devoted to preparation and dissemination of outlook and situation type information. This includes both on-campus and off-campus conferences and meetings, radio and television programs, and published materials. A number of colleges and universities have a long history of work in livestock marketing and regularly issue (monthly or weekly) publications devoted in full or in part to analyses designed to assist producers in both short- and long-run marketing decisions.

Commodity Futures Trading Commission

The Commodity Futures Trading Commission (CFTC) was established by the U.S. Congress in 1974, initially on a temporary basis, with the responsibility of supervising and regulating activities in the futures markets. CFTC replaced the Commodity Exchange Authority which had been under USDA. Although the basic functions of CFTC are supervisory and regulatory, it collects a substantial amount of data on futures trading which are made available to market analysts.

PRIVATE FIRMS, AGENCIES, AND ORGANIZATIONS

Newspapers, Radio, and Television

The management of news media usually respond to the interests of subscribers, listeners, and viewers. But their response is conditioned by the degree of interest shown, the availability of materials with which to satisfy the interest, and personal considerations. Livestock producers continually request more and improved market information—there is no question about producer interest, but that message is not always conveyed to the management. News media, generally, are not organized to collect market information; they depend upon other sources. The Federal-State Market News Service has attempted to fill this need and releases are made available through news wire services.

The available materials are accepted and utilized in all degrees of coverage (from zero to full coverage) by various newspapers, radio, and television stations. Many areas lack local newspapers which give adequate livestock market news, and some of the more remote areas have problems with radio and television reception, yet these media often are listed by producers among the most important sources of information (Bohlen and Beal 1967; Roberts 1959). In most important livestock-producing areas, certain newspapers, radio stations, and television stations have developed a rather respected reputation for livestock market reporting. These sources usually are known among producers and market personnel or can be easily determined by inquiry from experienced stockmen or market representatives.

Farm and Trade Magazines

Farm magazines have been cited as one of the most important sources of livestock marketing information, and a large number of magazines cater to the livestock clientele (Roberts 1959). The coverage among farm magazines ranges all the way from general farm magazines to those devoted to a single aspect (such as feeding), a single specie, or even to a particular breed. The latter, however, ordinarily are less market oriented than others. Some magazines regularly devote a section to livestock marketing, and some utilize the services of special contributing editors or correspondents for preparing the materials. Others include special articles at irregular intervals.

Trade magazines are oriented to problems and issues of a particular sector of the industry, which may be commodity groups such as wool or meat, or functional groups such as wholesalers or retailers.

The following magazines are cited simply as examples with orientation toward a variety of interest groups: *Beef, Farm Journal, Farm Quarterly, Feedlot Management, National Hog Farmer, National Livestock Producer, Beef Digest, National Provisioner, Successful Farming, Western Livestock Journal, Western Meat Industry, Progressive Grocer, Super Market News, Farm Futures,* and *Commodities.*

Market Agencies

Many market agencies—i.e., commission firms, auction markets, and related organizations (such as livestock market foundations found at some terminals)—prepare and distribute market information. In most cases emphasis is placed upon prices and market conditions of the particular market where the agency is located. A common form of presentation is a weekly market newsletter or card. However, auctions and foundations often sponsor radio and television broadcasts.

Market agencies are frank in acknowledging a dual purpose in their market reporting activities, as follows: (a) a desire to provide informational service, and (b) its use as a competitive tool in market promotion and solicitation of business. From a business standpoint both objectives are perfectly legitimate and tend to increase the degree of competition in the marketing system as long as the information is accurate and methods of presentation are fair.

Personnel of progressive market agencies also personally visit farms and ranches of customers and potential customers and participate in field programs. Again, these may have promotional and solicitation aspects, but at the same time they can be informational.

Market Analysis and Price Reporting Firms

There are a number of market analysis firms which make a business of selling market information—usually on a subscription basis. This class of agency tends to be more specific in forecasts and recommendations than most public agencies. A wide variety of services and reports are offered, ranging from monthly letters to daily private wire reports and continuous toll-free telephone advisory service. The frequency of report, extent of market coverage, and degree of analytical sophistication generally are reflected in the price charged. The following is in no sense a complete listing. The firms named are simply examples of this type of service: *Doanes Agricultural Digest, Kiplinger Agricultural Letter, National Provisioner's Daily Market and News Service* (the Yellow Sheet), *Meat Sheet* (the Pink Sheet), *Livestock Business Advisory Service, Farmers Grain and Livestock, Inc., Texas Agricultural Research Associates, Top Farmer Market Services, Professional Cattle Consultants, Commodity News Service,* and *Reuters Limited.* The Yellow Sheet and Pink Sheet differ from others listed here in two major respects. First, their coverage is devoted primarily to wholesale meat prices, while the others are concerned primarily with livestock. Second, they basically report market prices rather than perform analyses. They are unique examples of private firms in the livestock-meat industry complex in the area of price reporting—a field almost exclusively in the domain of public agencies.

Industry Organizations

Producer Associations.—The interest of producer organizations in livestock marketing dates back many decades. The Grange considered marketing problems a major issue shortly after the Civil War. Efforts of the Corn Belt Meat Producers Association, and the American National Livestock Association in promoting the establishment of market news reporting were mentioned earlier in this chapter. Every major farm organization has long been concerned with various aspects of marketing. Until recently the efforts of producer organizations have been directed almost exclusively at public policy issues. Several such issues were mentioned in earlier chapters—monopolistic trade practices, and the establishment of federal grade standards. Within recent years, however, several producer organizations have taken a more direct approach to marketing problems. This is an attempt on the part of producers to shift the balance of bargaining power to producers.

Two of these organizations merit special attention. The two organizations prominent in this approach are both utilizing privately-developed market intelligence systems as a basic tool. They are the National Cattlemen's Association, and the National Farmers Organization. The National Cattlemen's Association set up a separate corporation, Cattle Marketing Information Services, Inc. (CMIS) to operate their service, appropriately dubbed CATTLE-FAX. Both CMIS and NFO have developed systems for determining, on a relatively local basis, numbers of livestock on feed, potential supplies at future dates, together with current prices, market conditions, and other market factors. In addition to data collected from its own members, each has access to and compiles a great amount of market information made available by many other sources. This, however, is about as far as the similarity goes.

NFO's use of its information is entirely a centralized operation. Information is funnelled into the central office where it is used by this office in contract negotiation with packers and other selling operations for the membership.

CMIS assembles, analyzes, and interprets its information centrally, but transmits it by phone and by mail back to its members. Members then utilize the information individually (or in local pooled arrangements) in negotiating direct sales with packers. CATTLE-FAX subscribers pay a monthly fee—a fixed minimum plus a charge per head up to a stipulated maximum total fee. The CATTLE-FAX service would be rather expensive for a single farm-sized cattle feeder, but pooled arrangements can be set up where several small operators share the cost of an installation which may be located in a bank or business establishment convenient to the group. NFO's cost is included in membership dues.

A number of other producer organizations are active in marketing work and have varying degrees of market information systems, but generally on substantially smaller scales than the two mentioned above. The following are cited as additional examples without elaboration on their programs: The American Farm Bureau Federation (which sponsors the American Marketing Association), the American Sheep Producers Council, the Mid-West Wool Association, the National Swine Producers Council, and the National Livestock Producers Association.

A variety of other industry organizations with interests in market information exists at different levels of the total livestock-meat marketing system. Only four will be cited here. The American Meat Institute is mentioned as representative of the packer-processor sector of the industry, the Super Market Institute, Inc., and the National Association of Retail Grocers as representative of retailers, and the National Livestock and Meat Board, which cuts across all levels of the marketing system.

American Meat Institute.—A group of meat packers met in 1906 and organized the American Meat Packers Association. The name was changed to Institute of American Meat Packers in 1919, and to American Meat Institute (AMI) in 1940. Membership in AMI is composed of packers, processors, and associate members. Its major activities include sponsorship of research, and educational and promotional programs. While the AMI may be classed as a trade organization with primary allegiance to its members, its declared objectives and multidimensional program of activities are of direct and indirect benefit to the entire livestock-meat economy. The central office of this organization maintains a section devoted to continuous development of market information. Among its releases are an annual, *Financial Facts About the Meat Packing Industry*, and a *Weekly Report*. The former draws heavily from data collected directly by AMI—data not available from any other source. These original data are supplemented by government information and this publication is available to the public. The weekly release primarily is an AMI staff summary and interpretation of currently relevant government releases. Its distribution is limited to members of the Institute. In 1966 the AMI, under authorship of its then Vice President and Director of Marketing, J. Russell Ives, published a book length treatise entitled *The Livestock and Meat Economy of the United States*. This is a comprehensive treatment of economic aspects of livestock production, meat packing, processing, wholesaling, and retailing.

Super Market Institute and National Association of Retail Grocers.—These are trade organizations oriented primarily to interests of their members. The National Association of Retail Grocers has a long history with food retailers, having been organized in 1893. The Super Market Institute was organized in 1937, concurrently with the rapid expansion in super markets. Both carry an extensive informational program covering all phases of retail operation. Super Market Institute is somewhat unique in that it maintains a well-stocked and up-to-date library of informational materials on all aspects of food retailing.

National Livestock and Meat Board (NLMB).—The NLMB is an industrywide organization. Its directorate includes representatives (through organizations) from all sectors of the livestock-meat economy— livestock producers, market agencies, packers, processors, wholesalers, retailers, and eating establishments. Producer groups, however, hold about half the directorate positions. The NLMB was organized in 1923 as a nonprofit, nongovernmental, service association. Its major activities are (1) sponsorship of research, (2) assembling of information, (3) carrying out of educational programs, and (4) promotion of meat, meat products, and merchandising techniques. Over the years, this organization has financed a wide variety of research ranging from highly technical characteristics of meat to applied techniques in marketing.

From an organizational standpoint, NLMB operates separate beef, pork, lamb, and sausage promotion programs under their respective Beef Industry Council, Pork Committee, Lamb Committee, and Sausage Council. The program of NLMB is financed through voluntary contributions of livestock producers in cooperation with participating marketing firms and by participating meat packers. Marketing firms and packers, in the case of direct sales, assemble the funds from producer sales and forward them to NLMB. However, upon a producer's request the money is returned to him. It is estimated that contributions are made on about 45% of all livestock marketed and slaughtered. The Board also receives some direct contributions from organizations.

The Board's informational and educational activities are directed primarily at promotion of meat, and generally is consumer oriented. Demonstrations in meat cutting, merchandising, and cooking are performed by permanently-employed specialists. A wide variety of charts, pamphlets, brochures, and visual aids are prepared by the Board's own experts (marketing specialists, nutritionists, home economists). These are made available to communications media, professional groups (doctors, food specialists, food journalists), educational institutions, the meat trade, etc. Considerable effort is directed toward meat merchandising and toward professional groups whose members are in a position to recommend meat as a food item or influence consumer attitudes toward meat.

BIBLIOGRAPHY

ANON. 1965. A half century of Federal Market News. Am. Cattle Producer June, 15.

BOHLEN, J. M., and BEAL, G. M. 1967. Dissemination of farm market news and its importance in decision making. Iowa Agr. Expt. Sta. Res. Bull. 553.

DIETRICH, R. A. 1967. Price information and meat marketing in Texas and Oklahoma. USDA Econ. Res. Serv. (in cooperation with Texas Agr. Expt. Sta. and Oklahoma Agr. Expt. Sta.) Agr. Econ. Res. Rept. 115.

DOWELL, A. A., and BJORKA, K. 1941. Livestock Marketing. McGraw-Hill Book Co., New York.

ENGELMAN, G. 1956. The role of livestock market news in a free pricing economy. Paper presented to Beef Cattle Breeders' and Herdsmen's Short Course, Univ. Florida, Gainesville.

NEWELL, S. R. 1954. Reporting supplies and markets. USDA Yearbook Agr.

PHILLIPS, V. B. 1961. Price formation and pricing efficiency in marketing agricultural products: the role of market news and grade standards. Proc. 26th Ann. Conf. Natl. Assoc. Sci. Teachers, Washington, D.C., April.

PIERCE, J. C. (undated) The Livestock Division—its programs—its aims. USDA Consumer Marketing Serv. (unnumbered).

ROBERTS, W. P. 1959. Economic information for stockmen; where to find it. Wyoming Agr. Expt. Sta. Bull. 363.

SEUFFERLE, C. H. 1960. Livestock marketing information. Nevada Agr. Expt. Sta. Circ. 28.

SMEBY, A. B. 1961. History of livestock, meat, wool market news. USDA (unnumbered).

STRASZHEIM, R. E. 1968. The how and why of government crop and livestock reports. Econ. Marketing Inform. Indiana Farmers, June 28, Agr. Staff, Purdue Univ.

USDA. 1969A. The story of U.S. agricultural estimates. USDA Statist. Reporting Serv. Misc. Publ. 1088.

USDA. 1969B. Market news keeps pace with livestock industry. USDA Consumer Marketing Serv., Agr. Marketing July, 4–5.

USDA. 1970A. The market news service on livestock, meat and wool. USDA Consumer Marketing Serv. Marketing Bull. 50.

USDA. 1970B. Federal-State market news reports—a directory of services available. USDA Consumer Marketing Serv. 21, Revised Nov.

USDA. 1975. Glossary of terms. USDA Agr. Mktg. Serv. AMS-566.

UVACEK, E., Jr., and GOODWIN, J. (undated) Interpreting livestock and meat market news. Southern Coop. Ext. Serv. Marketing Publ. 69–3.

WASSON, C. R. 1954. The market's nervous system. USDA Yearbook Agr.

13

Regulatory and Inspection Measures

The livestock-meat industry is faced with a rather wide array of governmental regulations. Not all were set up specifically for the control of marketing problems (in fact, most were not), but all affect either directly or indirectly various aspects of the marketing system. As in other governmental regulations, whether they be traffic control, drug control, antilittering, registration of firearms, property zoning, licensing, etc., the basic justification for control (or restriction) of individual liberties is enhancement of the general welfare. General welfare is a broad term construed to include such concerns as health, economic well-being, social conditions, aesthetic values, humane consideration, etc. This is not a static situation as human values change over time, technological and economic developments introduce new factors, and new knowledge is gained through study and research. Thus, the establishment of regulatory measures is not a once-and-for-all event. Further changes may be expected. As stated by Brooks (1954), "The regulation of marketing is designed to promote the public welfare and, therefore, is to be modified or altered from time to time to meet the changing needs of the nation."

MEAT INSPECTION

Meat inspection is concerned with the wholesomeness, cleanliness, and truthfulness in labeling of meat and meat products. During the agrarian era when people slaughtered their own livestock, processed the meat, and consumed it themselves, there was little public concern about wholesomeness and sanitation; responsibility was left to individuals. The philosophy of individual responsibility carried over to commerce as the economy moved into the commercial era. The prevailing doctrine was "caveat emptor" or "let the buyer beware."

Some states and cities acted on behalf of the public with food inspection laws long before the federal government. But forces were underway which led to action by the federal government.

The first federal move with respect to meat was prompted by problems in the export trade. By the late 1800's the United States had developed a substantial export market in both pork products and beef. But there was stiff opposition from foreign countries on economic and political grounds, plus alleged presence of diseased beef and incidence of trichinosis from the consumption of imported U.S. pork. A number of countries imposed import restrictions which had a crippling effect on the U.S. livestock-meat industry (Clemen 1923).

Loss of these markets aroused producers and packers to the extent that Congress enacted an act in 1890 that required "the certification of certain meats processed for exportation. In 1891 and 1895 Congress extended meat inspection to partial protection of domestic consumers by requiring inspection of animals for disease before slaughter" (Crawford 1954).

This act, however, proved inadequate for full protection and was followed by the Meat Inspection Act of 1906. One of the most influential factors at that time was a description by Upton Sinclair of conditions in Chicago packing plants. This appeared in his 1906 novel *The Jungle* and received wide publicity in newspapers and magazines. Whether the report was fact or fiction, it aroused intense public attention and emotion.

The 1906 act kept the ante-mortem inspection and added a post-mortem inspection along with other features which substantially strengthened its effectiveness. The standards and procedures of the 1906 Act still set the pattern for meat inspection. A major change, however, was made in 1967. Prior to 1967, authority for federal meat inspection was limited only to meat which moved in interstate or foreign commerce. This left authority for inspection of meat moving in intrastate trade up to the individual states. The result was a hodgepodge of state and municipal regulations with adequate inspection in some states, and none at all in others.

Even though nearly 85% of the nation's meat supply was produced in federally inspected plants, Havel (1968) reported that only 5555 of the 14,832 meat facilities operated on an intrastate basis were currently subject to state or local inspection, leaving 9277 plants exempt from any inspection controls.

Congress, over much opposition, passed the Federal Meat Inspection Act, commonly known as the Wholesome Meat Act, in December, 1967. This act specified that by Dec. 1, 1969 each state was to have in operation, inspection standards and procedures for red meat at least equal to those of federal requirements, or submit to federal inspection of meat and plants in which meat was slaughtered for sale within that state's borders.[1] Farm slaughter of meat for home consumption was excluded. Subsequently, an additional year's extension was granted for compliance.

---

[1]The Wholesome Meat Act of 1967 covered inspection of all red meats but did not apply to poultry, fish, or certain wild game animals. Poultry inspection was covered in the Poultry Products Inspection Act of 1968.

The purpose of meat inspection is to safeguard health by (1) eliminating diseased and otherwise unwholesome meat from human consumption, (2) maintaining sanitary conditions during slaughtering and processing, (3) preventing the addition or use of harmful ingredients, and (4) preventing false or misleading labeling of meat and meat products. This is an important and immense assignment, requiring a large and well-trained group of personnel. The basic core of employees are licensed veterinarians who are assisted by trained food inspectors.

The entire regulatory process includes examination of plants before granting approval as an FI plant, ante-mortem and post-mortem inspection of animals and carcasses, inspection of plant operations, checks on ingredients used in processing, checks on the truthfulness of labels, inspection of imported meats, and inspection of plants in foreign countries which export meat to the United States. Operators desiring federal inspection are required to submit plans and specifications of their plants to USDA. Examination of the plans and specifications is followed by examination of the plant. If the plant meets requirements, the operator is granted an official number which appears on the stamp used to mark the meat and meat products produced. The number identifies that particular plant. Shipments of meat bearing that number can easily be traced if the need arises.

Inspection regulations prescribe that animals be examined on the premises before they are brought to the killing floor (i.e., the ante-mortem inspection). This applies to all species. Any animal which shows obvious signs of disease or disorders which are cause for condemnation is immediately tagged "U.S. CONDEMNED." It must be killed, if not already dead, by an official of the establishment and placed in a disposition tank in the presence of an inspector. As the animal is placed in the tank, the inspector removes the tag and reports the disposition. Animals which show some signs of cause for suspicion of disease or disorder that cannot be confirmed during the ante-mortem inspection are tagged "U.S. SUSPECT." These animals are slaughtered separately and are given an extra thorough post-mortem inspection which determines whether they are condemned or passed.

Animals which pass the ante-mortem inspection are cleared for slaughter. At the time of slaughter each carcass and viscera is carefully examined. If evidence of abnormal conditions exist, the carcass is tagged "U.S. RETAINED," moved from the regular line, and kept separate until a more detailed examination can be made. "U.S. Retained" meat is kept in a locked compartment pending further examination. Depending upon the results of the examination it either is condemned or passed. Carcasses which pass inspection are stamped with a circular brand bearing the plant number and the abbreviated inscription "U.S. INSP'D &

P'S'D" (see below). The imprint is made with a harmless (edible) indelible fluid. This mark is imprinted at various places on a carcass so that after breaking, the mark will show on various wholesale cuts. Under certain

conditions parts of a carcass may be condemned and the remainder passed.

Processed meats are marked by a circular brand on the container or package, showing the establishment (plant) number and the wording "U.S. INSPECTED AND PASSED BY DEPARTMENT OF AGRICULTURE" (see below).

Inspection continues beyond the killing floor. In the processing departments inspectors check to see that sanitary conditions are maintained, that added ingredients meet specifications, and that labeling is truthful. Periodically, samples of processed products, ingredients, containers and wrapping materials are selected for laboratory analysis.

The meat inspection service also covers imported meats. Before any country is permitted to ship to the United States that country's own inspection system must rate at least equal to that of the United States. The export country designates and certifies which of its plants may ship to the United States. These plants then are inspected at least once a year by a USDA foreign review officer. Inspectors at U.S. ports of entry check inshipments (on a sample basis) for wholesomeness and cleanliness. Meat inspection is carried out under the Food Safety and Quality Service of USDA.

Packers sometimes buy slaughter animals from producers "subject to inspection." This is a practice used with animals which show some reason for possible condemnation by the inspectors. The agreement between packer and producer usually stipulates a price that will apply if the carcass (or part of it) passes inspection. Payment usually is withheld pending inspection. This practice has features which are beneficial to both producers and packers. Without the practice, packers undoubtedly would reduce the offering price enough to compensate for the risk factor involved, or simply refuse to buy such animals. Producers are afforded a market for animals they might not be able to sell otherwise and which they probably would hesitate to slaughter for home consumption.

Almost 93% of all red meat is produced under federal inspection. This percentage increased substantially following the Wholesome Meat Act of 1967, but has tapered off in recent years. Most plants now have USDA or equivalent inspection so that further increases will be largely from new plants.

Under provisions of the Meat Inspection Act a series of regulations was issued in 1978 to effect a lowering of the level of sodium nitrite (or equivalent of potassium nitrite) in the production of bacon. These regulations were prompted by concern over the reported relationship between nitrosamines in fried bacon and the occurrence of cancer. Nitrites have been used for many years in bacon production for control of botulism organisms.

The cost of federal meat inspection is borne by the federal government—except that when plants operate overtime, plant owners reimburse the federal government for overtime worked by inspectors. It is estimated that the cost of inspection is about ¼¢ per lb of red meat. By any reasonable standards that must be considered a very nominal cost for insurance of wholesome and clean meat.

FOOD, DRUG, AND COSMETICS ACT

Public concern over wholesomeness of food during the early period mentioned in the previous section was not limited to meat. The separate Meat Inspection Act attests to the particular importance attached to meat, but concurrently with passage of the Meat Act, Congress also enacted a Food and Drug Act. Both of these Acts were signed on the same day, June 30, 1906 by President Theodore Roosevelt. This action dealt a *coup de grace* to the doctrine of *caveat emptor*. The major purpose of the Food and Drug Act was to assure the public of wholesome, sanitary, unadulterated and truthfully labelled foods. Changing conditions, new products and new technology necessitated changes and strengthening in the law. This was accomplished by the Food, Drug, and Cosmetics Act of 1938. The 1938 law as amended provides the framework of today's pure food and drug regulations. Administration of these Acts was under the Department of Agriculture from 1906 to 1940 at which time it was transferred to the Federal Security Agency. In 1953, the Federal Security Agency was reorganized and became the Department of Health,

Education and Welfare, which currently administers the Act under its Food and Drug Administration (FDA).

The concurrent enactment of a separate Meat Act, and Food and Drug Act, might appear to indicate Congressional intent that the Meat Act cover all aspects of regulation with respect to meat. However, changing circumstances have introduced new problems. The two agencies cooperate in some areas. One of the areas of recent mutual concern is residues in meat resulting from the use of feed additives, drugs, and pesticides. USDA inspectors test for such residues on a random sampling basis and turn findings over to the FDA for enforcement. FDA has the authority to initiate criminal prosecution of producers who market livestock containing illegal drug and hormone residues.

ANIMAL DISEASE CONTROL

Early U.S. history recorded numerous costly outbreaks of animal disease. Effective veterinary treatments were not available. Quarantines were applied in some areas to restrict the spread of disease and counter quarantines were applied in other areas as retaliatory measures. As a result, "Our system of interstate and export animal transportation was denounced as a disgrace and an outrage on the first principles of humanity. There was a growing demand for protection of the public health in connection with the meat supply" (Van Houweling 1956). A Veterinary Division was established in the USDA in 1883 and was given responsibility of developing information on the prevalence of animal disease and means of controlling and eradicating disease. In 1884, Congress enacted *An Act for the Establishment of the Bureau of Animal Industry.*

Currently, authority and responsibility at the federal level for control and eradication of animal disease is lodged in the Animal and Plant Health Inspection Service, USDA. Its programs are carried out in cooperation with states.

An interaction exists between disease control and meat inspection programs. The more effective the degree of control, the fewer animals and carcasses will be condemned during inspection. And disease detection during inspection often assists in disease control. By use of purchase records, brands, tags, etc., authorities often can trace the origin of a disease back to a particular owner or locality where control measures can be applied. Disease control activities, of course, are not limited to connections with meat inspection. Detection of disease also is effected by federal inspectors stationed at designated public markets, by state inspectors at local markets, and by veterinarians in their regular practice.

Detection is followed up by appropriate measures, depending upon the circumstances. With some diseases, e.g., foot and mouth disease, this calls for immediate disposal. In other cases, movement of the animals is restricted—quarantine may be imposed, treatment or immunization may be required, or immediate sale for slaughter may be required (as is the case with brucellosis reactors). Generalization about state restrictions is difficult due to variation among states. At federally inspected public markets, diseased animals are segregated and disposed of under supervision of an inspector. Contaminated vehicles, pens, and premises are disinfected under supervision of an inspector and the state of origin is notified. Outshipments of animals which pass inspection are certified for interstate shipments.

At the international level, a complete embargo is enforced on the importation of live animals or fresh meat from countries infected with especially dreaded diseases, such as rinderpest, and foot and mouth diseases. Livestock may be imported from so-called clean countries, but under strict inspection. Each year substantial numbers of feeder cattle are imported from Mexico and Canada. These cattle must come in through designated ports of entry where they are examined by U.S. inspectors.

Disease control measures of foreign countries still are a factor in U.S. livestock and meat exports. In relation to total U.S. production, the exportation of livestock is relatively small, yet it has increased substantially in recent years. Breeding cattle is by far the major item, with lesser numbers and value of sheep, lambs, hogs, and horses. "The major obstacle that U.S. cattle face is in meeting certain foreign requirements for Blue Tongue disease, once thought to occur only in sheep but now known to affect cattle as well" (Dobbins 1968). The U.S. exports only a minor tonnage of meat, but overseas sales of certain products (lard, tallow, variety meats, hides, and skins) are very important to the industry. It is impractical to attempt to list here detailed restrictions imposed by particular countries on the importation of U.S. products. A number of restrictions are based on presence of animal disease, some are based on hormone residues. Ostensibly, these regulations are enforced in the interest of public health, but in some cases there are strong suggestions that they also are being used to control imports for economic and political reasons.

FEDERAL TRADE COMMISSION

The Federal Trade Commission (FTC) was established in 1914 by enactment of the Federal Trade Commission Act. Conditions which led to establishment of FTC have been discussed in earlier chapters. This was just one step in a series of governmental actions designed to curb unfair trade practices and a tendency toward monopolization of trade by big business. Prior to FTC, there had been the Interstate Commerce Commission Act (1887) and the Sherman Antitrust Act (1890). The Interstate Commerce Commission (ICC) was the first federal regulatory agency empowered to control railroad rates and services. Its purpose was to eliminate rate discrimination among shippers and cut-throat competition among carriers. The Sherman Antitrust Act was designed to curb industrial monopoly. Both had a common purpose of restraining monopoly, but the approaches were substantially different. The ICC Act in effect made public utilities of the railroads. It might be noted parenthetically that a Special Studies Subcommittee of the U.S. House of Representatives, House Government Operations Committee, in a 1969 study, recommended the creation of a public utility type of commission to regulate the U.S. cattle industry. Opposition in both legislative branches killed that proposal.

The approach under the Sherman Act was an effort to make competition effective by outlawing (1) certain restraints to trade, (2) monopoly, and (3) attempts to monopolize. Due largely to judicial difficulties in defining and interpreting terms, and to administrative apathy, Congress in 1914 tried to clarify the situation by enacting the Federal Trade Commission Act and the Clayton Act. The Clayton Act listed a series of practices which specifically were declared to be unfair and illegal. The FTC was established as the watchdog agency to determine violations. Its major role now is the investigation of unfair trade practices and the issuance of rulings. Major actions are turned over to the courts. One of FTC's early actions was an investigation of practices in the meat packing industry in 1917. This was mentioned in Chap. 7 as a major factor in the Packers Consent Decree of 1920—an agreement whereby the five major packers of that time consented to abide by a number of stringent trade and operational specifications. Those packers have tried periodically to gain modifications of the Consent Decree but the courts and Department of Justice stood firm on all original specifications until 1971. At that time, the Department of Justice agreed to an amendment freeing the remaining four concerned packers (Swift & Co., Armour & Co., Wilson & Co., and Cudahy Co.) from a restriction on production and distribution of more than 100 specified product lines. The amendment allows manufacturing and wholesaling but maintains the original restriction on retailing.

PACKERS AND STOCKYARDS ACT

The Packers and Stockyards Act was enacted in 1921 in an attempt to provide relief for livestock producers who charged packers and market agencies with a variety of unfair and monopolistic practices. The Act originally was administered by the Packers and Stockyards

Administration—an agency directly responsible to the Secretary of Agriculture, USDA. Administrative reorganization in 1977 placed the agency under the Agricultural Marketing Service of the USDA and dropped the word "Adminstration" from its title. Major provisions of the act, as they apply to particular types of markets, were discussed in Chap. 6. In general, the act is designed to regulate business practices of those who engage in the buying and selling of livestock and meat which enter interstate and international trade. Regulations cover the activities of stockyards, companies, market agencies, dealers, packer buyers, and meat packers.

With respect to livestock marketing (USDA 1969B), the act specifically prohibits any stockyard owner, market agency, or dealer subject to its jurisdiction, from engaging in any unfair, unjustly discriminatory, or deceptive practice or device in connection with the receiving, marketing, buying, selling, feeding, watering, holding, delivery, shipment, weighing, or handling of livestock.

By definition, the term meat packer is rather broad. In addition to those who buy livestock across state lines for slaughter, "those who manufacture or prepare meats or meat food products for sale or shipment in interstate commerce are also packers, including wholesalers, fabricators, restaurant and hotel suppliers, and food chains who purchase meat in commerce for fabricating or other processing or preparation" (USDA 1969B). Packers may not (USDA 1969B): (1) Engage in or use any unfair, unjustly discriminatory, or deceptive practice. (2) Make or give any undue or unreasonable preference or advantage to any person or locality, or subject any person or locality to any undue or unreasonable prejudice or disadvantage. (3) Agree or arrange with any other packer to apportion purchase or sale territories or supplies for the purpose or with the effect of restraining commerce or creating a monopoly. (4) Engage in any act for the purpose of or with the effect of manipulating or controlling prices, creating a monopoly, or restraining commerce.

An amendment in 1968 required that payment for livestock purchased on grade and yield basis be made on hot carcass weight. This was a much debated issue. Prior to the 1968 amendment a wide variety of arrangements prevailed in terms of adjusting hot weight to a cold equivalent, with resulting dissatisfaction among producers. A general feeling prevailed that shrinkage adjustments in calculating cold weight tended to favor the packers. This amendment included provisions for tare adjustments for the weight of rollers, gambels, hooks, and other equipment on a uniform basis. Presumably, carcass grade and weight arrangements are now more uniform than previously was the case.

Producers who have reason to believe that they have suffered loss or damage from violations of the P&S Act may petition for reparations. This

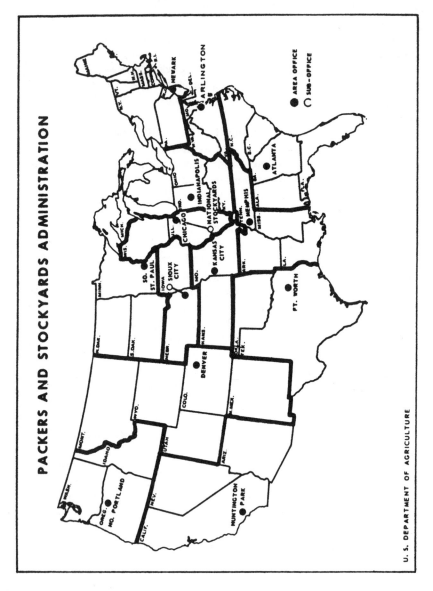

PACKERS AND STOCKYARDS ADMINISTRATION

U. S. DEPARTMENT OF AGRICULTURE

FIG. 13.1.   AREAS AND LOCATIONS OF P&S AREA OFFICES

can be done by filing a written complaint with an area supervisor or directly with the Secretary of Agriculture within 90 days of the wrongful action. Areas and the location of area offices are shown in Fig. 13.1. Some of the more commonly mentioned unfair trade practices which would justify filing a complaint are: dishonest weights, failure to pay for livestock, failure to rectify a mix-up in ownership of livestock, misrepresentation of livestock, etc. Upon receiving a complaint, P&S carries out an investigation without charge.

Under administrative authority the Packers and Stockyards Administration issued a regulation in 1974 prohibiting dual ownership of packing plants and custom feedlots. Packers may own and operate feedlots for their own livestock and may place livestock which they own in custom feedlots, but are prohibited from owning a custom feedlot.

An important amendment in 1976 assures prompt payment to producers from the sale of slaughter livestock. Producers had advocated such a law for years, but Congress did not act until the bankruptcy of a leading beef packer left Midwest cattle feeders with some $20 million of unpaid cattle sales. Under this amendment packers who buy $500,000 or more of livestock a year must carry a bond equal to a 2-day kill. Buyers of slaughter livestock must pay sellers by the close of the business day following the sale, unless buyer and seller mutually agree to do otherwise. The law also provides that if a packer pays by check, the account receivable proceeds therefrom, and product inventory of the packer is held in trust to cover value of the producer's sale until packer's check is cleared by the packer's bank. The law applies to interstate meat wholesalers, brokers, dealers, and distributors, as well as packers.

HUMANE TREATMENT OF LIVESTOCK

The 28-Hour Law

One of the earliest federal laws related to the marketing of livestock was a statute enacted in 1873 intended to alleviate cruelty to livestock while in transit on railroads (Dowell and Bjorka 1941). The Act stipulated that livestock could not be kept in continuous transit more than 28 hr without a rest stop. At a maximum of 28 hr, the livestock were to be unloaded, fed, watered, and rested for at least 5 consecutive hours. Transportation agencies were made responsible for furnishing the necessary facilities and services. In practice, however, the facilities and services actually furnished were, in many cases, inadequate. Due to aroused sentiment, the 1873 Act was repealed and another, known as the 28-Hour Law, was passed in 1906. The purpose of this Act was similar to the earlier one, but it was more specific with respect to minimum quantity of feed, cleanliness of water, and methods and procedures for handling stock during loading and unloading. In cars that were loaded with sufficient space for each animal to lie down, feeding and watering was permissible without unloading. In addition, the 1906 Act provided for an 8-hr extension, or a total of 36 hr, upon request. This law is still effective.

A limitation of time in continuous transit has economic as well as humane implications. Extended deprivation of feed, water, and rest undoubtedly results in discomfort and suffering. Prior to the Act, periods of 60 hr or more in transit were reported. However, associated costs also are important. Shrinkage, bruising, crippling, and death increase with time in transit. This law was effective in accomplishing both the intended objective of alleviating cruelty and reducing above-mentioned related costs. Need for such a law is considerably diminished now compared to earlier years. Improved, faster freight service makes it possible for many shipments to reach their destination within 36 hr, and with expeditious service shipments from mid-country can reach either coast with 1 rest stop. Trucks were not included in the provisions of the original law. But most buyers of livestock insist on a rest stop on an extended trip, for economic reasons—whether the haul be by rail or truck.

Humane Slaughter Act

Public sentiment over alleged cruelty to livestock during slaughter reached a peak during the 1950's. Methods of handling, shackling, stunning, and killing were considered inhumane and abhorrent by some people. A campaign for modification of slaughter methods culminated in the Humane Slaughter Act of 1958. Provisions of the Act instructed the Secretary of the U.S. Department of Agriculture to prescribe humane methods and made that office responsible for administration. Unlike many laws this Act did not prescribe criminal penalties for non-compliance. However, it did stipulate that government purchases of meat must be from plants using humane methods and this is a persuasive feature. Among other features, this Act provides for keeping animals relatively calm prior to stunning, and using electrical, mechanical, or chemical methods for rendering the animals unconscious prior to killing.

As was mentioned with respect to the 28-Hour Law, humane slaughtering has economic as well as humanitarian aspects. Capital requirements for complying with the recommended mechanical methods are negligible consisting usually of a captive bolt stunner, a small caliber gun device. These devices are faster and use less labor than the previous methods. The cost of electrical stunners and carbon dioxide tunnel installations (the common chemical method) are offset by elimination of shackling and associated equipment, but probably the most important economic plus is in fewer bruises and damaged parts of pork carcasses that formerly resulted from shackling and hanging prior to sticking of hogs.

Provisions of the Act exempted killing methods used in religious rituals such as kosher slaughtering.

Even though compliance with the law is not mandatory per se, it is reported that most meat is produced under approved methods.

Commodity Futures Trading Commission Act

Trading in livestock and livestock product futures is regulated under the Commodity Futures Trading Commission (CFTC), an autonomous agency of the federal government. Trading in commodity futures has a long history. As mentioned in Chap. 10, the Chicago Board of Trade was organized in 1848. While the market served a useful economic purpose, a number of abuses was evident from time to time in price manipulation and undue price fluctuations associated with speculative maneuvering. Bills were introduced in Congress as early as 1884 in attempts to provide federal supervision and regulation of trading. However, it was not until 1922 that the first law was passed—the Grain Futures Act. This Act was amended and renamed the Commodity Exchange Act in 1936.

The 1936 Act, with subsequent amendments, remained the basic law for the maintenance of fair and honest trading in commodity futures markets until the CFTC was established by congressional mandate in 1974. The commission became effective in 1975. A major responsibility of the commission is the prevention of price manipulation and unfair trading practices. This involves, among other things, the regulation of registration requirements, financial standards, and trading practices. The definition of hedging was broadened from the rather narrow concept specified in the Commodity Exchange Act.

Major emphasis is placed on the detection of attempted speculative price manipulation and trading activities which distort competitive prices. Some publications on early efforts at "cornering the market" depict an action-filled, romantic situation, but to a hedger or legitimate speculator who may be caught in a distorted market, the situation is anything but romantic.

MEAT IMPORT LAW OF 1964

The question of regulating (either restricting or encouraging) the supply of meat, like any other product, raises extremely complex issues. Conflicts of interest are inherent between producers and consumers, between producers and importers, and within the producing sector itself. At the national level the interest of a particular group may be incompatible with the overall national interest. This point is discussed further in the next chapter on "International Trade." Cattle producers during the early 1960's became deeply concerned over rising imports of beef. Concern had been expressed on previous occasions, but the problem had seemed to ebb and flow without reaching major proportions. Occasionally, imports would increase to about 5% of total beef production, usually in response to relatively high U.S. prices, and then drop back to about 1% as U.S prices declined. A nominal tariff had been imposed on livestock and meat for many years and this, along with tariffs in general, had been reduced from a high point of the 1930's. As mentioned in an earlier section, a complete ban had been enforced for many years against the importation of fresh meat from countries infected with rinderpest, and foot and mouth disease. But the situation was different in the late 1950's and early 1960's.

Imports of beef in the early 1960's increased to more than 10% of total U.S. production—about twice the level of previous high points (see Fig. 13.2). A number of factors were involved, but a major one was the discontinuance of an agreement in 1958 between Great Britain and Australia whereby Australia no longer was required to ship stipulated quantities of low grade beef to Great Britain. The U.S. market was the most lucrative outlet in the world and substantial quantities were coming in, not only from Australia, but from New Zealand, Ireland, and other countries. The situation was compounded by relatively high production in the United States.

Attempts were made during the late 1950's and early 1960's to control the meat import situation through voluntary agreements with major exporters. More stringent types of regulation were viewed unfavorably by State Department officials because an effort was under way to implement a broad tariff liberalization policy. However, spokesmen for the cattle industry did not view voluntary control as a satisfactory solution to the problem. After a series of hearings Congress enacted the Meat Import Law (Public Law 88–482) in August, 1964, to become effective Jan. 1, 1965. This bill provided for import quotas, based on a formula, for fresh, chilled, and frozen beef, veal, mutton, and goat meat. This included both carcass and boneless meat. It did not include pork, lamb, or canned meats. The purpose of the law was to limit annual imports of the specified meats to a level comparable to a selected base period (1959–1963) with an annual adjustment ("growth factor") based on changes in domestic production relative to the base period.

The base quota was established as the average annual quantity imported during the base period (1959–1963) which was 725,400,000 lb. Each year the growth factor is determined by calculating the percentage by which the estimated U.S. commercial production of the specified meats in the current calendar year and the 2 preceding years (i.e., a 3-yr average) exceeds (or falls short of) average annual U.S. production dur-

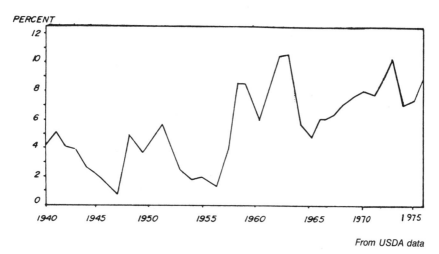

From USDA data

FIG. 13.2.   BEEF IMPORTS AS PERCENTAGE OF TOTAL BEEF PRODUCTION

ing the base period. Thus, to determine the growth factor at the beginning of a given year, it is necessary to estimate production for that year. The calculated growth factor then is multiplied times the base quantity to determine the amount of increase (or decrease) in the base. This increase (or decrease) added to the base gives an adjusted base quota. With U.S. production on an increasing trend, the annual adjustment typically has been upward. The Act allows a 10% leeway above the adjusted base quota before quotas would be applied to individual countries. Thus a "quota trigger point" is determined at 110% of the adjusted base quota.

In this process, the Secretary of Agriculture is required to estimate at the beginning of each quarter-year the quantity of prospective imports. If the quantity of prospective imports exceeds the trigger point, the President is required to invoke a quota on imports of these meats. In case quotas are imposed, the total import quota would be allocated among exporting countries on the basis of shares supplied by those countries during a representative period.

But the law also contained provisions for extraordinary conditions under which the President could suspend or increase quotas. These were (1) if he determines it necessary because of overriding economic or national security interests, (2) if the supply of meat is inadequate to meet domestic demand at reasonable prices, or (3) if international agreements are entered into after enactment of the Act which have the same effect as the Act.

It is seen that the Meat Import Law provides for a flexible quota. As U.S. production increases, the quota increases. At the time it was passed there was a considerable body of opinion that imports would never reach the trigger point. In the late 1960's, however, it became apparent that imports would reach that point unless some action was taken. In an effort to avert the necessity of invoking mandatory individual quotas, the Secretary of Agriculture, in the fall of 1968, began a program of negotiating voluntary quota agreements with exporting countries, which action is based on provisions of the Agricultural Act of 1956. Among the major exporters, only Canada refused to go along with this program. Most countries abided by the agreements on direct shipments[2] but some exported meat to Canada where it was then transhipped to the United States.

When it appeared in 1970 that the trigger point would be exceeded, President Nixon, under requirements of the act, issued a proclamation on June 30, ". . . to place a limitation on imports of meats covered by the Act. At the same time, the President suspended the limitation, as authorized by law, after determining that this action was required by the overriding interest of the United States" (USDA 1970B). Thus, it may be said that quotas were triggered, but immediately rescinded. This same sequence has occurred several times in subsequent years. On two occasions, when domestic meat prices increased substantially from previous levels (1973 and 1978), the President authorized increased imports above the trigger point. That was done, ostensibly, in the public interest.

Prior to 1977 substantial quantities of beef and veal were brought into the U.S. by way of Foreign Trade Zones in circumvention of import quotas. That loophole was closed by administrative regulation in late 1976 which specified that, beginning January 1, 1977, any foreign beef or veal processed in Foreign Trade Zones, possessions, or territories and shipped into the U.S. must be included in the quota of the country of origin.

Provisions of the Import Quota Law cause imports to increase when domestic production increases, and decrease when domestic production decreases—with a slight lag due to the 3-yr moving average feature. Since price movements bear an inverse relationship to available supplies, the act tends to accentuate price swings associated with production cycles. This has provoked several proposals to amend the law to make effects of imports contra-cyclical. In most respects, such an amendment would be beneficial to both producers and consumers, and its enactment may be expected.

While specific quotas have been imposed only to a minor degree, the existence of the law has a considerable psychological effect on the negotiating process.

---

[2]Several Central American countries were exceptions and were issued mandatory quotas.

POLLUTION CONTROL

A widespread public concern with environmental degradation reached major proportions during the 1960's. Livestock feedlots, with large numbers of animals in highly concentrated circumstances give rise to air pollution from feedlot odors and water pollution from waste materials carried in water run-off.

Pollution problems are indirectly related to marketing in that regulations designed to control pollution influence the location of feedlots which, in turn, affects transportation costs and competitive position with respect to both feed and feeder livestock procurement and sale of the finished livestock.

A number of states have enacted legislation designed to protect the interests of both the public and feedlot owners. One common form of regulation is the requirement of a license, or permit, for operation of a feedlot. Requirements for obtaining a permit vary, but generally include provision for adequate drainage and control of run-off. One of the most vexatious problems arises where housing developments expand near the vicinity of a long-established feedlot and then the occupants complain of feedlot odors. Protestations of both water pollution and air pollution have resulted in the closing of some commercial feedlots, notwithstanding valid permits and licenses. Concern with ecological values undoubtedly make feedlot pollution an increasing problem.

The federal government, under authority of the U.S. Environmental Protection Agency, has extensive pollution control requirements, many of which apply to livestock feedlot operations.

BIBLIOGRAPHY

ACKER, D. 1963. Animal Science and Industry. Prentice-Hall, Englewood Cliffs, N.J.

BROOKS, N. 1954. The wide range of regulation. USDA Yearbook Agr.

CLEMEN, R. A. 1923. The American Livestock and Meat Industry. Ronald Press, New York.

COMMODITY FUTURES TRADING COMMISSION. 1978. Report on farmers' use of futures market and forward contracts. Commodity Futures Trading Commission, (unnumbered), Feb. 15.

CRAWFORD, C. W. 1954. The long fight for pure foods. USDA Yearbook Agr.

DOBBINS, C. E. 1968. Past and prospects: U.S. livestock in export markets. USDA Foreign Agr. 6, No. 46, 1–3, Nov. 11.

DOWELL, A. A., and BJORKA, K. 1941. Livestock Marketing. McGraw-Hill Book Co., New York.

FOWLER, S. H. 1961. The Marketing of Livestock and Meat, 2nd Edition. Interstate Printers & Publishers, Danville, Ill.

GAST, L. L. 1969. USDA's watchdog is added protection. USDA Agr. Res. Serv., Agr. Marketing, Sept. 4.

GUNDERSON, F. L. et al. 1963. Food Standards and Definitions in the United States. Academic Press, New York.

HAVEL, J. T. 1968. Inspection of intrastate meat in Kansas. In Your Government. Univ. Kansas, Sept. 15.

JOHNSON, A. C. Jr. 1977. Public regulation of futures trading in the United States. University of Wisconsin, Dept. of Agr. Econ. Economic Issues, No. 9.

KAUFFMAN, R. R. 1965. Recent developments in futures trading under the Commodity Exchange Act. USDA Commodity Exchange Authority Agr. Inform. Bull. 155.

LEE, R. F., and HARPER, H. W. 1966. Meat and poultry inspection. USDA Yearbook Agr.

USDA. 1952 The inspection stamp as a guide to wholesome meat. USDA Agr. Inform. Bull. 92.

USDA. 1956. The little purple stamp. USDA Agr. Res. Serv., Agr. Research, Vol. 4, No. 12. 4–10.

USDA. 1969A. His mission: consumer protection. USDA Agr. Marketing 14, No. 11, 3.

USDA. 1969B. The Packers and Stockyard Act, what it is and how it operates. USDA P&S Admin. PA-399, Revised Aug.

USDA. 1970A. Commodity Exchange Act as amended. USDA Commodity Exchange Authority, Revised Feb. (unnumbered).

USDA. 1970B. Livestock and meat situation. USDA Econ. Res. Serv. LMS-174.

USDA. 1971. Accurate weights, guidelines for weighing livestock. USDA Packers and Stockyards Administration, PA-986.

USDA. 1974. Regulations and statements of general policy issued under the packers and stockyards act. USDA Packers and Stockyards Administration, (unnumbered), June.

USDA. 1975. Packers and stockyards act. USDA Packers and Stockyards Administration, PA-1019.

USHEW, 1970. Requirements of the United States Food, Drug and Cosmetics Act. U.S. Health, Education, Welfare, FDA Publ. 2.

VANHOUWELING, C. O. 1956. Our battle against animal diseases. USDA Yearbook Agr.

WILLIAMS, W. F., and STOUT, T. T. 1964. Economics of the Livestock-Meat Industry. Macmillan Co., New York.

ZIEGLER, T. 1965. The Meat We Eat. Interstate Printers & Publishers, Danville, Ill.

14

International Trade in Livestock, Meat and Wool

International trade in meat involves only approximately 8% to 10% of world production. But this relatively small proportion disguises significant economic and political implications. Two particularly important economic aspects are as follows: (1) at any given time the import-export position of a given country may deviate substantially from the average of the aggregate world situation, and (2) over the years the import-export position of a country can (and does) change.

With regard to the former, the United Kingdom exports virtually no beef, but in recent years has imported nearly 40% of its consumption. On the other hand, Australia and New Zealand import no beef but export approximately 50% of their production. The United States exports substantial quantities of some products and imports others. Many of the less developed countries subsist almost entirely on their own production. Thus, to say that only 8% to 10% of the world's meat moves in international trade ignores the fact that some countries are highly dependent upon imports, some are highly dependent upon exports, some are dependent upon both, and others are virtually independent.

The United States furnishes a good example of a country whose import-export position has changed drastically over time. As may be seen in Fig. 14.1, U.S. exports of both pork and beef expanded rapidly following the Civil War, reached a peak about World War I, then dropped precipitously, have remained low since then with the exception of a brief rise during World War II and have shown a slightly increasing trend since the early 1960's. An equally dramatic change is apparent in imports of beef (Fig. 14.2) which have increased from virtually nothing prior to World War I to 2 billion pounds in recent years. Imports of pork, although much less than beef, have picked up significantly since 1950.

Such changes over time, and differences among countries in import-export balances, prompts the question: What gives rise to international trade?

BASES FOR INTERNATIONAL TRADE

International trade has individual as well as general welfare implications. At the national level, trade policy is governed by a mix of economic

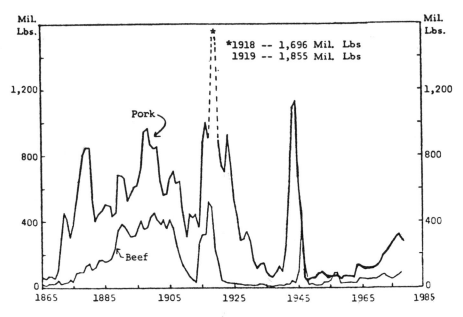

FIG. 14.1. U.S. EXPORTS OF BEEF AND PORK (PRODUCT WEIGHT BASIS), 1865–1977. 1865–1970 reproduced from Ives (1966); 1964–1977 data from USDA

and political considerations which may vary from country to country. From the standpoint of individuals, the motive which generates trade is essentially the same whether it be local, regional, or international in scope. That motive is profit. In the absence of artificial restrictions, an incentive exists to move goods internationally if the price in an importing country is high enough to cover production costs as well as cost of moving the product to that country. From the consumer's standpoint, an incentive exists to buy imported goods when such goods can be purchased cheaper than domestically produced products (of equal quality).

But these conditions simply are reflections of more basic, underlying economic conditions. Why are some countries able to produce certain products more cheaply than others? Why doesn't a single low cost country produce the entire world's supply of a given product?

The Principle of Absolute Advantage

The United States can produce cameras, but the cost is higher than the price of cameras shipped in from Japan, for example. Japan can raise hogs, but if they want pork, it can be purchased from the United States at a price less than the cost of producing it in Japan. These blunt statements

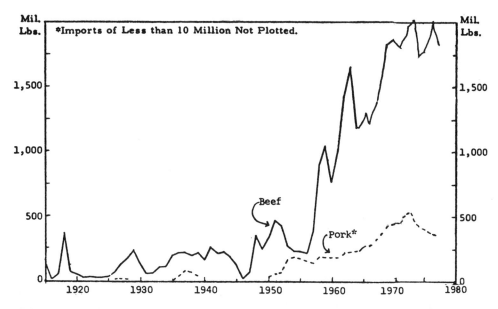

FIG. 14.2.   U.S. IMPORTS OF BEEF AND PORK (CARCASS WEIGHT EQUIVALENT),
1915–1977. 1915–1964 reproduced from Ives (1966); 1965–1977 data from USDA

of apparent fact gloss over some difficult problems in comparing absolute
costs, but serve to provide an introduction to basic factors behind
specialization and trade.

The following hypothetical example illustrates the gain for each country by specializing and trading. Assume that with 1 unit of labor and
capital the United States can produce either 1000 lb of pork or 12
cameras. With an equivalent expenditure of labor and capital, Japan can
produce 800 lb of pork or 20 cameras. Now if the United States and Japan
each devoted 2 units of labor and capital to production, and if each produced both items, the total output would be:

	Pork Production (Lb)	Camera Production
United States	1000	12
Japan	800	20
Total	1800	32

On the other hand, if each devoted 2 units of labor and capital to
production of the product in which it has an absolute advantage, the
result would be:

	Pork Production (Lb)	Camera Production
United States	2000	—
Japan	—	40
Total	2000	40

In total production the gain is 200 lb of pork and 8 cameras. By trading at an exchange of 1000 lb of pork for 20 cameras, each could have pork and cameras as follows:

	Pork (Lb)	Cameras
United States	1000	20
Japan	1000	20
Total	2000	40

It is apparent that in the absence of trade restrictions, each would specialize, trading would follow, and each would benefit. But what makes it possible for one country, or one area to gain an absolute cost advantage over another? Generally the contributing factors can be classified as follows: (1) natural phenomena, such as climate, geographic characteristics, or natural resources; (2) particular attributes or characteristics of the people; (3) capital accumulation and industrial development; and (4) differing proportions of available resources relative to the demands for them.

Many examples could be given where geography, climate, or natural resources give an area an absolute cost advantage in producing certain products compared to other areas. Natural phenomena are particularly important in the production of feed supplies for livestock, and feed is the primary factor in livestock production. Natural forage generally is the cheapest feed for maintaining breeding herds of cattle and ewe flocks, and in producing grass-fed beef. Thus, the vast range areas of the United States, Argentina, Australia, and New Zealand have a decided absolute cost advantage over such areas as central Europe in producing grass-fed cattle or maintaining breeding stock.

In addition to natural phenomena, human skills and capabilities play an important role in costs of production. Education, training, and experience vary greatly around the world. This is true in all industries, including livestock production. Custom and tradition also can be important modifying forces. In Nigeria, for example, the Fulani tribesmen are superb cattle husbandrymen. Yet the century-old tradition of accumulating cattle as a symbol of wealth tends to limit the off-take of their herds. Human characteristics, customs, traditions, and religious beliefs also have varying effects on the demand for goods, and particularly the demand for meat. In India, beef is not eaten by a large segment of the population. In other areas, Moslems and people of the Jewish faith do not eat pork.

Capital accumulation and associated industrial development affect both costs of production and, indirectly (through effects on personal income) the demand for meat and other products. The more highly developed countries have a tremendous head start on less developed countries in capital accumulation. Without this, large-scale enterprises and economies of size would be impossible. Producers in countries with this possibility hold a competitive cost advantage in this factor over their counterparts in less developed countries.

The relative proportions of available resources also influence the use to which those resources may be most profitably utilized. An example in contrast would be Belgium compared to New Zealand. Belgium may be better adapted to lamb production than is New Zealand from the standpoint of soil and climate. But Belgium has a much higher human population density. She must specialize in higher valued land use—high-yielding grain and forage crops, industrial production, etc. New Zealand has relatively less pressure on the land and can utilize it for lamb production.

Cost differences arising from such factors are evident for many goods in many countries. And in many cases, the differences are obvious. Kansas can hardly compete with Honduras in producing bananas. Niger cannot compete with the United States in producing heavy construction equipment. Great Britain cannot produce grass-finished beef as cheaply as Argentina, and so on. Yet a comparison of costs in different countries is a tenuous undertaking, especially in the not-so-obvious cases. Differences exist in monetary units and exchange rates. The applied technology differs greatly, as does the available mix of resources and other factors. Fortunately, it is not necessary to depend upon analyses of absolute costs. Another approach is available: the determination of "comparative costs" or "comparative advantage."

The Principle of Comparative Advantage

Let the following assumption hold for an elementary case of just 2 countries—the United States and a foreign country—and 2 products—grain-finished beef and watches. There are no artificial barriers to trade (such as tariffs, quotas, exchange controls, etc.). Both countries have the natural resources, manpower, skills, and capital to produce either or both products. The cost structure is such that with 1 unit of labor and capital, the United States can produce 1800 lb of fed beef or 6 watches. The foreign country, with 1 unit of labor and capital can produce 600 lb of fed beef or 4 watches. Clearly, the United States has an advantage in the production of both products, but it is apparent that its advantage in the production of beef is greater than in watches. The foreign country is at a disadvantage in both products, but its disadvantage is least in the production of watches.

Suppose each country uses 2 units of resources in producing beef and 1 unit in producing watches. The total quantity produced would be:

	Beef Production (Lb)	Watch Production
United States	3600	6
Foreign country	1200	4
Total	4800	10

Now let the United States specialize and devote all three units of resources to beef production and the foreign country specialize in watch production. Total production would be:

	Beef Production (Lb)	Watch Production
United States	5400	—
Foreign country	—	12
Total	5400	12

The United States would not be interested in trading unless it could get a watch for something less than 300 lb of beef, for it can produce a watch at that ratio of resource use. The foreign country would not be interested in trading unless it could get more than 150 lb of beef per watch. Assume the bargaining settles at 200 lb of beef for 1 watch. If 1400 lb of beef were exchanged for 7 watches, the distributions of the two goods would be:

	Beef (Lb)	Watches
United States	4000	7
Foreign country	1400	5
Total	5400	12

Both countries in this case would benefit by specialization and trading even though the United States was shown to have an advantage in the production of both products.

A generalization referred to as the "principle of comparative advantage" emerges from this set of circumstances. It may be stated as follows: countries gain by producing the products in which they have the greatest comparative advantage, or those in which they have the least comparative disadvantage.

Does this mean that certain countries would produce the entire world's supply of some products and other countries the entire supply of other products? The answer is qualified. Many countries are at such a decided disadvantage in the production of some goods that no output is feasible. However, it will be advantageous for most countries to produce a variety of goods, particularly when transfer costs are taken into consideration. The relative cost situation also changes as any given country attempts to expand production and exploit its advantage. As production increases, unit costs eventually rise. It is advantageous for countries to expand trade until the cost ratios of producing the commodities are equalized in the respective countries. This is likely to happen before any one country produces the entire world's supply of most products.

Conflicting Interests in International Trade

The above are theoretical concepts—explanations of economic pressures behind international trade in a freely competitive situation. The real world, however, deviates considerably from the perfectly competitive model. For one thing, the competitive model assumes mobility of resources, and resources do not move freely within a country, let alone between countries. In spite of possible gains, governments around the world have instituted a wide array of obstacles to international trade such as tariffs, licenses, quotas, arbitrary control of exchange rates, etc. Among the reasons given for restrictions are the protection of new industries, the shielding of established industries from the rigors of foreign competition, retaliation to another country's restrictions, and the desire to encourage domestic production (in an effort to avoid dependence upon imports in time of war or to avoid a situation of weakness in international politics). Most countries encourage exports through such things as domestic subsidies, overseas promotion, etc. Countries occasionally tax exports in an effort to discourage outmovements and keep the product available for domestic use; and on occasion, countries encourage imports to bolster domestic supplies. Usually, however, governments tend to restrict imports as a measure of protection for home industries. As noted in the previous chapter, the U.S. Meat Import Law of 1964 provided for import quotas as a protective device for the U.S. cattle industry.

Arguments for and against free trade center around three different points of view: (1) its effects on particular industries—or more precisely, individuals within particular industries; (2) its effects on the general national interest, irrespective of the implications to individuals, and (3) its worldwide effects, irrespective of the implications to particular countries or individuals. If "world welfare" were the goal, all barriers to trade and migration of people would be removed. Some workers in traditionally low-wage countries would be expected to move to high-wage countries. The resulting tendency for wages to even out would raise average world wages—to the benefit of some individuals, but to the detriment of others. If national welfare were the goal it might be suggested that relatively cheap beef be imported from Australia. The usual argument is that consumers in general would benefit from the resulting decrease in cost of living. But another result would be a decline in cattle prices to the detriment of cattle producers. Many cattle producers also produce wheat and the export market is vital for wheat. A given individual may find it advantageous to favor import restrictions on beef, but oppose a foreign country's restrictions on wheat imports. Thus, conflicts of interest arise within an industry, between individual and national interests, and between national and world interests. These are exceedingly complex issues, involving political and social as well as economic implications.

Governments have responded over the years to a multitude of pressures and considerations. Broad, general policy objectives set the guidelines within which specific regulations are applied to particular products. And the general policy objectives can, and do, change. In 1930, the Smoot-Hawley Tariff Act culminated a highly protectionist policy—a policy of high tariffs. This policy drew retaliation by trade restrictions of other countries and international trade was choked off. The United States changed its policy in 1934 under the Reciprocal Trade Agreements Program which was aimed at reducing tariffs. Liberalization of trade has been the general policy since, and tariffs on many products have been substantially reduced. The policy, however, has met opposition where individual industries have felt a particular disadvantage. As mentioned, the beef cattle industry is a case in point. Swine producers also are seriously questioning the rising importation of pork.

In spite of numerous governmental impediments, large quantities of meat and substantial numbers of livestock do in fact move in international trade. Countries with advantageous cost conditions and various programs of government encouragement maintain constant pressure to export meat to favorably priced markets. Changing cost structures continually alter the production cost ratios among countries. This, together with supply-demand structures, continually alters the price levels. The U.S. beef industry provides an example of such changes and their effect on trade. With the opening of vast range areas during the last half of the 1800's and early decades of the 1900's, the United States possessed both an absolute and comparative advantage in grass-finished beef production. American exporters, in spite of numerous health and tariff restrictions, opened a sizeable market in European countries. Following World War I, trade wars, largely in the form designed to protect home industry and combat world-wide depression unemployment, reduced most international movements. Economic recovery during and following World War II brought relative affluence in the United States, and with it, a growing demand for grain-finished beef. The intensity of this demand dried up the supply of grass-finished beef, resulting in a relative shortage and increased price for lower quality beef used in hamburgers and processed meats. In the meantime, producers in the low cost range areas of Australia, New Zealand, Argentina, and other countries not only took over the European market, but began extensive shipments of grass-finished beef into the United States. Partially instrumental in these developments were various subsidies and incentive programs which encouraged production in some of the exporting countries, but the main factors were relative costs of production and meat prices in the respective countries. Export-import activity then and now attests to the importance of economic forces behind international trade.

Meat Trade

As may be seen in Table 14.1 beef and veal comprise more than ½ of total world red meat production, and 8% of production enters international trade. Pork comprises 40% of world red meat production, and 7% of

TABLE 14.1.   RED MEATS: PERCENTAGE OF TOTAL WORLD PRODUCTION AND PERCENTAGE MOVING IN INTERNATIONAL TRADE, 1974

Item	Beef and Veal	Pork	Lamb, Mutton and Goat Meat	Total
Percentage each specie is of total world red meat production	54	40	6	100
Percentage of production moving in international trade	8	7	15	10

Source: Calculated from USDA (1975).

it enters world trade. In contrast, lamb, mutton, and goat meat amounts to only 6% of world red meat production, but the proportion entering world trade is about double that of either beef or pork. International trade in red meat bears little resemblance to the free trade envisioned in the competitive model. Virtually every country has restrictions of one sort or another on meat trade.

Beef.—Table 14.2 shows the import-export movements of beef and veal for important producing and consuming countries. Beef exports of a country are influenced by production capacity, current weather conditions, the stage of the cattle cycle, and national economic and political policies. For many years Argentina was the leading exporter, but in the early 1970's Australia emerged as the leader. Argentina's decline was related to national policies which discouraged production and trade, and to unfavorable weather. Argentina has the capacity to produce and could regain the lead if economic incentives were provided.

The rank of various countries in beef exports changes some from year to year, but typically the same 6 or 8 countries are found near the top. In addition to Australia and Argentina these include New Zealand, Ireland, France, and Uruguay. The Netherlands, West Germany (the Federal Republic of Germany), and Denmark export substantial quantities of beef. Mexico and Central American countries individually export only relatively minor quantities, but their shipments have increased in recent years and collectively they are assuming a more important role in world trade. For the continent of Africa, data are available for only a few countries. Currently, African trade is relatively insignificant. However, if or when African countries gain control over disease and improve sanitation in packing plants, Africa will make its presence known in the beef trade.

The United States is the major importer of beef—a position she gained from the United Kingdom during the 1960's. The United Kingdom is a strong second, followed by Italy, West Germany, and Spain. West Germany, like several other European countries, has a sizeable trade in both imports and exports of beef and veal. Generally, these are not identical products. Some countries import wholesale cuts of fresh meat which are processed and reshipped in the processed form. In recent years Japan and the USSR have joined the top 10 as importers of beef.

Pork.—Denmark, long renowned for superior quality bacon and hams, is the chief exporter; and her neighbors, the Netherlands, Belgium, and Luxembourg, are important exporters (Table 14.3). These countries have for many years carried out effective breeding, feeding, and processing programs with rigidly enforced specifications designed to produce bacon, hams, and other pork products that meet the highly discriminating British consumer standards. Three eastern European countries, Poland, Yugoslavia, and Hungary, are surplus pork producers and move substantial quantities in the international market. Ireland is one of the few European countries which has a substantial net surplus of both pork and beef. The determination of these European countries to participate in international meat trade, coupled with the powerful economic forces alluded to

TABLE 14.2.   BEEF AND VEAL[1]: INTERNATIONAL TRADE IN SELECTED COUNTRIES, AVERAGE 1967–71, ANNUAL 1972–75 IN THOUSANDS OF METRIC TONS

Continent and Country:	Average 1967–71		1972		1973		1974		1975[2]	
	Exports	Imports	Exports	Imports	Exports	Imports	Exports	Imports	Exports	Imports
North America:										
Canada	38.2	67.6	42.4	98.9	40.9	104.0	26.8	83.5	21.3	90.7
Costa Rica	22.9	.3	34.2	.7	29.9	.2	41.5	.2	39.8	.2
Dominican Republic	4.7	.3	9.4	.4	10.0	.4	9.2	.4	5.3	.4
El Salvador	—	.1	5.2	.2	6.2	.1	7.7	—	3.4	—
Guatemala	16.1	—	20.5	—	22.8	—	18.3	—	15.7	—
Honduras	14.2	.1	24.5	.2	26.7	.2	17.3	.2	22.8	.2
Mexico	45.2	.6	58.3	.5	38.5	.5	19.6	.4	14.0	.3
Nicaragua	27.9	—	40.7	.1	34.1	.1	21.7	.2	29.6	.2
Panama	2.2	—	3.3	.4	1.4	.4	1.7	.4	3.6	.4
United States	19.0	731.0	28.2	905.4	41.2	916.3	26.7	746.6	24.2	808.3
Total North America	190.4	800.4	267.1	1,006.8	251.6	1,022.1	192.5	831.9	179.6	900.7
South America:										
Argentina	641.7	—	674.4	—	499.8	—	263.2	—	270.3	—
Brazil	97.0	1.4	228.0	1.5	170.2	1.6	88.8	53.2	84.4	30.0
Chile	.1	13.8	—	38.0	—	17.6	—	43.8	—	5.0
Colombia	7.3	—	28.0	—	36.2	—	22.2	—	15.1	—
Paraguay	34.3	—	40.0	—	44.7	—	24.0	—	34.0	—
Peru	—	8.9	—	10.4	—	8.3	—	7.0	—	8.8
Uruguay	106.0	—	145.5	—	109.1	—	107.6	—	100.9	—
Total South America	886.4	24.2	1,115.9	49.9	860.1	27.5	505.9	104.1	504.8	43.9
Europe:										
Western:										
EC:										
Belgium-Luxembourg	23.4	31.1	31.2	43.9	34.6	48.1	32.7	28.7	36.5	36.6
Denmark	102.8	1.5	84.2	1.0	105.8	1.1	121.5	1.0	144.6	1.6
France	151.7	66.9	159.3	171.2	177.2	186.1	310.0	129.7	346.7	170.1
Germany, Federal Republic of	45.1	207.4	48.5	341.1	79.1	329.6	121.2	243.1	142.8	271.3
Ireland	162.0	.3	144.0	.9	165.9	.7	248.3	1.3	332.2	2.6

TABLE 14.2 (Continued)

Italy	3.8	308.3	6.9	349.9	4.3	453.7	4.0	305.9	5.6	329.1
Netherlands	98.6	51.9	116.8	75.2	116.8	80.1	133.3	45.1	134.3	51.2
United Kingdom	8.7	446.0	54.4	482.2	70.7	472.0	67.9	398.6	123.9	349.5
Total EC	596.1	1,113.2	645.3	1,465.4	754.4	1,571.4	1,038.8	1,153.5	1,266.6	1,212.1
Austria	3.7	9.2	6.2	12.7	6.9	13.8	4.6	2.3	6.1	2.2
Finland	7.4	.2	6.0	1.7	—	10.1	8.5	—	—	2.4
Greece	—	56.2	—	51.5	—	74.8	—	33.7	.3	49.1
Norway	1.9	2.9	.9	5.9	.8	4.7	1.1	2.0	.4	5.4
Portugal	.1	16.9	.9	32.1	.2	20.4	.1	36.4	.1	23.7
Spain	.3	96.3	—	84.4	1.3	81.3	1.9	19.9	1.1	31.9
Sweden	21.9	12.0	11.4	9.8	4.4	9.4	3.7	11.1	.9	16.0
Switzerland	3.2	35.5	2.4	42.9	2.6	41.3	2.5	20.1	2.1	14.9
Total Western Europe	634.7	1,342.4	673.3	1,706.5	770.7	1,827.2	1,061.3	1,279.1	1,277.6	1,357.6
Eastern:										
Bulgaria	9.7	9.6	8.6	1.6	9.2	1.7	3.2	10.2	3.0	10.1
Czechoslovakia	18.4	63.8	13.5	33.1	13.4	80.5	4.2	35.2	4.0	35.0
German Democratic Republic³	1.7	61.2	4.4	42.4	1.1	46.2	.3	19.7	.2	20.0
Hungary	33.6	8.7	32.4	8.8	35.7	13.2	33.1	4.1	73.4	10.0
Poland	35.8	5.6	31.2	19.0	47.7	17.4	70.0	2.0	71.0	2.0
Yugoslavia	74.8	1.6	56.3	3.1	69.8	7.8	40.9	3.2	43.7	—
Total Eastern Europe	173.9	150.4	146.4	108.0	176.9	166.8	151.7	74.3	195.3	77.1
Total Europe	808.6	1,492.8	819.7	1,814.5	947.5	1,993.9	1,212.9	1,353.5	1,472.9	1,434.7
USSR³	72.1	33.6	32.1	32.1	30.4	15.9	26.4	292.8	23.0	371.5

TABLE 14.2. (Continued)

Africa:										
Rep. South Africa	30.0	22.4	74.6	20.0	74.5	37.3	26.4	33.7	26.0	30.3
Total Africa	30.0	22.4	74.6	20.0	74.5	37.3	26.4	33.7	26.0	30.3
Asia:										
China, Rep of (Taiwan)	.1	.1	—	1.1	—	1.2	—	1.2	—	25.5
Iran	—	.2	—	.2	—	.4	—	5.4	—	15.0
Israel	—	35.6	—	13.6	—	14.1	—	16.0	—	16.5
Japan	—	34.0	—	87.4	—	194.1	—	83.6	—	70.3
Korea, Rep of	—	.5	.2	.7	1.7	2.2	1.4	.5	.1	.2
Philippines	—	12.6	—	4.9	—	.7	—	2.4	—	10.1
Turkey	.1	.2	.5	—	.1	—	.1	—	.1	—
Total Asia	.3	83.2	.7	108.0	1.8	212.8	1.6	109.0	.2	137.8
Oceania:										
Australia	451.9	—	749.3	—	883.8	—	503.3	—	751.9	—
New Zealand	221.2	—	264.8	—	287.1	—	259.4	—	300.7	—
Total Oceania	673.1	—	1,014.1	—	1,170.9	—	762.7	—	1,052.6	—
Total Selected Countries	2,660.9	2,456.6	3,324.2	3,031.2	3,336.9	3,309.5	2,728.3	2,725.0	3,259.0	2,918.9

Note: A Dash denotes no trade or trade less than half the unit shown.
Source: USDA (1977C).
[1]Carcass weight equivalent basis: excludes fat, offals, and live animals.
[2]Preliminary.
[3]Estimate based on trading partner data.
Foreign Agricultural Service. Prepared or estimated on the basis of official statistics of foreign governments, other foreign source materials, reports of U.S. agricultural attachés and foreign service officers, results of office research, and related information.

TABLE 14.3 PORK¹: INTERNATIONAL TRADE IN SELECTED COUNTRIES, AVERAGE 1967–71, ANNUAL 1972–75 IN THOUSANDS OF METRIC TONS

Continent and Country	Average 1967–71		1972		1973		1974		1975²	
	Exports	Imports	Exports	Imports	Exports	Imports	Exports	Imports	Exports	Imports
North America:										
Canada	32.3	21.4	52.6	21.3	56.9	24.9	41.7	32.7	40.8	46.2
United States	40.4	192.7	48.3	230.9	77.6	232.2	47.4	213.2	95.7	194.6
Total North America	72.7	214.1	100.9	252.2	134.5	257.1	89.1	245.9	136.5	240.8
South America:										
Argentina	2.4	—	1.7	—	11.3	—	1.6	—	3.3	—
Brazil	.9	.1	.6	.1	3.2	.2	3.0	.1	5.7	.1
Chile	—	.2	—	6.3	—	2.5	—	11.5	—	1.0
Peru	—	.6	—	.3	—	.5	—	.5	—	.5
Total South America	3.4	.9	2.2	6.7	14.5	3.2	4.6	12.1	8.9	1.6
Europe:										
Western:										
EC:										
Belgium-Luxembourg	109.0	19.5	194.5	14.0	223.0	16.1	247.7	16.9	231.3	21.8
Denmark	540.2	1.5	570.4	2.4	541.0	1.1	558.8	.9	549.6	.6
France	16.0	173.4	19.1	185.7	22.7	194.9	25.7	206.0	24.8	208.7
Germany, Federal Republic of	20.8	109.6	18.4	256.1	13.7	263.1	16.6	309.7	18.8	319.8
Ireland	53.7	.2	68.9	1.4	47.8	.6	37.5	1.9	19.4	2.9
Italy	9.1	95.2	15.8	151.5	13.4	188.2	17.5	227.1	22.8	239.8
Netherlands	304.8	9.6	373.3	12.4	430.1	10.7	454.8	23.5	433.3	11.6
United Kingdom	10.9	658.2	6.0	664.9	13.3	565.5	24.0	522.5	9.1	439.2
Total EC	1,064.6	1,067.2	1,266.6	1,288.5	1,305.2	1,240.1	1,382.7	1,308.5	1,309.1	1,244.4
Austria	1.9	2.4	.2	7.8	.1	13.8	.5	3.9	.3	3.7
Finland	12.6	.3	20.1	.1	9.2	.8	7.7	—	.1	4.1
Greece	—	5.2	.1	4.8	—	10.9	—	6.3	—	.8
Norway	.6	5.2	3.6	3.3	4.0	4.6	.8	7.8	.3	8.1
Portugal	.4	5.3	.5	12.8	1.1	5.6	.8	11.8	.7	6.5
Spain	1.2	5.9	.8	82.3	1.7	44.4	2.5	14.5	1.6	49.2

TABLE 14.3. (Continued)

Sweden	37.6	18.1	66.5	23.2	51.3	19.1	45.3	24.5	34.4	19.2
Switzerland	2.8	9.9	2.7	7.9	3.8	8.5	3.4	5.2	2.6	3.4
Total Western Europe	1,121.7	1,119.6	1,351.0	1,430.6	1,376.3	1,347.2	1,443.7	1,382.6	1,349.1	1,339.5
Eastern:										
Bulgaria	14.5	5.3	10.9	6.6	9.3	11.8	3.8	25.5	8.0	25.0
Czechoslovakia	1.8	20.1	1.4	3.6	1.2	4.1	.6	5.4	.6	5.0
German Democratic Republic	19.8	9.8	15.6	11.3	23.9	—	24.8	5.5	25.0	3.0
Hungary	29.0	24.8	51.4	2.9	16.8	13.0	64.4	14.6	51.8	2.0
Poland	146.3	64.9	148.4	48.2	159.7	38.5	145.9	3.8	126.1	14.4
Yugoslavia	51.4	8.2	61.4	4.8	33.4	36.6	27.6	12.9	53.4	3.2
Total Eastern Europe	262.8	133.0	289.2	77.3	244.2	104.0	267.1	67.7	264.9	52.6
Total Europe	1,384.4	1,252.6	1,650.2	1,507.9	1,620.6	1,451.2	1,710.8	1,450.4	1,614.0	1,392.1
USSR[3]	51.4	11.5	49.9	47.8	68.1	63.1	52.7	67.9	40.5	46.1
Africa:										
Rep South Africa	5.0	2.6	2.0	1.0	4.9	1.7	5.8	2.1	6.0	1.3
Total Africa	5.0	2.6	2.0	1.0	4.9	1.7	5.8	2.1	6.0	1.3
Asia:										
China, Rep of (Taiwan)	2.8	—	12.5	—	35.7	—	13.6	—	7.8	—
Japan	.1	21.6	.1	76.2	—	152.5	—	60.2	—	134.7
Korea Rep of	.2	.1	4.3	.6	3.4	1.4	5.4	.8	8.5	.7
Philippines	—	1.1	—	1.2	—	.9	—	1.1	—	.4
Total Asia	3.1	22.8	17.0	77.9	39.1	154.8	19.1	62.1	16.3	135.8
Oceania:										
Australia	3.1	—	14.5	—	18.5	—	2.6	—	5.7	—

TABLE 14.3. (Continued)

New Zealand..........:	1.0	1.2	.7	.2	.1				
Total Oceania........:	4.0	15.7	19.3	2.9	5.8				
Total Selected Countries....:	1,524.1	1,504.4	1,837.8	1,893.5	1,901.1	1,931.2	1,884.9	1,840.5	1,828.1

Note: A dash denotes no trade or trade less than half the unit shown.
Source: USDA (1977C).
[1]Carcass weight equivalent basis; excludes fat, offals, and live animals.
[2]Preliminary.
[3]Estimate based on trading partner data.
Foreign Agricultural Service. Prepared or estimated on the basis of official statistics of foreign governments, other foreign source materials, reports of U.S. agricultural attachés and foreign service officers, results of office research, and related information.

earlier, is reflected in the fact that their canned bacon, ham, and other meat products may be found on the shelves of food stores throughout the United States. The United States also exports significant quantities of pork products. These consist primarily of lard, variety meats, casings, etc.

Not included in Table 14.3 are lard and other fats. The United States is the major producer, user, and exporter of these products.

The United Kingdom is a clear leader in pork imports. West Germany, the United States, Italy, and France are substantial importers, but at a significantly lower level than the U.K. Japan has emerged as an important importer in recent years. Even though the U.S. is listed as a leader in both exports and imports of pork, she imports 3 to 4 times more than she exports.

Lamb, Mutton, and Goat Meat.—Import-export movements of lamb, mutton, and goat meat are shown in Table 14.4. Goat meat is a minor fraction of the total, but it is an important meat in certain countries—particularly in some Near Eastern and African countries not shown in these tabulations. New Zealand is in a class by itself in the exportation of lamb and mutton—accounting for over 50% of all international shipments. Australia is the next most important exporter. Together, these two "down-under" countries move approximately 85% of the lamb and mutton. However, Argentina also exports sizeable quantities.

The chief importers of lamb, mutton, and goat meat are the United Kingdom and Japan. Additional important importers are France, Greece, Canada, and the United States. The USSR has increased imports substantially in recent years.

Horsemeat.—Horsemeat is important in the United States only as a source of pet food, but it is commonly used as human food in northern European countries and in Japan. Japan is the leading importer followed by Belgium, Luxembourg, Netherlands, and France. Argentina and Brazil are the outstanding sources of horsemeat—accounting for more than one-half of international shipments. Other South American countries, Mexico, Canada, and Poland also are major suppliers.

Cattle Hides, Calf and Kip Skins

Often overlooked in discussions of the livestock economy is the extensive and important trade in hides and skins. The production and availability of hides and skins parallel the production of live animals. Expanding world cattle production has increased the availability of these

TABLE 14.4. LAMB, MUTTON AND GOAT MEAT[1]: INTERNATIONAL TRADE IN SELECTED COUNTRIES, AVERAGE 1967-71, ANNUAL 1972-75 IN THOUSANDS OF METRIC TONS

Continent and Country	Average 1967-71 Exports	Imports	1972 Exports	Imports	1973 Exports	Imports	1974 Exports	Imports	1975[2] Exports	Imports
North America:										
Canada	.2	45.1	.3	48.0	.1	34.6	.1	21.9	.1	25.3
Mexico	—	.3	—	.9	—	.6	—	.4	—	.3
United States	1.1	58.6	.9	67.1	1.2	23.6	1.8	11.8	1.8	12.2
Total North America	1.2	104.0	1.2	116.0	1.3	58.8	1.9	34.1	1.9	37.9
South America:										
Argentina	52.2	—	16.8	—	28.4	—	24.0	—	22.0	—
Brazil	.4	—	1.6	—	.5	—	.5	.9	2.5	.5
Chile	.4	—	—	—	—	.1	—	9.8	—	1.0
Peru	—	8.2	—	6.7	—	3.6	—	5.0	—	3.9
Uruguay	12.9	—	2.9	—	2.7	—	3.7	—	9.5	—
Total South America	65.9	8.2	21.4	6.7	31.6	3.7	28.2	15.7	34.0	5.4
Europe:										
Western:										
EC:										
Belgium-Luxembourg	2.1	4.6	.3	7.4	.4	8.0	.1	8.1	1.3	.1
Denmark	.3	.4	.1	.8	.1	1.0	.1	.9	.1	1.5
France	.1	25.6	.2	39.2	.2	46.9	.4	43.9	.4	51.6
Germany, Federal Republic of	2.0	4.8	2.1	6.6	1.7	12.4	2.1	10.0	5.6	19.6
Ireland	11.9	—	11.8	—	11.7	—	11.1	—	11.7	—
Italy	.2	6.8	.4	10.4	.2	11.6	—	6.3	—	12.1
Netherlands	8.0	1.4	9.6	2.1	8.7	1.7	12.9	1.0	14.4	1.3
United Kingdom	9.0	357.5	23.5	355.9	27.9	271.0	27.7	215.2	34.2	244.7
Total EC	33.5	401.0	48.1	422.5	51.0	352.6	54.5	285.3	67.8	330.9
Greece	—	40.8	—	46.8	—	41.0	—	7.8	—	12.0
Iceland	4.1	—	1.9	—	3.3	—	3.2	—	2.9	—
Norway	—	.8	—	2.2	.1	2.3	—	3.3	—	4.3
Portugal	—	—	—	.4	—	.2	—	.6	—	.5

TABLE 14.4. (Continued)

Spain	.5	.3	.7	.8	1.7	2.6	.6	3.5	.4	1.2
Sweden	—	.4	—	.9	—	1.2	—	.7	—	.6
Switzerland	—	3.1	—	3.8	—	4.7	—	4.3	—	3.0
Total Western Europe	38.2	446.6	50.7	477.3	56.1	404.5	58.3	305.5	71.1	362.6
Eastern:										
Bulgaria	3.9	1.3	5.4	1.0	4.6	1.0	8.0	1.0	8.0	1.0
Czechoslovakia	—	1.3	—	.9	—	.7	—	1.0	—	1.0
Hungary	1.8	1.8	2.8	.4	5.2	.6	3.8	.2	4.0	.2
Poland	.1	.5	—	—	—	—	—	—	—	—
Yugoslavia	3.9	.4	3.4	.4	2.9	—	2.4	1.5	3.3	—
Total Eastern Europe	9.6	5.3	11.5	2.8	12.7	2.3	14.1	3.7	15.3	2.2
Total Europe	47.9	451.8	62.2	480.1	68.9	406.8	72.4	309.2	86.3	354.8
USSR³	.2	18.7	—	8.0	—	15.4	—	95.0	—	36.6
Africa:										
Rep South Africa	.5	.2	.5	—	.4	1.3	.7	.1	.7	.1
Total Africa	.5	.2	.5	—	.4	1.3	.7	.1	.7	.1
Asia:										
Iran	—	5.8	—	7.4	—	12.2	—	18.8	—	35.0
Japan	—	148.1	—	211.9	—	187.0	—	126.0	—	183.1
Korea Rep of	.1	.2	3.2	6.3	6.0	10.0	9.3	16.3	10.4	17.2

TABLE 14.4. (Continued)

Philippines	—	.2	—	.2	—	.1	—	.1	—	.1
Turkey	2.8	—	4.5	—	6.4	—	5.3	—	5.0	—
Total Asia	3.0	154.3	7.7	225.8	12.4	209.2	14.6	161.1	15.4	235.4
Oceania:										
Australia	250.0	—	373.2	—	229.9	—	139.6	—	242.1	—
New Zealand	457.7	—	470.9	—	436.0	—	395.9	—	405.3	—
Total Oceania	707.7	—	844.1	—	665.8	—	535.6	—	647.4	—
Total Selected Countries	826.3	737.3	937.1	836.8	780.4	695.2	653.3	615.1	785.7	670.1

Note: A dash denotes no trade or trade less than half the unit shown.
Source: USDA (1977C).
[1]Carcass weight equivalent basis; excludes fat, offals, and live animals.
[2]Preliminary.
[3]Estimate based on trading partner data.
Foreign Agricultural Service. Prepared or estimated on the basis of official statistics of foreign governments, other foreign source materials, reports of U.S. agricultural attachés and foreign service officers, results of office research, and related information.

products and in spite of stiff competition from synthetics, a relatively strong world-wide demand exists for hides and skins.

Prior to the 1960's, Argentina was the leading hide exporter, but in recent years that role has been taken over by the United States. Together, the United States and Argentina dominate exports. Other leading suppliers are Canada, Australia, West Germany, Spain, Yugoslavia, Poland, Czechoslovakia, and the Soviet Union.

In recent years the leading producers of calf and kip skins have been the United States, France, Argentina, West Germany, New Zealand, Australia, and Italy. However, in exports the leaders have been the Netherlands, France, the United States, New Zealand, and West Germany.

A major change among importers of skins has been the recent rapid rise to leadership by Spain. Other important importers are Japan, West Germany, and Italy.

Wool Trade

World wool production reached a peak about 1970, then turned downward (Fig. 14.3). U.S. production has been on a decline for many years (Table 14.5) despite a government incentive program designed to encourage domestic production. Congress enacted a law in 1954 providing for incentive payments to wool growers with a stated purpose of maintaining U.S. production at least 300 million pounds a year. This action was prompted by a belief that wool is a strategic war material. But, producers have not responded to the incentives. Lack of labor on the big ranches is given as a chief cause. On family farms, sheep, while consistently returning more income per dollar invested than other livestock, have not been able to stem the drift to other livestock and crop enterprises.

In the period following World War II domestic demand for wool was weakened by wide acceptance of man-made fibers (domestic and imported) and from imported wool and wool fabrics. The importation of raw wool remained at a relatively high level until the late 1960's, but then a precipitous decline set in. The same pattern applied to the foreign trade balance of wool textile products, especially for apparel wool (Table 14.5).

Wool produced in the United States is used chiefly in the fabrication of clothing, blankets, etc. Wool used for rugs is all imported. Australia and New Zealand are the chief suppliers of wool, followed by South Africa and countries of South America.

U.S. MEAT ANIMAL IMPORTS AND EXPORTS

The economics of transporting meat compared to live animals limits the international movement of slaughter animals to trivial numbers. There

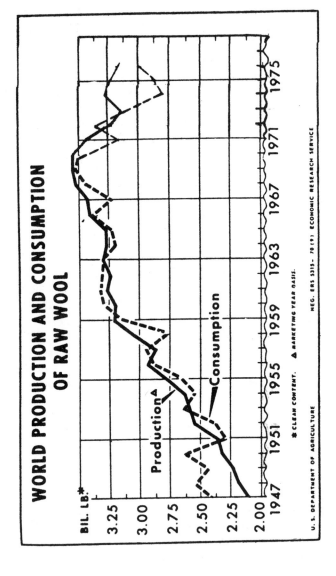

FIG. 14.3.   WORLD PRODUCTION AND CONSUMPTION OF RAW WOOL
(1971 and later years added by author from USDA data)

TABLE 14.5.   WOOL: DOMESTIC PRODUCTION, IMPORTS OF RAW WOOL AND THE FOREIGN TRADE IMPORT BALANCE OF WOOL TEXTILE PRODUCTS, CLEAN BASIS, U.S. 1976

| Year | Domestic Production[1] | | | Imports of Raw Wool[2] | | | Foreign Trade Import Balance of Wool Textile Products[3] | | |
	Shorn	Pulled	Total	Duti-able	Duty-free	Total	Ap-parel Wool	Carpet Wool	Total
				(Million Lb)					
1960	119.4	25.2	144.6	74.3	153.9	228.2	98.9	28.5	127.4
1961	116.6	25.9	142.5	90.3	157.3	247.7	95.2	27.7	122.9
1962	111.0	22.4	133.4	125.8	143.5	269.2	112.3	29.0	141.3
1963	104.6	21.6	126.2	109.2	168.0	277.2	125.4	21.5	147.0
1964	101.3	18.3	119.6	98.4	113.9	212.3	107.1	27.0	134.1
1965	96.1	17.0	113.1	162.6	108.9	271.6	122.6	21.4	144.0
1966	93.0	17.6	110.6	162.5	114.6	277.2	117.9	16.2	134.2
1967	90.1	16.3	106.5	109.1	78.2	187.3	105.4	9.4	114.8
1968	84.6	14.9	99.6	129.7	119.6	249.3	128.1	8.5	136.6
1969	79.1	12.5	91.5	93.5	95.7	189.2	112.6	8.1	120.8
1970	77.1	11.1	88.2	79.8	73.3	153.1	102.2	6.9	109.1
1971	76.4	8.7	85.1	42.7	83.9	126.6	69.7	8.0	77.7
1972	75.8	7.1	82.9	24.8	71.8	96.6	50.8	11.2	62.0
1973	69.3	5.8	75.1	18.0	39.9	57.9	45.0	11.6	56.6
1974	63.4	4.2	67.6	11.8	15.1	26.9	38.3	10.0	48.3
1975	57.3	3.9	61.1	16.6	17.0	33.6	37.5	9.5	47.0
1976[4]	58.4	2.9	61.3	38.4	19.1	57.5	71.6	11.8	83.4

Source: USDA (1977B) and earlier issues.
[1]Production as reported converted on basis of 45% yield of shorn wool for 1960 through 1963, 47.7% yield 1964 to 1975 and 53% for 1976.
[2]Imports of raw wool for consumption.
[3]Raw wool content of semiprocessed and manufactured wool textile products.
[4]Preliminary.

are, however, small but significant movements of breeding stock and as far as the United States and her neighbors are concerned, varying but often substantial movements of feeder cattle.

Beef Cattle

Imports of beef cattle are shown in Table 14.6. In the past 10 yr imports of breeding cattle have ranged from 6,000 to almost 25,000 head. This includes dairy as well as beef stock. Compared to the approximately 50 million head of breeding cows in the United States, it can be seen that from the standpoint of total numbers, breeding cattle imports are relatively insignificant. Yet for particular herd owners the importation of select stock undoubtedly is of ultimate importance. Were it not for the ban on importation of cattle from foot and mouth disease areas, the importation of breeding cattle would be larger.

Table 14.6 also shows the number of "other" cattle imported in recent years. A majority of these are feeder cattle and came almost entirely from Canada and Mexico, with a substantial predominance from Mexico

TABLE 14.6.   IMPORTS OF MEAT ANIMALS, UNITED STATES, 1960–1976

Year	Breeding Cattle[1] (Hd)	Cattle Other Cattle[2] Canada (Hd)	Mexico (Hd)	Other (Hd)	Total (Hd)	Sheep and Lambs[1] (Hd)	Hogs[1] (Hd)
1960	18,624	233,415	390,517	—	623,932	49,76	6,162
1961	19,922	454,717	543,064	—	997,781	979	3,151
1962	17,773	464,856	751,885	—	1,216,741	20,84	3,277
1963	18,562	236,122	585,342	376	821,840	3,091	4,323
1964	17,735	183,307	330,901	700	514,908	12,680	5,094
1965	17,647	560,105	535,260	—	1,095,365	19,073	14,453
1966	18,876	475,587	584,085	327	1,060,002	8,310	22,698
1967	11,387	227,042	500,418	40	727,500	12,403	34,926
1968	14,915	306,117	702,308	27	1,008,452	26,579	21,678
1969	20,855	187,733	810,387	58	998,178	22,805	13,430
1970	24,762	170,947	936,583	219	1,107,749	11,716	67,832
1971	21,624	180,721	752,209	215	933,145	5,454	77,532
1972	17,441	227,850	915,767	250	1,143,867	13,765	89,032
1973	15,541	330,340	672,654	2,125	1,005,119	9,514	87,615
1974	12,082	111,226	434,700	721	546,687	900	196,347
1975	6,391	183,390	196,043	1,189	380,622	3,497	29,768
1976	11,225	448,156	507,768	869	956,793	4,607	45,577

[1]Source: USDA (1977A) and earlier issues.
[2]Source: USDA (1978) and earlier issues.

in most years. When U.S. slaughter cattle prices are attractive, Canada and Mexico are induced to ship in some slaughter animals.

Exports of cattle are shown in Table 14.7. The exportation of "other" (i.e., feeder) cattle is negligible. Most of the breeding cattle exported from the U.S. go to Canada and Mexico. However, it is not unusual for annual shipments to go to 30 different countries.

Sheep and Hogs

Imports of live sheep and lambs, and hogs are somewhat variable and relatively small in numbers (see Table 14.6). In the early 1960's wide publicity was given to the importation of slaughter lambs from New Zealand, but this soon dwindled. Most imports of sheep and hogs are breeding animals.

Exports of sheep, though small, have increased substantially during recent years (Table 14.7). World trade in live hogs is of minor importance—virtually all are breeding stock. The U.S. is a leader in exportation of hogs and, while the numbers are small, for particular breeders this is an important business.

MAJOR SOURCES OF U.S. MEAT AND LIVESTOCK PRODUCT IMPORTS

Table 14.8 shows the chief suppliers of meat to the United States. In recent years Australia alone has furnished about 40% of the U.S. imports

TABLE 14.7.   MEAT ANIMAL EXPORTS: NUMBER OF CATTLE, SHEEP AND HOGS EXPORTED, UNITED STATES, 1962–1976

Year	For Breeding (Hd)	Cattle Other Cattle (Hd)	Total (Hd)	Sheep[1] (Hd)	Hogs (Hd)
1962	18,039	1,273	19,312	37,336	3,330
1963	22,428	727	23,155	31,493	3,899
1964	28,164	33,467	61,631	22,809	16,567
1965	32,380	21,791	54,171	25,315	12,180
1966	26,946	8,371	35,317	59,054	9,649
1967	31,749	23,573	55,322	120,733	12,932
1968	31,917	3,808	35,725	117,677	13,714
1969	34,063	5,123	39,186	106,237	18,620
1970	26,323	61,714	88,037	132,856	25,654
1971	33,271	59,685	92,956	213,806	17,347
1972	40,115	63,805	103,920	159,428	12,316
1973	79,939	192,642	272,581	204,339	16,802
1974	88,509	115,871	204,380	290,659	15,801
1975	71,421	124,339	195,760	339,246	15,960
1976	58,961	145,583	204,544	244,450	9,900

Source: USDA (1978) and earlier issues.
[1]Sheep only for 1962–1964. Beginning 1965 includes sheep, lambs, and goats.

of beef and veal. New Zealand supplied an additional 18%. Argentina, Brazil, Mexico, and Canada are all important suppliers, and the Central American countries have come up rapidly in recent years. Imports from Ireland, formerly a chief supplier, have dropped substantially since Ireland was admitted into the European Community trading bloc in 1973.

The bulk of imported beef is in frozen, boneless form (see Table 14.9) and is used primarily for processing purposes—including hamburgers. Not shown in either Table 14.8 or Table 14.9 are sizeable quantities of beef produced from imported feeder cattle. In some years this amounts to 1 million head. While that is only about 2½% of total cattle slaughter, beef from these cattle is 18% to 20% of total beef imports and the bulk is block beef rather than processing beef.

Lamb and mutton imports into the United States come almost exclusively from Australia and New Zealand (Table 14.8). As shown in Table 14.9 the importation of lamb has remained fairly stable over recent years, but incoming mutton and goat meat has declined sharply—down from 104 million lb in 1965–1969 to less than 2 million lb in 1977.

Denmark, Poland, Netherlands, and Canada supply most of the U.S. pork imports (Table 14.8). This principally is canned bacon, hams, and shoulders (Table 14.9).

MAJOR DESTINATIONS OF U.S. EXPORTS

As may be seen in Table 14.10, U.S. exports of meat go chiefly to its territories, but Canada is a large and expanding outlet. In recent years

TABLE 14.8. MEAT IMPORTS: UNITED STATES, BY COUNTRIES, 1970–1976

Product and Year	Imports, by Country of Origin, Product Weight in Million Pounds												Total imports in Million Pounds	
	Canada	Mexico	Argentina	Brazil	Denmark	West Germany	Poland	Netherlands	Ireland	Australia	New Zealand	All Other	Product Weight	Carcass Weight Equivalent
Beef and veal														
Avg 1965–1969	51.2	56.7	101.2	23.7	0.1	[1]	[1]	[1]	49.9	414.5	169.3	118.9	983.6	1327
1970	80.6	78.6	141.1	28.8	0.4	[1]	[1]	[1]	69.0	535.8	241.6	174.2	1350.1	1816
1971	80.1	79.1	88.4	63.0	2.2	[1]	[1]	[1]	64.0	505.4	241.8	186.7	1310.7	1756
1972	59.6	81.9	94.1	48.0	2.4	0.2	—	[1]	31.1	674.7	266.4	222.5	1480.9	1996
1973	56.3	67.0	81.5	46.2	2.2	1.2	—	[1]	22.0	697.9	291.3	231.1	1496.7	2022
1974	36.9	38.8	89.0	39.5	2.7	0.7	—	[1]	44.0	514.3	259.9	193.1	1217.9	1646
1975	21.4	29.8	56.2	34.9	2.9	0.1	—	—	6.8	681.2	276.8	204.6	1314.7	1782
1976[3]	84.4	52.8	95.0	73.0	3.0	0.3	—	—	4.5	679.0	270.9	223.2	1486.1	2005
Lamb and mutton														
Avg 1965–1969	0.3	—	—	—	—	—	—	—	—	57.4	11.2	0.2	73.5	126
1970	0.6	—	—	—	—	—	—	—	—	60.1	22.2	0.1	83.0	122
1971	[1]	—	—	—	—	—	—	—	—	58.0	12.4	0.1	70.5	103
1972	0.3	—	—	—	—	—	—	—	—	72.4	20.1	0.1	92.9	148
1973	0.1	0.3	—	—	—	—	—	—	—	17.8	21.9	0.2	40.3	53
1974	[1]	1.6	—	—	—	—	—	—	[1]	6.6	13.4	0.1	21.7	26
1975	—	—	—	—	—	—	—	—	—	5.5	19.5	0.7	25.7	27
1976[3]	[1]	—	—	—	—	—	—	—	—	7.3	27.2	0.7	35.2	26

TABLE 14.8. (Continued)

Pork														
Avg 1965–1969	52.9	¹	¹	—	105.0	1.5	54.1	70.7	0.9	0.1	¹	16.8	301.4	386
1970	63.2	¹	¹	—	120.6	1.4	56.0	86.7	0.1	0.3	¹	19.3	347.6	449
1971	69.4	—V¹	—	128.1	1.7	54.9	82.5	0.1	0.3	—	19.5	356.5	458	
1972	67.5	4.0	—	—	151.8	1.2	66.6	75.3	0.2	0.4	0.1	27.6	394.7	508
1973	68.2	—	—	—	138.7	1.2	61.4	93.9	0.2	2.2	0.1	32.6	398.5	514
1974	53.7	—	—	—	122.0	1.0	64.2	78.8	0.5	0.2	0.2	41.3	362.0	470
1975	37.3	¹	¹	—	91.1	0.7	80.3	70.0	0.2	0.1	¹	47.4	327.1	429
1976³	28.8	—	—	0.1	87.2	1.1	82.3	54.8	0.6	¹	—	62.7	317.5	420
Total²														
Avg 1965–1969	101.9	56.7	101.3	23.7	105.1	1.5	56.1	70.8	50.8	471.9	180.5	135.9	1361.8	1838
1970	144.6	78.6	141.1	28.8	144.3	2.3	56.2	88.1	69.1	597.3	264.0	195.3	1809.7	2387
1971	149.5	79.1	88.5	63.0	148.8	2.4	55.0	83.4	64.1	564.3	254.2	207.7	1760.0	2317
1972	127.5	85.9	94.2	48.0	172.1	2.2	66.7	75.7	31.3	747.9	286.7	251.7	1989.9	2653
1973	125.1	67.3	81.5	46.2	155.2	2.9	61.8	94.0	22.2	718.5	313.5	265.8	1953.7	2589
1974	91.0	40.4	89.0	39.6	136.4	2.2	64.6	79.0	44.6	521.5	273.5	235.4	1617.2	2142
1975	59.1	29.8	56.2	35.1	101.0	1.1	84.4	70.3	7.0	687.0	296.5	254.2	1681.7	2238
1976³	113.8	52.8	95.0	73.0	98.0	1.7	84.1	55.0	5.1	686.7	298.1	289.0	1852.3	2461

Source: USDA (1977A) and earlier issues.
¹Less than 50,000 pounds.
²Includes quantities of other canned, prepared or preserved meat n.e.s.
³Preliminary.

TABLE 14.9. U.S. IMPORTS AND EXPORTS OF LIVESTOCK AND LIVESTOCK PRODUCTS, AVG 1965–1969, ANNUAL 1972–1977

Item	Imports						
	Avg 1965–1969	1972	1973	1974	1975	1976	1977[1]
	Million Pounds						
Meat (carcass weight)							
Beef:							
Boneless, fresh or frozen	1,082.1	1,714.5	1,771.2	1,416.7	1,611.0	1,685.1	1,621.9
Fresh or frozen	21.6	12.3	18.9	10.7	7.5	21.0	20.5
Canned	143.9	139.8	131.3	131.8	89.3	214.7	172.4
Pickled or cured	1.1	0.7	0.5	0.7	1.2	2.1	1.1
Other processed	57.8	92.9	68.9	56.4	48.5	155.7	122.9
Total	1,306.6	1,960.2	1,990.8	1,615.8	1,757.5	2,078.6	1,938.8
Veal							
Fresh or frozen	19.8	36.1	31.2	30.5	24.3	22.0	23.9
Total (includes canned)	—	36.1	31.2	30.5	24.3	22.0	23.9
Pork							
Fresh or frozen	45.7	68.2	64.7	49.4	33.7	24.7	26.9
Hams and shoulders, not cooked	2.0	1.3	1.2	1.6	0.8	0.7	0.6
Hams and shoulders, canned	282.0	427.2	430.5	404.3	384.2	403.5	379.8
Other	56.6	41.4	36.7	32.3	20.6	40.2	32.1
Total	386.3	538.1	533.1	487.6	439.3	469.1	439.4
Lamb	21.3	37.3	27.3	17.8	24.6	34.6	21.1
Mutton and Goat	104.4	111.2	35.8	7.8	2.2	1.7	1.4
Total red meat	1,838.5	2,682.9	2,608.2	2,159.5	2,247.9	2,606.0	2,424.6
Variety Meats (product weight)	3.7	7.9	7.2	5.9	5.6	5.3	5.7
Animal Fats:							
Lard	[7]	0.3	[5]	0.3	0.1	0.1	—

TABLE 14.9. (Continued)

	Avg 1965–1969	1972	1973	1974	1975	1976	1977
Inedible tallow and grease[2]	3.4	3.7	4.3	4.0	2.2	1.7	3.5
Edible tallow and greases[3]	—	2.1	10.8	11.9	15.5	8.8	3.3
Mohair (clean content)	—	—	—	—	—	—	—
Wool (clean basis))							
Dutiable	131.5	24.8	19.6	11.8	16.6	38.4	36.2
Duty-free	103.4	71.8	40.7	15.2	17.0	19.1	16.8
Total wool	234.9	96.6	60.3	27.0	33.6	57.5	53.0
Hides and Skins, 1,000 pieces							
Cattle	305	292	709	520	950	962	932
Calf	409	87	216	200	68	46	29
Kip	404	174	222	239	154	48	51
Sheep and lamb	25,910	16,852	12,894	15,732	15,520	16,615	15,468
Live animals, number							
Cattle[4]	995,968	1,169,035	1,023,444	556,189	382,928	972,619	1,127,639
Hogs	21,437	89,032	87,615	196,347	29,768	45,577	43,030
Sheep and lambs	17,834	13,765	9,514	900	3,497	4,607	8,546

Source: USDA (1978) and earlier issues.

	Exports						
Item	Avg 1965–1969	1972	1973	1974	1975	1976	1977
Meat (carcass weight)							
Beef:							
Boneless, fresh or frozen							
Fresh or frozen	19.7	41.3	68.0	43.9	42.7	77.1	85.3

TABLE 14.9. (Continued)

Canned	2.5	2.0	1.5	2.7	0.7	1.0	1.0
Pickled or cured	12.9	10.5	9.1	4.3	0.9	0.9	1.0
Other processed	5.7	5.7	8.1	8.3	6.3	8.5	10.3
Total	40.7	59.5	86.7	59.2	50.6	87.5	97.6
Veal							
Fresh or frozen	0.7	2.0	3.4	3.1	2.2	2.4	5.0
Total (includes canned)	1.3	2.6	4.2	4.0	2.8	2.4	5.0
Pork							
Fresh or frozen	40.0	72.8	120.4	59.9	142.5	220.7	207.0
Hams and shoulders, not cooked	15.1	13.6	30.0	21.9	45.8	63.5	49.3
Hams and shoulders, canned	3.2	3.4	3.2	2.4	1.8	2.8	4.6
Other	25.4	22.8	23.8	24.4	26.2	29.2	32.9
Total	83.7	112.6	177.4	108.6	216.3	316.2	293.8
Lamb							
Mutton and Goat	2.4[6]	2.0	2.7	4.0	3.9	3.8	4.6
	—	—	—	—	—	—	—
Total red meat	128.2	176.7	271.0	175.8	273.6	409.9	401.0
Variety Meats (product weight)	224.8	254.1	281.9	295.9	293.6	379.7	381.5

TABLE 14.9. (Continued)

Animal Fats:							
Lard	206.8	164.4	113.3	161.4	87.7	180.6	182.0
Inedible tallow and grease[2]	2,086.5	2,349.3	2,297.0	2,619.7	2,009.0	2,330.7	2,885.4
Edible tallow and greases[3]	15.0	18.9	22.2	65.5	10.5	41.6	22.8
Mohair (clean content)	11.7	19.3	10.6	8.0	9.0	7.4	6.6
Wool (clean basis)							
Dutiable	—	—	—	—	—	—	—
Duty-free	—	—	—	—	—	—	—
Total wool	0.3	11.2	3.7	4.	7.7	1.1	0.4
Hides and Skins, 1,000 pieces							
Cattle	13,423	17,578	16,866	48,429	21,269	25,270	24,488
Calf	1,803	1,621	1,608	1,802	2,051	1,657	2,052
Kip	454	451	279	365	352	505	456
Sheep and lamb	3,266	5,872	5,792	4,897	4,571	3,409	3,273
Live animals, number							
Cattle[4]	43,948	103,920	272,581	204,380	195,926	204,544	106,997
Hogs	13,419	12,316	16,802	15,801	15,960	10,768	10,212
Sheep and lambs	85,803	159,428	204,339	290,659	339,246	244,450	205,149

shipments of beef and pork to Japan have increased substantially. Historically, the largest dollar export item among livestock products has been tallow, but in recent years hides and skins, variety meats, and red meats have made significant gains (Table 14.11). The value of lard, casings, and mohair exports has been on a declining trend for many years.

Quantities of export items are shown in Table 14.9. While the major items of U.S. export do not constitute a large fraction of the total quantity or value of livestock and meat produced, they nevertheless constitute an important fraction. Generally, these are items with a low demand preference in the United States. Without foreign outlets their value would be considerably less and this would have an adverse effect on U.S. livestock prices. In recent years the value of livestock and livestock products exported has equalled that of imports.

EFFECT OF IMPORTS ON U.S. LIVESTOCK PRICES

As pointed out in Chap. 13, rises in beef imports during the late 1950's and early 1960's were blamed for depressed cattle prices and led to enactment of the Meat Import Quota Law of 1964. Consumer concern over high meat prices in 1973 prompted the President of the United States to cancel import restrictions in view of "over-riding economic interests." Voluntary quotas which were negotiated during 1974–1977 allowed imports essentially at the maximum level provided under the Meat Import Quota Law. That was a period of cattle liquidation, high beef production, and disasterously low cattle prices. Cattlemen reacted with deep concern. Remedial proposals ranged from complete prohibition of imports to contra-cyclical quotas. The latter would liberalize quotas during periods of cyclically declining domestic beef production, and impose increasingly stricter quotas during periods of cyclically increasing domestic beef production

The major item in imported meat is frozen, boneless beef (see Table 14.9) which competes directly with U.S. boned beef from cull beef and dairy cows and bulls. It competes directly also with trimmings and increasingly larger portions of lean cuts from steer and heifer carcasses. Most imported beef is used in processed meats—including hamburger. Indirectly, imported beef competes with grain-fed beef, pork, lamb, and other foods. It is well known that one type of meat can be substituted for another meat or for another food.

In terms of demand concepts discussed in Chap. 3, we are concerned here with the effect that increased quantities of imported beef have, not only on identical of comparable U.S. products, but also on prices of sub-

TABLE 14.10. MEAT EXPORTS: UNITED STATES EXPORTS AND SHIPMENTS BY COUNTRIES, AVG 1966-1969, ANNUAL 1970-1976

Product and year	Exports, by destination, product weight											Shipments to territories[1]	Total exports and shipments	
	Canada	Mexico	France	Bahamas	West Germany	Jamaica	Japan	Netherlands	Venezuela	All other	Total		Product weight	Carcass weight equivalent
	Million pounds													
Beef and veal:														
Avg 1965-1969	12.5	0.3	0.5	5.5	0.2	1.1	0.3	0.3	[2]	10.5	31.3	33.5	64.8	92
1970	11.6	0.4	0.3	7.5	[2]	1.y	1.1	0.2	[2]	6.6	29.3	45.9	75.2	104
1971	24.5	0.2	0.3	7.0	0.2	1.8	1.7	0.2	—	6.1	42.0	50.3	92.3	121
1972	34.3	0.2	0.3	6.6	0.3	1.9	1.6	0.2	[2]	6.8	52.2	38.8	91.0	124
1973	34.6	0.3	0.6	7.0	0.2	1.4	24.8	0.2	[2]	10.0	79.1	45.4	124.5	152
1974	15.5	0.7	0.4	6.7	[2]	1.5	13.4	0.6	—	11.8	50.7	30.7	101.3	130
1975	7.9	0.9	0.1	6.0	0.2	1.3	17.7	0.8	[2]	10.7	45.6	57.1	102.7	124
1976[4]	19.0	1.2	0.1	5.2	0.5	0.9	34.3	1.3	[2]	18.1	80.6	62.9	143.5	171
Lamb and mutton														
Avg 1965-1969	0.3	[2]	[2]	0.6	—	[2]	[2]	—	[2]	0.7	1.6	1.5	3.0	6
1970	[2]	0.1	—	0.5	—	[2]	—	—	0.1	0.4	1.1	3.3	4.4	7
1971	0.1	0.1	—	0.6	[2]	[2]	—	—	[2]	0.5	1.3	3.5	4.8	8
1972	[2]	0.1	[2]	0.5	—	[2]	—	[2]	0.1	0.6	1.3	2.0	3.3	7
1973	0.2	0.1	—	0.9	—	[2]	[2]	[2]	[2]	0.5	1.7	1.0	2.7	6
1974	0.8	0.4	—	0.7	—	0.1	—	[2]	0.1	0.4	2.5	1.4	3.9	8
1975	1.2	0.4	—	0.7	—	0.1	—	[2]	0.1	0.4	2.9	2.2	5.1	8
1976[4]	1.2	0.3	[2]	0.7	—	[2]	[2]	[2]	0.1	0.8	3.1	1.7	4.8	8
Pork														
Avg 1965-1969	36.7	2.6	0.8	3.3	0.2	2.8	16.7	0.2	2.4	10.4	75.9	69.0	144.8	168
1970	23.5	2.7	0.1	3.5	0.1	1.3	16.2	[2]	1.1	12.7	61.2	85.5	146.7	177
1971	13.6	2.1	0.2	3.5	0.1	2.2	25.7	0.4	0.9	16.6	65.3	93.4	158.7	183
1972	31.6	1.2	0.1	3.6	0.1	1.5	46.3	0.2	0.7	14.0	99.3	94.4	193.7	223
1973	43.4	1.5	0.4	4.5	[2]	1.0	96.8	0.2	0.8	12.1	160.7	83.1	243.8	268
1974	51.0	1.6	0.3	4.6	—	1.3	21.5	0.1	1.1	13.1	94.6	77.4	172.0	196
1975	74.5	1.7	0.2	4.2	[2]	1.6	101.1	0.3	1.0	16.4	201.0	84.8	285.8	310
1976[4]	160.1	4.0	0.1	4.5	[2]	0.5	118.1	0.2	2.2	13.4	303.1	87.6	390.7	416
Total[3]														
Avg 1965-1969	52.1	3.8	1.4	10.8	0.8	4.6	17.5	0.6	2.5	31.0	125.2	136.8	259.9	265
1970	38.9	3.8	0.8	12.9	0.3	3.7	17.8	0.3	1.2	28.0	107.7	180.3	288.1	288
1971	42.6	2.8	1.7	12.5	0.7	4.2	28.3	0.6	0.9	30.7	125.0	190.8	315.8	312
1972	70.7	2.1	1.4	12.2	0.5	3.8	48.7	0.6	0.8	28.0	168.8	187.9	356.7	354
1973	84.3	2.5	2.6	14.0	0.3	3.0	124.3	0.5	1.0	30.5	263.0	168.9	431.9	426
1974	72.0	3.1	4.6	13.9	0.1	3.5	37.5	1.0	1.2	35.9	172.8	168.4	341.2	334
1975	88.6	3.6	0.8	12.8	0.3	3.3	120.6	1.2	1.2	34.5	266.9	182.8	449.7	442
1976[4]	187.5	6.3	0.2	12.8	0.6	1.5	154.5	1.6	2.4	41.9	409.3	199.2	608.5	595

Source: USDA (1977A) and earlier issues.

TABLE 4.11.   LIVESTOCK PRODUCTS: ANNUAL U.S. EXPORTS (MILLION DOLLARS), 1960–1976

Year	Red Meats (Excluding Offals)	Live Animals	Animal Fats Lard	Animal Fats Tallow	Hides and Skins	Other Animal By-Products Variety Meats (offals)	Other Animal By-Products Casings, Mohair, etc.	Total
1960	36.3	13.0	60.6	115.2	76.4	25.2	25.9	352.6
1961	35.7	11.6	46.7	134.6	86.1	27.2	29.8	371.7
1962	34.1	10.7	40.6	106.9	82.9	25.5	25.5	326.2
1963	53.0	13.5	48.5	123.7	75.0	31.9	26.7	372.3
1964	65.8	21.0	69.8	179.8	92.7	47.9	12.5	489.5
1965	46.9	21.8	30.2	193.2	108.5	56.0	15.9	472.5
1966	47.6	19.6	19.6	167.9	154.8	58.5	18.0	486.0
1967	47.7	28.3	18.6	157.3	127.4	57.0	13.9	450.2
1968	61.9	21.7	14.2	134.4	120.8	55.0	18.1	426.1
1969	93.9	26.3	25.4	139.0	151.7	61.7	19.9	517.9
1970	61.3	40.5	45.5	198.6	144.4	69.5	21.0	580.8
1971	71.9	46.1	33.3	232.7	155.1	78.2	19.0	636.3
1972	110.3	66.8	18.9	188.4	291.8	88.6	24.3	789.1
1973	22t.3	167.1	19.6	310.5	375.4	123.9	31.1	1252.9
1974	152.6	154.1	40.4	540.0	337.2	113.1	28.1	1365.5
1975	268.8	112.9	23.6	331.8	291.6	109.9	37.4	1176.0
1976	397.1	132.0	35.4	403.7	518.0	151.6	49.6	1687.4

Source: USDA (1977B) and earlier issues.

stitute products such as grain-finished beef and pork. This involves the concepts of price elasticity of demand, cross elasticity of demand, and cross price flexibility.

During the height of discussions over imported beef in 1963 the USDA (1963) published results of a study showing that imports of beef and veal have a greater effect on cow prices than on fed beef prices. More specifically the study indicated that a 20% increase in imports (from 1962 levels) would result in 80–90¢ per cwt reduction in utility grade cow prices. The same increase in imports would cause 50–60¢ decline in choice steer prices. It is generally recognized that, in the short run, changes in the supply of beef are the primary factor affecting prices. The USDA study showed that for the period 1948–1962 a 10% change in steer and heifer beef production was associated with a 13% change in the opposite direction of fed cattle prices, and a 23% change in the opposite direction of cow prices. Furthermore, a 10% change in cow beef production (domestic cow beef plus imported beef) was associated with a 3% change in the opposite direction of fed beef prices and a 7.5% change in the opposite direction of utility grade cow prices. These are average net changes after consideration of other factors which affect prices.

Purcell (1968) estimated the effect of imported beef and veal on prices and revenue for cattle and hogs on an annual basis from 1947 to 1966. Results are shown in Table 14.12. The United States was a slight net importer in 1947 but the effects were not measurable. Since 1948 the United States has been a net importer every year. According to Purcell,

TABLE 14.12.   ESTIMATED EFFECT OF NET IMPORTS OF BEEF AND VEAL ON PRICE
AND REVENUE FOR CATTLE AND CALVES, AND HOGS, UNITED STATES, 1947–1966

Year	Price ($/Cwt)	Cattle and Calves Revenue[1] (Million $)	(%)	Price[3] ($/Cwt)	Hogs Revenue[1] (Million $)	(%)
1947	–	–	–	–	–	–
1948	−1.01	− 201.3	− 3.5	−0.74	−106	−2.6
1949	−0.72	− 143.6	− 2.4	−0.43	− 65	−1.9
1950	−1.06	− 210.7	− 3.4	−0.61	− 97	−2.8
1951	−1.49	− 271.2	− 4.2	−0.98	−169	−4.5
1952	−1.25	− 247.6	− 3.8	−0.68	−119	−3.4
1953	−0.67	− 171.4	− 2.8	−0.36	− 55	−1.5
1954	−0.53	− 142.3	− 2.3	−0.26	− 39	−1.1
1955	−0.48	− 133.2	− 2.1	−0.16	− 27	−1.0
1956	−0.29	− 85.2	− 1.4	−0.07	− 12	−0.5
1957	−0.82	− 234.1	− 3.6	−0.29	− 47	−1.6
1958	−2.45	− 638.3	− 9.1	−1.04	−170	−5.3
1959	−2.78	− 714.3	− 9.7	−0.85	−160	−6.1
1960	−2.02	− 565.0	− 7.8	−0.62	−114	−4.1
1961	−2.69	− 770.0	−10.5	−0.87	−158	−5.3
1962	−3.70	−1,063.3	−13.7	−1.16	−221	−7.2
1963	−4.27	−1,292.1	−16.7	−1.17	−235	−7.9
1964	−2.59	− 880.3	−11.8	−0.67	−136	−4.6
1965	−2.21	− 770.3	− 9.8	−0.78	−142	−3.9
1966	−2.78	−1,004.2	−12.5	−1.05	−198	−4.9

Source: Purcell (1968).
[1] Estimated change in revenue to primary producers in the United States attributed to net imports with actual domestic output (slaughter).
[2] Based on estimated reduction in revenue to primary producers due to imports relative to actual revenue.
[3] Percentage increase in beef and veal due to imports (estimated cross price flexibility −0.91).

imports in 1963 pushed cattle and calf prices down $4.27 per cwt. This is an aggregate average price decline for all classes of cattle and calves. Evidence from the USDA and other studies would indicate that cow prices were depressed more than $4.27 and fed cattle prices less than that amount. Revenues of cattle and calf producers in 1963 were off almost $1.3 billion or 16.7% as a result of imports. In 1966, the effect on cattle and calf revenues again reached $1 billion. Cross relationships are shown in the effects of beef and veal imports on price and revenue of hogs (Table 14.12). The relatively heavy beef imports of the early 1960's depressed hog prices $1.16–$1.17 per cwt and reduced hog revenues to farmers by $235 million, or almost 8% in 1963. Thus, it is apparent why livestock producers were incensed as imports increased in the late 1950's and the early 1960's. Their concern has continued as imports continue to press upon quota levels provided by the Import Quota Act.

A series of studies during the 1970's (Graeber and Farris 1972, Jackson 1972, Ehrich and Usman 1974, Freebairn and Rausser 1975, and Folwell and Shapouri 1976) confirmed the direction of effect as shown in earlier studies, even though magnitude of effect was somewhat different. Davis (1977) summarized four of these studies as indicated in Table 14.13. All indicate a negative effect on cattle prices—ranging from $1.08 per cwt to $1.91 per cwt for cull cows. The effect on price of slaughter steers is shown to be $0.24 per cwt and $0.60 per cwt in respective studies, $1.16 per cwt for feeder cattle, and $1.41 per cwt for an aggregated "all cattle" price. These studies were essentially comparable in estimating price effect for a 200 million lb increase in imports—relative to average imports over the period studied. In this type of analysis results are not expected to be identical among studies. The particular statistical model and time period used can affect derived estimates. The results are consistent in indicating a substantial price impact, and as pointed out by Davis (1977), the impact may have been even greater during those years when U.S. production was highest (and prices were relatively low). During those years the market would have been operating beyond the point of average imports—in a relatively inelastic segment of the demand function where any incremental change in quantity would have a greater proportional

TABLE 14.13. ESTIMATED EFFECTS OF INCREASED BEEF IMPORTS ON CATTLE PRICES IN DOLLARS PER HUNDREDWEIGHT

Cattle classification	STUDY			
	Farris & Graeber*	Rausser & Freebairn**	Folwell & Shapouri**	Ehrich & Usman**
	Dollars per Hundredweight			
All cattle			−1.41	
Cull cows	−1.91	−1.09		−1.08
Slaughter steers	− .24	− .60		
Feeder calves		−1.16		

Source: Davis (1977).
*Estimated at one pound per capita (202 million pounds) increase in beef imports.
**Estimated at 200 million pounds increase in beef imports.

effect on price than the same change would have at the point of average imports. Two of these studies estimated the effects of imports on farm revenues. Folwell and Shapouri (1976) estimated that 1975 cattle revenues were reduced $1.6 billion compared to what would have prevailed if imports had been held to 1964 levels and that 1976 cattle revenues were reduced $1.8 billion. Freebairn and Rausser's (1975) estimates were $1.1 billion and $1.5 billion for 1975 and 1976 respectively.

Jackson (1972) estimated that elimination of meat quotas in 1975 would have resulted in a 1.2% decrease in high grade beef prices and a 2.5% decrease in low grade beef prices at retail. At the farm level his analysis indicated a 1.9% decrease in high grade beef and a 3.9% decrease in low grade beef. Jackson's general conclusion was, ". . . that the quota affords only very modest protection or benefits to the U.S. beef producer. When considered in the light of side effects the quota has on domestic consumption, in terms of higher retail meat prices, and on those countries as a means of improving the incomes of beef producers has very little in its favor."

TRADE PROMOTIONAL ACTIVITIES

International trade is not a passive proposition. On the contrary, it is a highly competitive, aggressive contest among nations and individual firms. While trade restrictions impede the international flow of meat and other products, most countries with exportable surpluses are actively engaged in efforts to stimulate trade. Among the tactics used are: (1) international trade fairs; (2) solo country exhibits; (3) trade missions; (4) in-store promotions—including demonstrations, prizes, etc.; (5) promotional activities by agricultural attachés or other governmental agents stationed in foreign countries; (6) direct contact and selling by representatives of individual firms; (7) governmental manipulation of import-export regulations; (8) export subsidies; and (9) the use of marketing boards or comparable agencies in expediting export movements.

Recipient countries also utilize a number of devices to encourage and/or discourage imports and exports. Among the approaches used are import tariffs, export taxes, import and export licensing, exchange controls, embargos and import trade missions. While exporting countries are busy courting prospective importers, the importers are busy attempting to diversify their sources of supply.

Following is a summary of promotional activities of a group of selected countries.

Australia

Australia's greatest promotional expenditures are in wool marketing. She is the major contributor to the International Wool Secretariat—an international pooled effort in promoting the sale and use of wool. Other principal contributors to this organization are New Zealand and South Africa. In addition, Australia maintains Trade Commissioners in a number of overseas posts and participates in trade fairs in a number of importing countries. One of its principal meat promotional agencies is the Australian Meat Board. Operational funds are derived from a levy on livestock slaughtered. The Board maintains offices in London, New York, and Tokyo, and works in conjunction with the Australian Overseas Trade Publicity Committee. The central government matches Board funds for promotional proposals approved by the Committee.

Canada

Canada's promotional efforts in agricultural products are directed primarily at wheat sales. However, livestock and meat receive some attention in trade missions, trade fairs, exhibitions, and advertising and other publicity.

Denmark

Denmark carries on an aggressive overseas marketing program. Although not all of the following agencies spend equal effort on the promotion of livestock products, the listing indicates a well-planned, comprehensive approach to overseas marketing. A Traveling Ambassador for Agriculture was appointed in 1967. A marketing office is maintained in Beirut covering a number of Near East countries. Danish Food Centers are operated in London, Manchester, and Glasgow. These centers are partially financed by the Danish Bacon Factories Export Association. A Danish Food Festival and Danish Food Fair have operated in Japan. Denmark also has participated in various food fairs in Europe. Use is made of television and press advertising in the United Kingdom, Germany, and Japan.

Netherlands

The Netherlands also maintains a relatively extensive export marketing program. Most of the funds are derived from Marketing Board levies, but the central Ministry of Agriculture also contributes. The Netherlands participates in numerous food fairs, carries out substantial TV advertising programs, and in-store promotions. To foster the sale of breeding stock in less developed countries, the Netherlands sends experienced herdsmen to acquaint local herdsmen with the proper care and feeding of imported stock.

New Zealand

Like Australia, New Zealand is a heavy contributor to the International Wool Secretariat. New Zealand has a Wool Marketing Board but, other than fund contributions to the Secretariat its activities are directed primarily to domestic affairs. A Meat Board is active in promoting overseas sales of meat and meat products. The Board maintains an office in London which covers the United Kingdom and Europe. Promotional activities also are carried on by the Board in Canada and in Far Eastern countries. Activities include press and TV advertising, retail store display materials, demonstrations, and participation in food fairs. An office titled the Meat Export Department Co., Ltd., is maintained in Canada.

South Africa

South Africa is an important exporter of wool, and major emphasis is placed on that commodity. Minor amounts are spent for promotion of mohair and meat. The principal activities include trade missions, trade fairs, TV advertising, and in-store promotions.

European Community

The European Community (EC), generally referred to as the Common Market, is an organization of European nations formed under the so-called Treaty of Rome of January, 1958. Under this agreement, Belgium, France, Italy, Luxembourg, Netherlands, and West Germany formalized a comprehensive arrangement for cooperation in economic, social and political activities. Great Britain, Denmark, and Ireland were admitted in 1973. Of particular interest to international trade, the EC plan provides for a common policy whereby barriers are to be removed (over a period of time) for trade among the signatories, but trade with outside countries is strictly regulated. A primary purpose of the regulation is the protection of EC producers and the encouragement of local production. Consumer interests also are considered through various subsidy arrangements.

Imports of livestock products are permitted with varying degrees of regulation depending upon the degree to which the particular product is competitive or complementary to the domestic production and needs of the EC. One of the most potent means of regulating meat and livestock imports is by variable levies. By a complicated series of formula calculations, internal target prices are set up, and so-called sluice-gate prices are calculated for imported products. If the selling price of a third country is less than the sluice-gate price, a levy is imposed to make up the difference. Thus, the lower the outside price, the higher the variable levy. Other restrictions include requirements for import licenses, prior deposits by importers, and quotas. In addition, health regulations are relatively strict on meats.

EC restrictions are more severe on fresh meats than on by-products. Since the United States does not compete actively in the fresh meat trade of EC, the restrictions affect the United States less than other countries. Generally, the EC considers by-products obtained from the United States to be complementary to their production and needs.

United States

The United States works actively through its agricultural attachés—a branch of the diplomatic corp stationed in most countries. These attachés not only promote U.S. products but are valuable in arranging contacts for representatives of private firms in their individual promotional activities. They also work in conjunction with private firms in arranging for displays at trade fairs and in-store displays.

The United States sends meat and livestock trade missions to foreign countries many of which are sponsored jointly by the government and private organizations such as the American Meat Institute, the National Cattlemen's Association, and various breed associations. It should be noted also that the USDA works actively with exporting countries in negotiating voluntary quotas on meat shipments to the United States.

In 1976 a group composed of beef producers, pork producers, meat packers, and others interested in the export market, organized the United States Meat Export Federation Inc. with a major objective of expanding overseas sales of U.S. meat and meat products. The Federation is designed to coordinate export activities. Following its organization the Federation and USDA's Foreign Agricultural Service signed an agreement to join efforts in a cooperative program of sales promotion.

BIBLIOGRAPHY

ALBAUGH, R. 1965. The livestock and meat industry of Australia. USDA Foreign Agr. Serv. *M-164*.

DAVIS, E. E. 1977. Impact of beef imports. Texas Agr. Ext. Serv. Food and Fiber, Vol. 6, No. 7, 1–3.

DECOURCY, J. 1967. World production and trade: tallow and greases. USDA Foreign Agr. Serv. *M-182*.

EHRICH, R. L. and USMAN, M. 1974. Demand and supply functions for beef imports. Wyoming Agr. Expt. Sta. Bull. *604*.

FOLWELL, R. J. and SHAPOURI, H. 1976. An econometric analysis of the U.S. beef sector. Washington State Univ., Dept. Agr. Econ. Mimeographed (unnumbered).

FREEBAIRN, J. W. and RAUSSER, G. C. 1975. Effects of changes in the level of U.S. beef imports. Amer. J. Agr. Econ. Vol. 57, 676–688.

GRAEBER, K. E. and FARRIS, D. E. 1972. Beef cattle research in Texas, 1973. Texas Agr. Expt. Sta. PR-3217.

IVES, J. R. 1966. The Livestock and Meat Economy of the United States. Am. Meat Inst., Chicago.

JACKSON, G. H. 1972, The impact of eliminating the quota on U.S. imports of beef. Cornell Univ., Dept. Agr. Econ. A.E. Res. 338.

LEGE, F. M. III. 1974. Livestock exhibits throughout the world. USDA Foreign Agr. Serv. FAS M-259.

LEGE, F. M. III. 1976. Guide for U.S. cattle exporters. USDA Foreign Agr. Serv., Agr. Handbook No. 217.

LEGE, F. M. III. 1977. Suggested procedures for exporting breeding cattle and swine. USDA Foreign Agr. Serv. FAS M-274.

LEIGHTON, R. I. 1970. Economics of International Trade. McGraw-Hill Book Co., New York.

PURCELL, J. C. 1968. Trends and relations in the livestock-meat sector affecting prices and revenue to primary producers. Georgia Agr. Expt. Sta. Res. Bull. *35*.

USDA. 1975. Foreign agriculture circular. USDA Foreign Agr. Serv. FLM. 12–75.

USDA. 1977A. Livestock and meat statistics. USDA Econ. Res. Serv. and Stat. Reporting Serv., Stat. Bull. *522*.

USDA. 1977B. Handbook of agricultural charts. USDA Agr. Handbook No. 524.

USDA. 1977C. Livestock and meat. USDA Foreign Agr. Serv., FLM 3–77.

USDA. 1978. Livestock and meat situation. USDA Econ., Stat., and Co-op. Serv. LMS-220.

15

Marketing Costs

The USDA maintains and publishes three series on marketing costs: (1) price spreads, (2) the market basket, and (3) the marketing bill.

MARKETING SPREADS—GENERAL COMMENT

"Marketing spread" is a general term referring to the difference in value of a product at one point in the marketing system, and the value of an equivalent quantity of that product at another point in the marketing system. Price spread may be viewed as the value added between these designated points in the marketing system, or alternatively, it is the sum of all costs (middlemen costs) in performing whatever functions are necessary including profits. In the case of meat the USDA calculates a farm—wholesale (carcass) spread and a wholesale (carcass)—retail spread. The sum of these two is the farm—retail spread. "Marketing spread" often is used synonymously with price spread.

The marketing spread for the bundle of farm products (i.e., the farm—retail spread) in the so-called "market-basket" has amounted to about 60¢ per dollar spent by consumers in recent years. Commodities which require relatively little processing and other marketing services have a smaller spread, while those which require extensive processing have an even larger marketing spread.

Farmers are prone to emphasize the residual element which shows that they receive 40% of the consumer's dollar spent for food. During 1945, the last year of World War II, the farmers' share was 54%, the highest of record dating back to 1913. The average for all nonwar years since World War II has been about 40%. During the past 10 yr it has ranged from 39% to 46%. As will be shown later, the farmers' share of consumers' expenditures for beef, pork, and lamb differ substantially from this average of all food products.

Farmers tend to equate their relatively low income position to what appears to them to be a slim part of the consumer dollar, and often assume that the larger share going to market agencies is indicative of inefficient marketing, excessive profits, or both. Consumers, generally, are far removed from the agricultural scene and most are not well acquainted with the intricacies of the marketing system. To many, it is simple logic that rising retail prices must be putting profits in farmers' pockets. This notion may be heightened by occasional publicity of government programs which give the impression that farmers are getting rich from federal handouts for not producing.

Obviously, a considerable amount of misunderstanding exists about marketing spreads, their relation to efficiency of marketing, excess marketing profits, retail prices, and producer profits. An understanding of the method of calculating spreads is a first step to their interpretation.

PRICE SPREADS FOR BEEF, PORK, AND LAMB[1]

Farm-retail price spread, by definition, is the difference between the retail *price per unit* of a commodity and the price received by farmers for an equivalent quantity.

Method of Calculation

The basic data needed for calculating spreads are prices and values at the retail and farm levels, but in the case of beef, pork, and lamb, interest also centers in an in-between level— i.e., carcass or wholesale value. Therefore, for meat price spreads a determination is needed for three levels: (1) retail price; (2) wholesale or carcass value equivalent, and (3) farm value equivalent. The word "equivalent" is used because 1 lb of retail meat necessitates somewhat more than 1 lb at wholesale and even more at the farm level. The determination of retail price itself is no small matter, for a carcass is broken down into many cuts by the time it reaches the retail counter. Each cut comprises a different volume of the carcass and each sells at a different price. Cutting methods can affect the volume of meat in various cuts so this also must be standardized for calculation purposes. What is needed at retail, then, is a weighted average price per pound based on standardized yields of the various cuts.

Retail prices for U.S. Choice, Yield Grade 3 beef are derived from 30 cuts. Prices are obtained weekly by the Meat Animals Program Area, Commodity Economics Division, Economics, Statistics and Cooperatives Service, USDA. The retail cut nomenclature was changed in 1978 to conform, in general, with Uniform Meat Identity Standards. A continuing survey is maintained with about 40 retail chains throughout the country in which regular and special prices are obtained. Prices so obtained are applied to standardized proportions of cuts as shown in Column 1,

---

[1]Lawrence A. Duewer, Agricultural Economist, Commodity Economics Division, Economics Statistics and Cooperatives Service, USDA, provided much of the material in this section.

Table 15.1. The percentage of total retail cuts from carcass shown in Table 15.1 are not equivalent to carcass cut-out or yield discussed in Chap. 9. The figures shown here are net of retail cutting loss and retail shrink. Cutting losses are accounted for in a conversion to wholesale values. Adjustments are made to take into account the effects of retail shrink and price specials.

Earlier calculations, without these adjustments, prompted criticism from the meat trade and others. Consequently, changes were made to improve accuracy of the calculations. Among the improvements made in 1969 at the retail level (Duewer 1970) were an allowance for retail shrinkage due to pilferage, spoilage, refacing, conversion to lower-valued uses, and the extra weight added to the package to allow for dehydration loss. The revised procedure allows a 5% loss for beef. Special prices affect the composite price in two ways. The reduction in price of the cut on "special" reduces the average price of all cuts sold. This decrease is termed the price effect. When the price is reduced, consumers tend to buy more of that cut than they ordinarily would. As a result, a store may sell several times more than the carcass proportion of a cut when it is on special (Duewer 1970). This is called the volume effect. It should be noted that the values and prices indicated in Table 15.1 are not equivalent to the composite price calculated for price spread purposes. Further adjustments are made for price and volume effects.

The determination of carcass value necessitates a conversion factor based on weight loss from carcass to retail. The conversion factor is associated with animal types and cutting methods. These change over time, and USDA makes adjustments as needed. A number of adjustments were made in 1978 (Duewer 1978A). In the case of beef the carcass-retail conversion factor is 1.48 (prior to 1978 it was 1.41). This means that the value of 1.48 lb of carcass beef is needed to make an equivalent comparison with 1 lb at retail. A carcass-to-retail byproduct value was added for beef in 1978 to reflect the value of fat and bone trim. Beginning in 1978 beef trimmings were adjusted for fat removal (i.e., "defatted") in determining the amount of ground beef produced. Prior to 1978 carcass values were derived by weighting Midwest and West Coast carcass prices and adding a transportation differential to reflect average U.S. price of carcass beef. A new procedure adopted in 1978, ". . . uses Choice, Yield Grade 3, 600 to 700 lb steer carcass prices at 5 markets (East Coast, Colorado, Midwest, Amarillo Area, and Los Angeles) and weights them by population and consumption data to derive a U.S. price of carcass beef delivered to the city where consumed" (Duewer 1978B).

Live cattle prices formerly were determined at the farm level, but beginning in 1978 the live animal price was taken at the first market level after leaving the feedlot. USDA's Agricultural Marketing Service pro-

TABLE 15.1.   BEEF RETAIL PRICE PER POUND: PROPORTION CUTS ARE OF TOTAL RETAIL CUTS, AND RETAIL VALUE PER CUT AND PER 100 LB RETAIL CUTS FROM CARCASS[1]

Item	Percentage of Total Retail Cuts from Carcass (%)	Price/Lb ($)	Value/100 Lb ($)
Steaks			
Loin, Porterhouse Steak, BI[2]	2.2	3.08	6.78
Loin, Top Loin Steak, BI	1.0	3.24	3.24
Loin, Top Loin Steak, BO[3]	0.8	4.02	3.22
Loin, T-Bone Steak, BI	3.7	3.05	11.29
Loin, Sirloin Steak, BO	6.9	2.93	20.22
Round, Round Steak, BI	3.7	2.05	7.58
Round, Top Round Steak, BO	3.6	2.36	8.50
Round, Bottom Round Steak, BO	3.0	2.15	6.45
Chuck, Blade Steak, BI	4.3	1.30	5.59
Chuck, Arm Steak, BI	3.3	1.65	5.44
Rib, Steak, Large End, BI	2.3	2.39	5.50
Rib, Ribeye Steak, BO	2.3	3.93	9.04
Flank Steak, BO	0.7	2.83	1.98
Roasts			
Rib Roast, Large End, BI	2.6	2.14	5.56
Rib Roast, Small End, BI	2.6	2.43	6.32
Chuck, Blade Pot Roast, BI	1.0	1.20	1.20
Chuck, Arm Pot Roast, BI	6.3	1.55	9.76
Chuck, Pot Roast, BO	6.8	1.62	11.02
Round, Tip Roast, BO	3.6	2.17	7.81
Round, Eye Round Roast, BO	1.6	2.69	4.30
Round, Bottom Round Roast, BO	1.9	1.96	3.72
Round, Rump Roast, BO	3.3	2.08	6.86
Other Cuts			
Rib, Short Ribs, BI	4.0	1.25	5.00
Plate, Short Ribs, BI	1.0	.95	.95
Brisket, whole, BO	2.3	1.83	4.21
Ground beef not less than 70% lean	15.1	1.16	17.52
Ground beef not less than 80% lean	2.6	1.49	3.87
Stew, BO	6.2	1.70	10.54
Shank, cross cuts, BI	1.0	1.22	1.22
Kidney	0.3	.54	.16
TOTAL	100.0	—	194.9

Source: Duewer (1978A).
[1]Prices used were for July, 1978.
[2]Bone in.
[3]Bone out.

vides prices for Choice, Yield Grade 3, 1000 to 1200 lb steers at 8 markets. Four terminal market prices are used (Omaha, Sioux Falls, Sioux City, and South St. Paul) along with 4 direct market quotations (Iowa, Texas, Colorado, and California). Quotations from these markets are weighted uniformly to determine a live steer price. Under procedures established in 1978 a factor of 2.40 lb of live animal are needed to obtain one lb of retail beef (formerly that factor was 2.28). This change was made to reflect changes in animal type and industry practices. Then, live animal price multiplied by 2.40 gives the gross farm (live[2]) value. However, gross farm value includes a valuation of farm by-products (primarily hide and offal) which is subtracted from gross value to arrive at net farm value. Price spreads are calculated each week and month, then aggregated into quarterly and annual price spreads.

A simplified hypothetical example of beef price spreads for a given month might be:

Assume the following prices and values:

	¢/lb
Composite retail price	190
Wholesale carcass price	84
Farm level price	53
Farm by-product value	15
Carcass by-product value	2

(1) Equivalent gross carcass value..................... $84.0 \times 1.48 = 124.3¢$
(2) Equivalent net carcass value....................... $124.3 - 2.0 = 122.3¢$
(3) Carcass-retail price spread........................ $190.0 - 122.3 = 67.7¢$
(4) Equivalent gross farm value....................... $53.0 \times 2.40 = 127.2¢$
(5) Equivalent net farm value .... $127.2 - 15.0 = 112.2¢$
(6) Farm-carcass price spread .... $122.3 - 112.2 = 10.1¢$
(7) Farm-retail price spread........................... $190.0 - 112.2 = 77.8¢$
(8) Farmers' share of beef dollar...................... $\dfrac{112.2}{190.0} \times 100 = 59.1\%$

The approach in calculating pork and lamb price spreads follows the same general procedure as described for beef. Pork prices and values are determined for barrows and gilts, i.e., an average of all grades. Lamb prices are based on U.S. Choice grade. As was mentioned in the case of beef, a number of changes were made in pork price spread measurements in 1978. These changes are reflected in the following discussion. The 15 retail pork cuts used and the standard percentage that each cut is of the carcass, are shown in Table 15.2. The 146.5¢ shown in Table 15.2 is not the official retail price used in price spread calculations. An adjustment is made for price effect and volume effect of price specials.

A 4.0% retail shrink is allowed for pork. Midwest prices, with a transportation differential, are used to estimate the pork wholesale value. This value is multiplied by a wholesale conversion factor of 1.06 to obtain a carcass value equivalent to 1 lb of pork at retail. The average monthly price of barrows and gilts at 7 Corn Belt markets is used to obtain the

---

[2]"Farm" value is the nomenclature used although the live value is determined at the first market level. Prior to 1978 a live animal transport cost was deducted from market value to arrive at a "farm gate" value.

TABLE 15.2.   PORK RETAIL PRICE PER POUND: PROPORTION CUTS ARE OF TOTAL RETAIL CUTS, AND RETAIL VALUE PER CUT AND PER 100 LB RETAIL CUTS FROM CARCASS[1]

Item	Percentage of Total Retail Cuts from Carcass (%)	Price/Lb ($)	Value/100 Lb ($)
Pork loin, rib chops	5.1	2.02	10.30
Pork loin, top loin chops	4.9	2.10	10.29
Pork loin, blade roast	5.3	1.39	7.37
Pork loin, sirloin roast	4.7	1.46	6.86
Pork loin, assorted pork chops	1.4	1.50	2.10
Pork loin, tenderloin, whole	1.0	2.92	2.92
Smoked Ham, rump portion	8.2	1.19	9.76
Smoked Ham, shank portion	10.6	1.08	11.45
Smoked Ham, center slices	5.9	2.25	13.28
Smoked Ham, whole	4.3	1.24	5.33
Pork Shoulder, blade boston, roast	8.4	1.36	11.42
Pork, spareribs	4.5	1.64	7.38
Ground pork sausage	6.5	1.29	8.38
Smoked pork shoulder, picnic whole	10.2	.96	9.79
Sliced bacon	19.0	1.57	29.83
TOTAL	100.0	—	146.5

Source: Duewer (1978A).
[1]Prices used were for July, 1978

farm price of hogs (Duewer 1978B). To this price is applied a conversion factor of 1.70 to obtain gross farm value equivalent of 1 lb of pork at retail. And from this figure is deducted an allowance for pork by-products. The allowance recently was adjusted to reflect the trend in decreasing yield of lard.

The carcass price for lamb is derived from quotations at New York, Los Angeles, San Francisco, and the Seattle-Tacoma-Portland area to which a conversion factor of 1.18 is applied to obtain carcass value. Live lamb prices are from quotations for U.S. Choice grade at 11 leading public markets. The average conversion factor is 2.45, but actual monthly factors vary seasonally to allow for changes in the fleece.

Characteristics of Marketing Spreads

As mentioned above, the total farm-retail spread is composed of two components: (1) the farm-carcass (wholesale)[3] spread, and (2) the carcass (wholesale)-retail spread. Over the years these spreads have exhibited

[3]Prior to 1970, the spread between farm level price and wholesale meat price was referred to as the "farm-wholesale" spread. In the case of beef and lamb this was changed to "farm-carcass" spread in 1970 to more accurately reflect the fact that prices and values at the wholesale level are for carcasses rather than primal wholesale cuts. The term "wholesale" still is used for pork, as prices and values used apply to wholesale pork cuts.

some common characteristics and some which are preculiar to particular species.

While the spreads have not remained constant, they have been more stable than farm value. This may be observed in Fig. 15.1 for beef.[4] Farm value is directly related to prices received by producers. As farm price rises and falls, farm value rises and falls, but the spreads tend to be more fixed. This is largely explained by the fact that the elements which comprise slaughtering, processing, wholesaling, and retailing costs (wage rates, rent, taxes, interest charges, freight rates, profits, etc.) simply do not vary as farm price for livestock varies. The market structure under which these items are determined is subject to more institutional control than is the market for live animals.

The tendency for the farm-retail spread to remain relatively fixed results in a decline in the farmer's share of the meat dollar when farm prices fall and an increase when farm prices rise. Assume the farm-retail spread for pork is a constant 30¢ per lb. If the farm value were also 30¢, retail price would be 60¢ and the farmer's share would be 50%. But, if farm value dropped to 25¢, retail price (if it dropped the same amount) would be 55¢ and the farmer's share would be 45%. This has been a matter of contention with producers; and what compounds their concern even more is a tendency for retail prices to be rather sticky. Retail prices do change, but usually with a lag, following farm price and wholesale price changes. In the above example, if retail prices had remained at 60¢ after farm value dropped to 25¢, the farmer's share would have been 42%, at least temporarily. The reverse happens during rising farm prices (i.e., the farmer's share tends to rise), but if farmers' reactions are an indication of their feelings, this does not seem to balance the situation. Part of the reason it does not balance out in the eyes of producers is that the farmer's shares, over a period of time, have not averaged out at a constant level. As shown in Fig. 15.2 they have tended generally downward, though some leveling off is apparent since about the mid-1960's.

Over the same period the farm-retail spreads (the costs of marketing) have trended upward for all three species (see Figs. 15.3, 15.4, 15.5). This is a reflection of the general inflationary trend in most market input items. Freight rates were an exception until the late 1960's when that trend also turned upward. Labor is the largest single item among marketing costs, but the pronounced upward trend in hourly wage rates has been partially offset by increasing labor productivity. However, in spite of productivity gains, unit labor costs have risen substantially.

The carcass (wholesale)-retail spreads have exhibited a generally increasing trend for all three species (Figs. 15.3, 15.4, 15.5). These, basically, are retail marketing costs, and as calculated by USDA, all cutting,

---

[4]Pork and lamb exhibit the same characteristics.

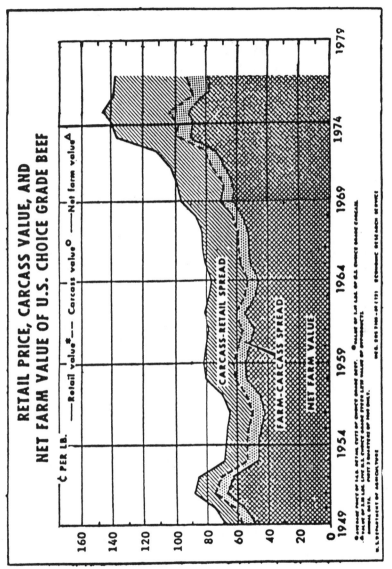

FIG. 15.1.   RETAIL PRICE, CARCASS VALUE, AND NET FARM VALUE OF CHOICE GRADE OF BEEF
(1970 and later years added by author from USDA data)

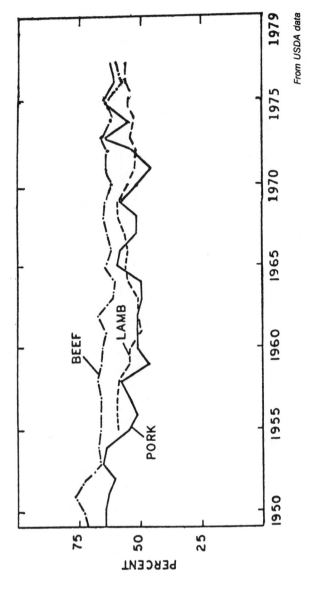

FIG. 15.2.   FARMER'S SHARE OF CONSUMER'S MEAT DOLLAR

processing, packaging, and merchandising costs of beef and lamb are attributed to the retail level. This is not entirely appropriate as more and more breaking, cutting, and processing is being done at the packer level. However, it does tend to explain why the carcass-retail spread for beef and lamb is greater than for pork. Pork, traditionally, has been broken into wholesale cuts and a considerable share of the processing done by packers which puts this cost in the farm-wholesale spread. Thus, the farm-carcass (wholesale) spread is greater for pork than for beef and lamb. The farm-carcass (wholesale) and the carcass (wholesale)-retail spreads for all species have been on an increasing trend—except the farm-carcass spread for beef (Fig. 15.3). This is attributed to improved technology and increases in efficiency of labor and equipment (Duewer 1970).

Implications with Respect to Marketing Efficiency

The farmer's share of the consumer's dollar spent for beef in 1969 was 65%, during the same time it was 7% for canned beets. There is a tendency for some persons to conclude, on the basis of this evidence, that the marketing of beets must be less efficient than the marketing of beef. This is an unwarranted conclusion. Neither the farmer's share, nor the absolute amount of marketing spread is adequate in itself for evaluating marketing efficiency—either operational efficiency or pricing efficiency. It has been pointed out earlier that price spreads consist of marketing and processing costs, plus profits to participating agencies. The matter of profits will be examined later. Some products simply require more marketing inputs relative to the value of the product itself than do others. This is apparent in the case of beets and beef where processing costs are substantially higher for beets relative to the value of the product.

Some might argue that the marketing costs of both are too high. It is assumed in a competitive economy that astute businessmen are continually looking for ways to reduce costs, but the possibility of further improvements should be a matter of continuing concern for both individual firms and the general public. The public concern is reflected in numerous governmental research and extension projects designed to reduce marketing costs.

It would be possible to reduce the farm-retail price spread to zero (i.e., farmers could get 100% of the consumers' meat dollar if they slaughtered, processed, and delivered meat to the consumers' doors). This, however, is not necessarily the most efficient system. In fact, it was discovered long ago that specialization and trade, based on comparative advantage (as discussed in the previous chapter) would result in a greater total and per capita real income. A severe price and income squeeze on farmers

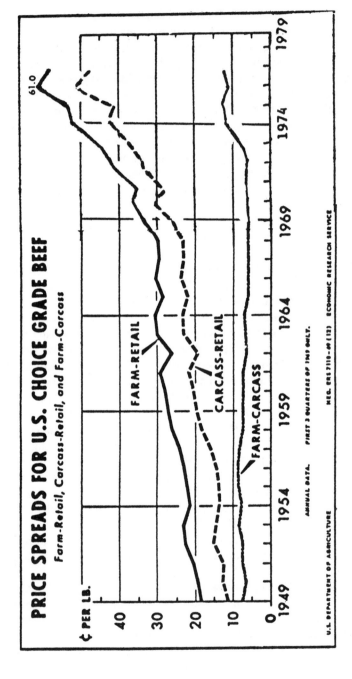

FIG. 15.3.   PRICE SPREADS FOR U.S. CHOICE GRADE BEEF
(1969 and later years added by author)

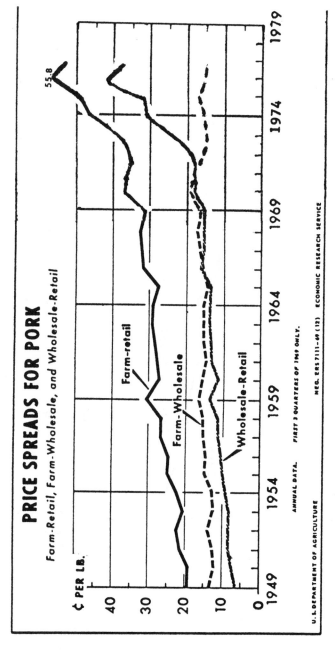

FIG. 15.4.   PRICE SPREADS FOR PORK
(1969 and later years added by author)

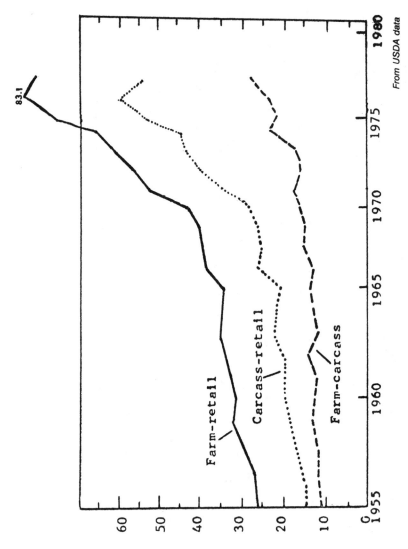

From USDA data

FIG. 15.5.   PRICE SPREADS FOR CHOICE GRADE LAMB

during the 1970's prompted several farm groups to slaughter, process, and retail their own meat. If there were excess profits in the system, this approach would prevail, but in a competitive situation economies normally accrue to specialization which is difficult to accomplish in a farm to retail operation.

The major utility aspects affecting marketing costs and the farmer's share are place, time, and form.

Place.—Transportation costs for meat could be reduced by raising beef cattle near the centers of population. This, however, would increase other costs—feed, labor, pollution abatement, etc.—and negate the comparative advantage of production in the Milo Belt and Corn Belt. It would be less efficient from an overall standpoint and consumer prices would increase.

Time.—Any product which has a seasonal variation in production is faced with a time problem if that product is to be made available to consumers uniformly throughout the year. Beets provide an excellent example of a product with an extreme seasonality of production—the entire crop is harvested within a relatively short period. In processed form, beets may be stored and made available for the remainder of the year. But processing and storage are costly and reduce the farmer's share. This cannot be classed as inefficiency.

Beef is produced more uniformly throughout the year as large commercial feedlots have grown in importance. If year-round production had been attempted when the industry was composed entirely of small family farms, costs undoubtedly would have increased. Net costs probably have not increased with commercial lots since cost gains from economies of scale and more complete utilization of facilities probably offset the cost-increasing tendency of reduced summer gains. However, there still is some seasonality in beef production. This means that slaughtering and processing facilities have unused capacity at times, which causes unit costs to increase and is reflected in marketing spreads and farmers' shares. To reduce this cost by completely leveling out production might increase production costs.

Cold storage is used to alleviate the time factor to a greater extent with pork than for beef or lamb. This increases the marketing spread and reduces the farmer's share, but again, cannot be classified as inefficiency per se.

Form.—Some products require more processing (change in form) than others. Pork is subjected to more processing than beef which goes far in explaining why the farmer's share is smaller for pork than for beef. Canned beets are an extreme example of a product which requires much more processing relative to value than either pork, beef, or lamb. Again, this is not an indictment of marketing efficiency.

Reduction in Farmer's Share Over Time.—As pointed out earlier, the farmer's share has been on a generally declining trend. Inefficiencies cannot be completely ruled out, but studies on market structure of the meat industry have failed to confirm any substantial degree of inefficiency. There is plenty of evidence that consumers are demanding (or merchandisers are selling them) more and more services. This includes such things as more trimming, more boning, better and more attractive packaging, more processing and preparation (e.g., fully cooked meats, TV dinners, etc.), more parking space, air-conditioned stores, carryout service, etc. The cost of all such items ends up in the marketing spread and reduces the farmer's percentage of the dollar spent for meat. The recent trend toward discount food stores with a reduction in some services may be partially responsible for the recent leveling off of farmer's share in recent years. However, there is no indication of a slackening in the demand for built-in convenience in meat items.

Implications with Respect to Profits

An increase in marketing profits would show up in the price spread, and, everything else being equal, would result in a decrease in the farmer's share. As has been pointed out, the food industry has been subjected to a number of investigations when monopoly and undue profits were suspected. The Consent Decree was an outgrowth of one such investigation and put some restrictions on the actions of packers involved. Other investigations have been conducted, the most recent being the 1965 National Commission on Food Marketing, which covered the entire food industry, including a complete investigation of the livestock and meat industry. The Commission was critical of some aspects, but ended with a generally favorable conclusion, expressed by Brandow (1966), its executive director, as follows: The Commission concluded its study believing that the contribution of the food industry to a high and rising level of living was fully comparable with that of other leading sectors of the economy. In broadest terms the industry is efficient and progressive. Supplied by a highly productive agriculture, manufacturers and distributors have provided consumers with varied, abundant, and nutritious array of goods at generally reasonable prices.

Some data are available on industry profits. These data are adequate for comparison with other industries, but the question of what constitutes a reasonable or adequate profit level cannot be settled—that requires a value judgment and differences of opinion are to be expected. Data were presented in Chap. 7 (Figs. 7.1, 7.2, 7.3) showing net income after taxes as a percentage of net worth for a number of U.S. industries, including chain food stores and meat packing. The meat packing industry generally had the lowest returns of the industries shown. "Profits to retail food chains were high relative to other industries during most of the postwar period. These high levels of profits resulted from a rapid rise in popularity of the supermarket. In response to this increase in demand, many thousands of supermarkets were built. As this rapid building program caught up with demand around 1960, profits for food retailers returned to levels comparable to other industries" (National Commission on Food Marketing 1966B). Typically, meat packing industry profits after taxes are about 1% of sales, which is at or near the bottom of the range for major U.S. industries. As a percentage of net worth, meat packing profits usually are less than average for major U.S. industries—and were 10% to 11% during the mid 1970's.

What does price spread and farmers' share of the consumers' expenditure reveal about producers' profits? The same generalization applies here as with packer's and retailer's profits discussed above. By itself, the farmer's share (either at a given time or a change over time) is not an adequate indicator of farmers' profit position. It tells nothing about the absolute amount of farm profit or its relation to that of other industries. A 30% share of $1.50 per retail pound of beef would be more dollars than 50% of 70¢ per lb beef.

The farmer's profit position can be better evaluated by examining his returns on net worth as was done with the other sectors. Farm management records show that livestock producers' returns are extremely unstable from year to year, and over the long period average about 3–4% of equity. This is substantially lower than other sectors of the meat industry. The effect of relatively low returns has been apparent for many years as livestock production is shifting into fewer and larger units.

Seasonality in Price Spreads

Seasonal variation in meat production leads to a seasonal variation in prices which, in return, affects price spreads. Production normally drops off during mid-summer and prices rise seasonally. During this period the farm value rises and the farm-retail spread declines slightly. With a lag of about one month, retail prices rise. It may be noted, however, that the farm-retail spread and retail prices are considerably more stable than production and net farm value. Beef and lamb also exhibit seasonal tendencies. The magnitude of seasonal variations associated with beef is considerably less than for either pork or lamb.

Price Spread versus Gross Margin and Profit Margin

Some confusion exists with respect to the definition of price spread, industry gross margin, and profit margin. On occasion they have been used interchangeably. They are not equivalent, and important errors are made in using them synonymously. Figure 15.6 illustrates the basic differences among the terms. "Price spreads, gross margin, and net profit margins measure different aspects and components of the spread between what farmers receive and consumers pay.

"Price spreads measure differences between price levels of subsequent stages in the marketing channel.

"Price spreads are normally greater than gross margins for any single marketing agency. Likewise, gross margins are greater than net profit margins" (USDA 1975A).

The farm-retail spread includes all costs and profits from assembly of the live animals (shown at the extreme left side of Fig. 15.6) to purchase of the meat at retail outlets (shown at extreme right side of Fig. 15.6). Within that system, packer gross margin (which includes packer costs and profits) and retail gross margin (which includes retailer costs and profits) are isolated diagramatically from other costs which are encompassed in USDA's spread computations.

"The farm-carcass spread includes approximate charges for marketing cattle, slaughtering, and transporting the dressed carcass to the cities where consumed.

"The carcass-retail spread includes not only the gross margin for retailing, but also the charges for other intermediate marketing services such as cutting carcasses into smaller portions, wholesaling, and local delivery to retail stores.

"Gross margins (the difference between dollars paid and dollars received) of packers and retailers, on the other hand, don't take into account all marketing functions. Rather, they represent the tab for a packer's or retailer's labor cost, packaging, overhead, other costs and any net profit. They exclude some items included in the spread, like charges for transportation and services performed by businesses other than meatpackers or retailers. Because such costs are included in what they pay for beef, gross margins of these firms are smaller than the overall USDA spreads.

"Profit margins, before and after taxes, are a relatively small component of the gross margin and total operating cost of a firm. They are

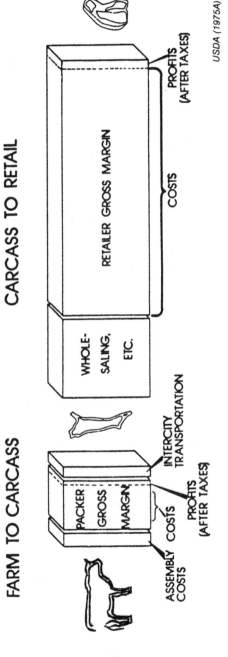

FIG. 15.6.   SCHEMATIC COMPARISON OF PRICE SPREAD, GROSS MARGIN, AND PROFIT

usually expressed as a percent of total sales or of stockholder's equity for a firm or group of firms, rather than for an individual product or group of products" (USDA 1975A).

MARKET BASKET

Market basket calculations are for a "basket" or a group of products, representative of average quantities of domestic, farm-originated food products purchased annually by wage earners and clerical worker families and single workers living alone. Market basket statistics have four components: (1) retail cost—which actually is somewhat less than an average sized family's expenditure for food since the market basket does not include cost of meals away from the home, imported foods, seafoods, and foods of nonfarm origin; (2) farm value—the gross returns to farmers for the quantity of farm products equivalent to those in the market basket; (3) farm-retail spread—the difference between retail cost and farm value; and (4) the farmer's share.

The farm-retail market basket spread is similar in concept to the price spread in that it is an estimate of costs of marketing (assembling, processing, transporting, and distributing). In the case of the market basket, the spread is the cost for the entire quantity in the basket, while the price spread discussed earlier was cost for one unit of the product—for meat it was the cost per retail pound.

Table 15.3 shows trends in components of the market basket. Retail costs have advanced at a faster rate than farm value. The farm-retail spread has increased for the same reasons explained under price spreads. The farmer's share is down from 1947–1949 levels. It partially recovered during the erratic price movements of the early 1970's, but then returned to its former trend.

In market basket calculations all meat is aggregated into one figure. In all three components, meat is by far the most important item. In recent years meat has comprised about 30% of retail sales, 40% of farm value, and 23% of the farm-retail spread. The next most important food group is dairy products, with about 18%, 22%, and 15% respectively.

Market basket statistics are published quarterly by the USDA.

THE MARKETING BILL

The marketing bill is the total cost of marketing the entire quantity of U.S. farm-originated foods purchased by civilians. It is the difference between consumer expenditures and farm value. The marketing bill statistics show the distribution of consumer expenditures between the marketing system and farmers, and the distribution of marketing costs among commodity groups and individual cost components such as labor.

Since marketing costs on a per unit basis have been on an increasing trend, it is no surprise that the total marketing bill also has increased. This is shown in Fig. 15.7. Out of a total marketing bill of $180 billion in 1977, the farm value was $56 billion; the marketing bill was the difference, or $124 billion.

In dollar expenditures by consumers the most important marketing bill item is meat—comprising about 30% of total expenditures.

During recent years about ⅓ of the increase in the marketing bill has been due to inflation. Table 15.4 illustrates why the marketing bill has increased. The cost of energy (fuel, power, and light) has increased at a spectacular rate. Wages, materials, and rent have increased substantially—and the profit rates have increased.

The marketing bill does not include all expenditures for food in the U.S. USDA started a new series in 1978 entitled Total Food Expenditures which includes a number of items not in the marketing bill. Among these items is food served in institutions, on airlines and dining cars, at recrea-

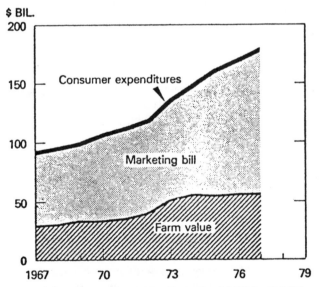

FOR DOMESTIC FARM FOODS PURCHASED BY CIVILIAN CONSUMERS FOR CONSUMPTION BOTH AT HOME AND AWAY FROM HOME.
1977 PRELIMINARY.

USDA (1977)

FIG. 15.7.   FARM-FOOD MARKETING BILL AND CONSUMER FOOD EXPENDITURES

TABLE 15.3.   THE MARKET BASKET OF FARM FOODS: RETAIL COST, FARM VALUE, FARM-RETAIL SPREAD, AND FARMER'S SHARE OF RETAIL COST, AVERAGE 1947–1949 AND 1957–1959, ANNUAL 1960–1977[1]

Year & Month	Retail Cost ($)	Farm Value ($)	Farm-retail Spread ($)	Farmer's Share (%)
Average:				
1947–1949	890	441	449	50
1957–1959	983	388	595	39
1960	991	383	608	39
1961	997	380	617	38
1962	1006	384	622	38
1963	1007	378	629	38
1964	1009	377	632	37
1965	1037	416	621	40
1966	1092	445	647	41
1967	1081	419	662	39
1968	1120	441	678	39
1969	1179	481	698	41
1970	1228	478	750	39
1971	1250	480	771	38
1972	1311	524	787	40
1973	1537	701	837	46
1974	1750	747	1002	43
1975	1876	784	1092	42
1976	1895	748	1148	39
1977[2]	1937	749	1188	39

Source: USDA (1978) and earlier issues.
[1] Retail cost of average quantities purchased annually per household in 1960–1961 by urban wage-earner and clerical worker families and single workers living alone, calculated from retail prices collected by the Bureau of Labor Statistics. Data for earlier years are published in Farm-Retail Spreads for Food Products 1947–1964, ERS-226, Apr. 1965.
[2] Preliminary.

tional facilities, at dormitories, through vending machines; donations; etc. Total food expenditures were estimated to be $219 billion in 1977 compared to consumer expenditures of $180 billion in marketing bill calculations.

FACTORS AFFECTING MARKETING COSTS AND SHARE TO FARMERS

The circumstances of an individual producer can vary considerably from the averages presented above, and the share accruing to individuals likewise can vary. The following factors affect a farmer's share: (1) Type of livestock program. A farmer who maintains a cow herd, retains and feeds out his own calves, sells direct to a local packer who, in turn, sells the meat to a local retailer would receive a considerably larger share of the retail dollar than a neighbor who buys yearling feeders, finishes them out and sells to a distant packer who ships the meat to a still more distant

TABLE 15.4.   FOOD MARKETING: SPREADS, COSTS, AND PROFIT RATES

Year	Farm-retail price spread [1]	Intermediate goods and services [1]			Hourly earnings [2]	Interest rate [3]	Profit rates after taxes			
		Total	Containers packaging	Fuel, power, and light			Food retailers [4]		Food manufacturers [5]	
							Sales	Equity	Sales	Equity
		1967=100			Dollars	Percent				
1970	113.4	113	108	108	3.03	8.48	—	—	2.5	10.8
1971	116.5	120	113	120	3.24	6.32	—	—	2.6	11.0
1972	118.9	126	117	126	3.45	5.82	—	—	2.6	11.2
1973	126.5	134	123	138	3.66	8.30	—	—	2.6	12.8
1974	151.5	159	151	202	3.99	11.28	—	—	2.9	13.9
1975	165.1	180	174	237	4.40	8.65	0.5	6.8	3.2	14.4
1976[6]	173.2	193	184	258	4.77	7.52	.8	10.0	3.4	14.9

Source: USDA (1978).

[1] Represents all goods purchased by food marketing firms except raw materials and plant and equipment, and all services except those performed by employees, calculated from wholesale price relatives. [2] Weighted composite of production employees in food manufacturing and nonsupervisory employees in wholesale and retail trade, calculated from data of the U.S. Department of Labor. [3] Bank rates on short-term business loans in 35 centers, Department of Commerce. [4] Federal Trade Commission. The data are based on reports from all food retailing corporations having more than $100 million in annual sales, and whose activities are at least 75 percent specialized in supermarket operations. Comparable data not available prior to third quarter 1974. [5] "Quarterly Financial Report," Federal Trade Commission. Data represent national aggregate estimates for corporations based upon a sample of company reports. Data since the fourth quarter of 1973 are imperfectly comparable with prior data because of changes in accounting methods. [6] Preliminary.

retailer. (2) Costs incurred and prices received, which will vary considerably among producers. Cost control is partially a matter of management expertise. Prices paid for feeder animals and received for finished stock are related to marketing ability, but vagaries of the market can affect profits. (3) Grade of livestock fed, particularly yield grade. This can affect the price and quantity of meat at retail.

Examples are available from studies which illustrate varying shares of returns to producers under several different types of production and marketing programs. The results obtained from these studies are not to be construed as suggestive of average returns which producers could expect from the indicated production or marketing programs. Neither are they suggestive of superiority of any production or marketing program over another.

Results from a S. Dakota report (Schulte undated) are shown in Fig. 15.8. In this study Example No. 1 assumes U.S. Choice grade steer calves were raised in southeastern S. Dakota, marketed as finished steers through the Sioux Falls terminal, slaughtered locally, and the

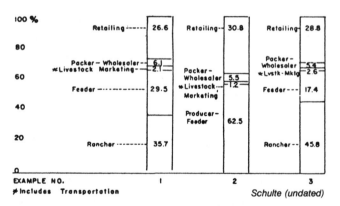

FIG. 15.8. BEEF: COMPARISON OF THE PERCENTAGE OF GROSS RETURNS TO PRODUCERS AND MARKETING AGENCIES

Choice carcasses sold to a New York retail firm. Example No. 2 assumes U.S. Good grade calves were raised and fed out on the same farm in southeastern S. Dakota, sold through the Sioux Falls terminal to a southwestern Minnesota packer who, in turn, sold the U.S. Good grade carcasses to a retailer in Omaha. Example No. 3 assumes U.S. Good grade yearlings in south central S. Dakota were sold through auction to a feeder in eastern S. Dakota, marketed as finished steers through the Sioux Falls terminal to a local packer who shipped the U.S. Good grade carcasses to a retail firm in Philadelphia. The major variation in shares is in the split between original producer (rancher or producer-feeder) and feeder. In Example No. 2, producer and feeder are one and the same person who gets the entire 62.5%. The combined rancher and feeder share in Examples No. 1 and No. 3 is approximately the same, but the division between rancher and feeder is considerably different in the two examples.

Figure 15.9 shows results of a USDA study (Duewer 1970). The beef example assumes U.S. Good grade feeder steers from near Casper, Wyo., were marketed through the Omaha terminal to a feeder near Lincoln, Neb.; sold as finished Choice grade steers through the Omaha terminal to a local packer who shipped the carcasses to a retailer in New York. The same program was assumed for 1967 and 1969. The pork example (again the same program at two different dates) assumes hogs were raised in western Iowa, sold through the Sioux City terminal to a local packer, and the wholesale cuts were shipped to a retail firm in Los Angeles. In the beef example, both the rancher and retailer shares were less in 1969 than in 1967. Most of the gain was in the feeder's share. In the pork example, both the retailer and packer-wholesaler shares declined from 1967 to 1969, with the gain going to the original producer.

These examples are presented, not to emphasize the magnitudes shown, but to illustrate the point that shares can vary with different

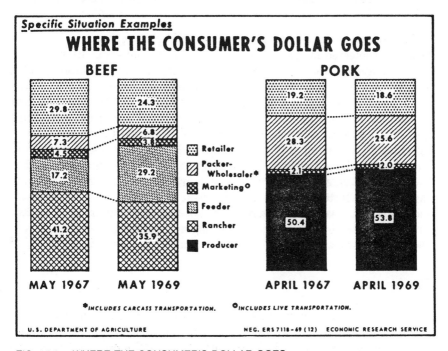

FIG. 15.9.   WHERE THE CONSUMER'S DOLLAR GOES

programs (the S. Dakota study), and that different shares can be obtained from the same program at different times (the USDA study).

Dissatisfaction exists among producers over the farmer's share, and various actions have been suggested to alleviate the situation. Farmers have, in some instances, set up farmer-owned cooperatives to capture a greater share of marketing profits. In a few instances farmers have organized cooperative packing plants. Most cooperative packing ventures failed. While several successful plants currently are operating, they are handling a negligible fraction of livestock slaughter.

Farmer organizations on occasion have indicated an interest in ownership and operation of a national retail food chain, but have not made the plunge. Cooperative ventures in retailing to date have been, in general, somewhat less than smashing successes. If the profit rates reported earlier in this chapter are accurate, it appears doubtful that any significant one-shot gain can be squeezed out in favor of farmers. However, this is no excuse for assuming a passive attitude toward marketing costs. Improvements undoubtedly can be made, and a series of small gains could be significant in improving the producer's position.

COMMENT ON SELECTED ITEMS OF MARKETING COST

Live animal shrinkage; transportation; and bruise, crippling, and death losses are selected for particular comment. In the determination of price spreads, account was taken of retail "store shrinkage." In addition, cooler shrinkage was recognized, along with trimming loss, in the conversion factors to carcass weight and live weight equivalent. Neither shrinkage in live animal weight from the feedlot to market place, nor bruise, crippling, and death loss were explicitly considered. Transportation costs were specifically accounted for in spread calculations, but merit additional comment. Rail and truck rate have more than doubled since 1969, and in view of the U.S. energy shortage, transportation will become an increasingly critical factor in livestock and meat shipments, as with all commodities. Transportation is significant not only in the number of dollars involved, but also as a factor in the changing structural characteristics of the livestock-meat industry. These two aspects are related.

Transportation

Early developments in transportation, and their impact upon the livestock and meat industry, were noted in previous chapters. Overland droving in eastern areas, the great cattle trails of the range area, and the era of canal building were all changed as the railroads pushed their steel rails throughout the country. The railroad system set the pattern for concentrated and centralized livestock marketing and meat packing locations. Then came the motor truck and highway system which were instrumental in reversal of the organizational structure (decentralization).

The impact of transportation development was not limited to the livestock-meat sector. Hulbert (1920), writing at the time some of these significant changes were occurring, said, "If the great American novel is ever written, I hazard the guess that its plot will be woven around the theme of American transportation, for that has been the vital factor in the national development of the United States. Every problem in the building of the Republic has been, in the last analysis, a problem of transportation." In view of subsequent concern with social and environmental problems, undoubtedly many currently would disagree with the attachment of this degree of importance to transportation. Even in the earlier days, concurrent developments in refrigeration, communications, and implementation of grade standards were teamed with transportation in changing the structure of the livestock and meat industry. But there is no question that the central importance of transportation and its significance lies in marketing cost reductions. Conceivably, droving could have continued as the mode of transport as production expanded to mid-continent. But, at what cost? Today, cattle are droved from the southern fringes of the Sahara Desert more than 1000 miles to Lagos, Nigeria, and other west African coastal cities. But consumption is only 8–10 lb per person and not many people can afford the price. Transportation alone would not solve the Nigerian problem, but it would help.

Innovations in Transportation.—It is not enough simply to refer to railroads and trucks as means of transportation. Many improvements have been (and are being) made within each system which sustain a continuous competitive contest between these two types of carriers. In this contest, trucks virtually have taken over all short hauls of both livestock and meat. It is estimated that more than 80% of all livestock is moved to market by trucks and trailers. The truck advantage in short hauls lies in the faster time, convenience, and flexibility including pickup and delivery at destination. Trucks are vying for long hauls of livestock by increasing capacity, controlling temperature, and improving designs that reduce bruising and crippling and expedite loading and unloading. Cross-country, triple-deck trucks haul thousands of hogs from the midwest to the west coast each year. Some of these are equipped with air conditioning units for use in hot weather. Others have sprinkling or fogging devices. Triple-deck trucks also are used for lambs; and double-deck cattle trucks have been in use for a number of years.

Railroads also have made improvements with multideck cars and improved construction for greater animal comfort and reduced bruising. These include such things as improved ventilation, suspension, coupling, and operational procedures that reduce starting and stopping jerks. Most railroads now operate fast, "express" livestock trains to speed delivery.

Some years ago railroads introduced "piggyback" hauling of loaded trucks as a counter measure to truck competition. Piggyback use has increased for meat in refrigerated trucks and is used to some extent for live animals. Significant improvements have been made in refrigeration capacity of rail cars, and in long hauls the railroads are competitive with trucks in meat transportation.

Containerization and palletization of meat shipments has progressed, but much research still is needed to perfect the techniques and maintain quality during shipment. These developments apply to both truck and railroads—and, in fact, extend also to airlines and shipping lines. Lack of standardization in container size is a problem not only for a given mode of transportation, but also especially where trans-shipment involves different modes of transportation.

Types of Motor Carriers.—A first, broad classification of motor carriers is "for-hire" and "private" truckers. Private truckers haul their own products and are subject only to state trucking regulations. For-hire truckers fall into two classes—regulated carriers which are subject to federal regulations and unregulated, or "exempt carriers" which, at the federal level, generally are subject only to regulations concerning the type of commodities that may be hauled, safety specifications, and maximum hour limitations for drivers. The crucial aspects of exemption are freedom in rates charged and routes of operation. Provisions of the Interstate Commerce Act placed regulation of transportation under the Interstate Commerce Commission. However, through a series of administrative rulings, court decisions, and amendments, the hauling of certain commodities by truck was exempt from federal regulation. The major classification of exempt commodities is unmanufactured agricultural products which includes livestock—with certain exceptions. Breeding, racing, and show animals are not exempt. Meat is not exempt on grounds that it has undergone a manufacturing process. The determination of what constitutes manufacturing as applied to agricultural and fishery products has been a matter of contention in establishing the list of exempt products. The exemption does not apply to railroads.

Within the group of regulated carriers a distinction is made between common carriers, which are available to the public generally, and contract carriers, which provide service to shippers under contract. Regulated carriers (railroads and regulated trucks) are subject to numerous federal restrictions including rate charges, routes used, and who may be issued permits. Certain states have regulations (taxes, rates charged, safety requirements, load limits, etc.) which apply to all carriers (exempt or regulated) operating in, or through, that state. However, as a rule, exempt carriers have greater flexibility in rates and in operation than do regulated carriers, and from a competitive standpoint, are in a position to make quicker adjustments to changing competitive situations. Thus, livestock truckers (operating under exempt status) can make quicker adjustments than can railroad officials (operating under regulated status) in the transportation of livestock.

Comparison of Truck and Rail Rates.—A regional study conducted by western states (Capener 1969) indicated that truck rates clearly were less than rail rates on hauls of up to 200–300 miles (Fig. 15.10). The rate for interstate trucks was greater than that for rail on both feeder and slaughter cattle on distances above 300–400 miles. And, rail rates on slaughter cattle were about 15% higher than rail rates on feeder cattle. Rates are important, but other factors must also be considered. Usually, some trucking is required to bring livestock to a rail loading point and take them away from the unloading point. Time in transit also is important and will be discussed later. The adequacy of rest stops is a factor in shrinkage. On this score, railroads have had better facilities than were available for trucks. However, in recent years improvements have been made in privately-operated rest stops.

The Back-Haul Problem.—A factor contributing to livestock transport costs is difficulty in obtaining back-hauls. Established truckers often have contacts for possible return freight, but timing is a problem. At best back-hauls are irregular, inconvenient, or seasonal. Railroads may be able to park cars in freight yards for use at their convenience, but many eventually return empty. Truckers normally must return to their base of operations within a relatively short time whether loaded or empty. The impact of empty returns can be illustrated by a hypothetical example. Assume the cost of operating a diesel powered semitrailer is $1.20 per loaded mile. If return hauls were never available, a 1-way charge of $2.40 per mile would be required to break even (ignoring the fact that costs would be slightly less on the empty return). However, if back-hauls were available ½ the time, a charge of $1.80 per mile would give a monetary return equivalent to $2.40 per mile without a back-haul.

An advantage of air-conditioned trucks is their adaptability to back-hauls of perishable products. Truckers operating under exempt status have a problem in that the return load also must be an exempt product if carried for hire. This difficulty may be avoided if the trucker buys (takes

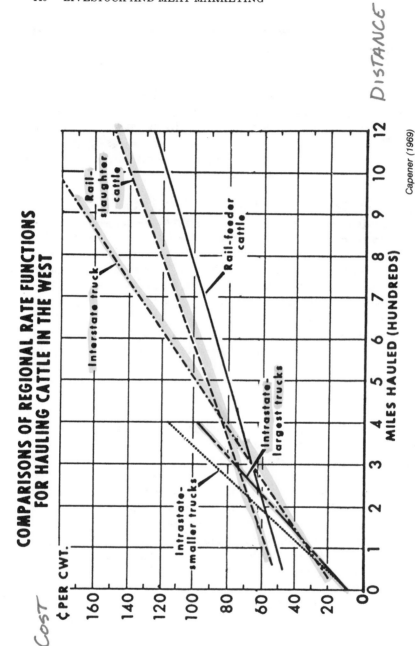

FIG. 15.10.   COMPARISONS OF REGIONAL RATE FUNCTIONS FOR HAULING CATTLE IN THE WEST

title) to the cargo for the return trip. This, however, can be a costly operation unless outlets are prearranged.

Industry Location Related to Transportation Costs.—The transportation rate structure plays a crucial role in industrial location. Other factors, of course, affect packing plant location, but everything else being equal, the relationship between costs of shipping live animals versus the cost of shipping dressed meat is a determining factor. The same situation holds for location of feedyards. The relationship between cost of shipping feed grain versus cost of shipping meat (or livestock) can determine where livestock will be fed.

After a period of relative stability in freight rates, an industry will develop a locational pattern compatible with that rate structure. Subsequent changes of an unequal geographical nature can disrupt the locational pattern. The flour milling industry in parts of the hard red winter wheat area experienced such a disruption during the early 1960's following a reduction in wheat rates to Southeastern states without equivalent reduction in flour rates. As a result, many flour mills ceased operation in the hard red winter wheat area and milling capacity was expanded in Southeastern states. Cattle feeding in the Milo Belt is somewhat vulnerable to the transportation rate structure. A reduction in feed grain rates to the Southwest or West Coast areas without equivalent changes in livestock and meat rates could disrupt cattle feeding and meat packing in the Milo Belt. Western interests have been active in recent years in attempts to effectuate such a change in the rate structure. This is just one aspect of the competitive game.

Shrinkage

Shrinkage is recognized as a loss of weight. Livestock typically shrink during the marketing process, and meat loses weight in marketing channels. Where livestock is sold on a liveweight basis, the number of pounds sold (or paid for) is equally as important as the price. Directly or indirectly, a reduction in weight represents a cost. When meat loses weight in marketing channels, that must be reflected as a cost. In marketing, a distinction is made between the following: (1) "actual" shrink, (2) "pencil" shrink, (3) "cooler" shrink, (4) "cutting" shrink, and (5) "store" shrink. Cooler shrink—a loss due to evaporation of moisture—generally ranges from 1.5–3% by weight. Cutting shrink is 2–3%, and store shrink is estimated to be 5–5.5%. Cooler shrink, cutting shrink, and store shrink were considered in the previous section. This section is devoted to live animal shrinkage, i.e., actual and pencil shrink.

Actual loss in body weight may arise from excretory shrink (the emptying of the digestive and urinary tracts) and tissue shrink (a loss of body tissue). "Tissue shrink occurs on long, extended shipments or during long periods of fast. These two types of shrinkage probably occur as two distinct phases in the shrinkage process. In the early part of shipment only excretory shrink occurs. At an unidentified stage in the movement both excretory and tissue shrinkage occur simultaneously. During the latter part of the shipment, tissue shrinkage is relatively more important" (Harston 1959A).

The economic importance of tissue shrink is readily apparent. A loss in body tissue reduces carcass weight and directly affects value. If buyers (and sellers) were able to precisely evaluate "fill" and if proper allowances were made in the agreed-upon price, excretory shrink would have no economic significance. In slaughter livestock this would involve a precise estimate of carcass yield, and in feeder livestock an estimate of deviation from normal fill condition (i.e., "excess fill" or the opposite, a "shrunk out" condition). Professional marketers are able to average out these estimates over a number of lots with a high degree of accuracy. The occasional buyer or seller usually lacks comparable ability and even the professional may be off on any given lot. In addition to impreciseness of knowledge, inequality of bargaining ability also may result in prices not adequately reflecting shrunk out or excess fill conditions. Therefore, excretory shrink cannot be dismissed as completely noneconomic.

Factors Related to Shrinkage.—A number of studies have been made over the years in attempts to measure the extent of shrinkage and determine the associated factors. Several factors have been isolated, but inconsistency in results points up the fact that shrinkage is a complex phenomenon. Efforts to derive predictive mathematical functions have, in general, been marked with a relatively low degree of accuracy. However, results of the studies do permit some generalizations on explanatory variables and provide some guidelines for manipulation of those variables which are controllable.

Major factors which affect shrinkage are time in transit, distance hauled, degree of fill, weather conditions, weighing conditions, sex, weight of the animals, class of animals, type of feed, mode of transportation, handling procedures, preconditioning, type of feedlot shelter, rate of gain near end of feeding period, and yield grade. Most of the research on shrinkage has been concerned with liveweight shrink. However, one study (Raikes and Tilley 1975) dealt with hot-carcass shrink as well as liveweight shrink.

Time in Transit and Distance Hauled.—Time in transit and distance hauled may be discussed together as these variables are highly interrelated and have the same general effect on shrinkage. Liveweight shrink consistently has been found to bear a positive relationship with distance (and time) to market (Abbenhaus and Penny 1951, Henning and Thomas 1962, and Raikes and Tilley 1975). Shrinkage, whether measured in pounds or percentage of body weight lost, increases with distance and time, but not at a constant rate throughout the trip. Shrinkage occurs at a faster rate during the early part of the trip than during the latter part, meaning that shrinkage increases at a decreasing rate with distance and time. Figure 15.11 shows the results of an Indiana study in which shrinkage of hogs increased throughout the distance, but the rate tapered off with distance. The same general type of relationship holds between shrinkage and time in transit, and is consistent also for cattle and sheep. Of these two variables, time in transit probably is the more realistic with which to deal. Two different truckers, hauling identical loads over the same route, might have different times in transit. Over-the-road speeds may differ, length of meal and coffee stops may differ, etc.

The amount of shrinkage to be expected with time or distance cannot be precisely specified because other factors also enter the picture. However, Tables 15.5 and 15.6 show the results of well-planned studies of cattle and hog shrinkage. It may be noted in Table 15.5 that average shrinkage for the group was 3.9% for the entire 200-mile haul and that nearly ½ of this occurred during the first 25 miles. Table 15.6 again illustrates a declining rate of shrinkage for hogs, reaching a maximum of 2.76% where the hogs were not allowed a fill at the market.

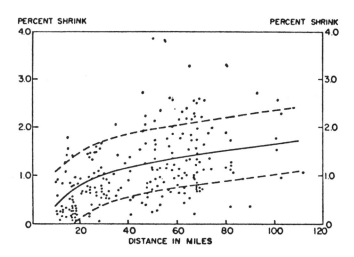

Stout and Cox (1959)

FIG. 15.11.   FARM-WEIGHT TO SALE-(OR CATCH) WEIGHT OF SELECTED HOGS SOLD AT LOCAL AND TERMINAL MARKETS IN RELATION TO DISTANCE TRAVELED, OCT. 31, 1956 TO FEB. 24, 1958

TABLE 15.5.   PERCENTAGE OF SHRINK FOR 60 FAT CATTLE BETWEEN EACH CHECK WEIGHING DURING 200 MILE TRUCK HAUL (TOTAL ANIMAL WEIGHT EQUALS 100 PERCENT)

Weight Classes (Lb)	No. Head in Each Class	Avg Weight (Lb)	0–25 (%)	25–50 (%)	50–100 (%)	100–200 (%)	Total Weight Lost (%)
					Miles Traveled Between Weighings		
Group avg	60	1122	1.8	0.7	0.8	0.6	3.9
Under 1000	11	954	1.5	0.7	0.9	0.8	3.9
1000–1099	10	1056	2.1	0.9	0.8	0.3	4.1
1100–1199	24	1139	1.8	0.8	0.8	0.7	4.1
Over 1200	15	1263	1.9	0.5	0.7	0.5	3.6

Source: Abbenhaus (1951).

Raikes and Tilley (1975) found that hot-carcass weight shrink was not significantly affected by distance.

Degree of Fill.—Degree of fill refers to the extent feed and/or water are denied, or made available, to the animals both prior to shipment and at the destination. Since excretory shrink results from elimination of the digestive and urinary tracts, it stands to reason that some of this loss will be quickly regained when the animals have access to feed and water. This is shown in Table 15.6. The degree of fill, however, is highly variable depending upon: whether the animals were taken off feed prior to shipping, quantity of feed and water consumed during rest stops, time allowed for fill, and degree of emotional disturbance during transit and during the fill period. Some shippers claim to regain all the shrink, but instances are known where emotional disturbance results in no fill at all.

TABLE 15.6.   RELATIONSHIP TO THE LENGTH OF HAUL TO SHRINKAGE OF HOGS (1132 LOTS; 38,303 HOGS)

Miles Hauled	Shrinkage When Not Fed at Market (%)	When Fed at Market (%)
0–5	1.06	[1]
6–15	1.12	1.03
16–25	1.39	1.24
26–35	1.75	1.51
36–45	2.06	1.79
46–55	2.50	1.99
56–65	2.68	2.03
66–75	2.76	2.08
76–85	[1]	2.14
86–95	[1]	2.16

Source: Wiley and Cox (1955).
[1] Sufficient data not available.

Raikes and Tilley (1975) found that withholding both feed and water from slaughter steers 12 hrs prior to slaughter had a significant effect on both liveweight shrink and hot carcass shrink. That degree of fasting resulted in an additional liveweight shrink of 2.128% over the control (non-fasted) animals, and hot carcass weight of the fasted steers was 7.762 lb lower than the control steers. Their work and that of others suggests that withholding of feed, but not water, will not increase hot carcass shrinkage, but that withholding both does increase hot carcass shrinkage.

Shrinkage is one variable which is at least partially controllable by a shipper. Generally, it is considered desirable to do any necessary sorting several days prior to shipping, withhold feed immediately prior to shipment, handle as quietly and gently as possible during loading (utilizing a pen and loading chute arrangement that facilitates speed with a minimum of excitement), and minimize time in transit consistent with road conditions and general safety considerations. Scheduled time of arrival at the market will depend upon length of haul. If a public market is used, schedule the time of arrival. On relatively short hauls, arrival 2–3 hr prior to market opening will give the animals time to settle down and partake of water. Then, if sale can be made shortly after the market opens, feeding probably is not economical (Wiley and Cox 1955). On longer hauls it is desirable to schedule arrival the evening prior to day of sale and give access to both feed and water.

Any attempt to gain excess fill is not recommended. Experienced buyers will insist on docking the price of such animals.

Weather Conditions.—Shrinkage tends to increase as temperatures move from moderate levels toward either extreme—heat or cold. This is illustrated in Fig. 15.12 in the case of hogs. "Extreme temperatures have an effect on fat cattle shrinkage during marketing. But other things—wind, rain, snow, humidity, and other weather conditions—seem to have more effect than temperature alone" (Harston 1959B). In the Raikes and Tilley (1975) study a significant relationship was shown between shrinkage and "feeding-period." Feeding period referred to different time periods,[5] and the authors reasoned that weather condition was the basic factor involved. The effects of temperature extremes can be alleviated by such things as properly adjusting ventilation; using straw bedding and covering open trucks in cold weather; using wet sand for bedding, and spraying and fogging during hot weather; and scheduling hauls during an optimum time of day (this will depend upon weather conditions).

---

[5]In this study one group of steers was marketed in May and another in October. These were the two time periods involved in the analysis.

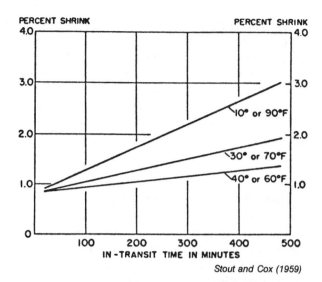

FIG. 15.12.   EFFECT OF VARIATION AROUND 50°F OVER TIME ON IN-TRANSIT SHRINKAGE OF SLAUGHTER-WEIGHT MARKET HOGS FROM 232 OBSERVATIONS

Weighing Conditions.—Weighing conditions include such arrangements as holding livestock off water and/or feed for a specified time prior to weighing, and in the case of feeder cattle or lambs, driving the livestock to a specified place for weighing. Weighing conditions affect shrinkage and hence value of the animals. Therefore, weighing conditions need to be considered as an integral part of the bargain, along with price. Harston (1959A) reported overnight shrink—12 hr stand—will vary with the type of feed. Cattle off green grass, wet beet pulp, or silage will generally shrink 4%, while fat cattle off concentrates will shrink about 2.5–3% if no feed and water are available. If feed and water are available, morning weights of fat cattle will be 2% less than evening weights. Range cattle not used to enclosures often shrink more than 5% when held in drylot overnight. In several respects, the considerations involved here are similar to those discussed above under degree of fill.

Shrinkage from driving livestock will depend not only on the distance, but also on weather conditions, speed of travel, and handling procedures. Harston (1959B) reported an 8.4% shrink on steers from a 50-mile drive under favorable handling conditions. Under the best of conditions such shrink probably amounts to 3–4% in a 6–8 mile drive.

Sex, Class, and Weight of Animals.—Most studies show that heifers shrink more than steers. However, Harston (1959B) reported the difference was not great except during the summer when heifers shrank 11.5 lb for each 10 lb steers shrank. Much variation occurs in comparison of sex. Bulls usually shrink a lot because of disturbing circumstances and strange animals in nearby pens. Calves also are heavy shrinkers largely because they are often weaned at market time. Finished cattle shrink more during the latter hours of transit (Table 15.7).

TABLE 15.7. COMPARISON OF SHRINKAGE BY FAT CATTLE AND FEEDER CATTLE

Time in Transit (Hr)	Rate of Shrink	
	Fat Cattle (%)	Feeder Cattle (%)
6	5.4	3.8
10–17	6.2	8.2
60	10.8	12.4

Source: Data from Harston (1959B).

Studies relating shrinkage to weight of hogs have been inconsistent (Bjorka 1938; Stout and Cox 1959). In the case of cattle, Harston (1959B) reported that shrinkage was not closely related to weight, except as weight is correlated with degree of fatness.

Mode of Transportation.—The comparative shrinkage by truck versus rail on short hauls is irrelevant since other considerations have given trucks a virtual monopoly on short hauls. The evidence in Fig. 15.13 indicates relatively little difference between rail and truck shrinkage in long hauls of cattle, at least up to 30–40 hr in transit. Factors other than mode of transportation easily could account for the apparent differences.

Preconditioning.—Studies of the effect of preconditioning on shrinkage have been somewhat inconclusive. It appears reasonable that weaning, vaccination, castration, and such operations cause stress. When accomplished as part of preconditioning, subsequent shrinkage would be reduced. However, preconditioning is not standardized and perhaps this has contributed to conflicting results. Additional research is needed to evaluate not only shrinkage, but also other economic aspects associated with preconditioning.

General Considerations.—To the extent that actual shrinkage can be predicted and controlled it becomes a bargaining issue (comparable to pencil shrink in the following section). Raikes and Tilley (1975) pointed out that packers have an incentive to require the withholding of both feed and water because the shrinkage in liveweight reduces their liveweight cost more than the reduced value due to carcass weight loss. However,

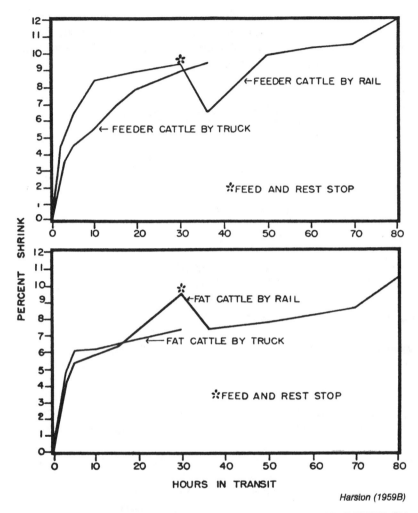

FIG. 15.13. HOW SHRINKAGE DIFFERS WHEN CATTLE ARE SHIPPED BY TRUCK AND RAIL (FEEDER AND FAT CATTLE)

producers with a knowledge of actual shrinkage under varying conditions may negotiate for an offsetting price increase.

Pencil Shrink.—Pencil shrink, as defined in Chap. 6, is a negotiated or an agreed upon deduction from scale weight in arriving at pay weight. This is a common practice, particularly in direct cattle sales. Normally, the deduction is specified as a percentage and may range from 2 to 5%, but more generally is 3–4%. The logic of pencil shrink is that some actual shrinkage will take place and the negotiated pencil shrink is an estimate or allowance to compensate the buyer for expected actual loss in weight accompanying the transaction. Everyone is aware that an adjustment could be made in the price and accomplish the same results without resorting to pencil shrink, but the custom is well established, and some buyers (and sellers) apparently believe that bargaining advantages accrue to them. The practice is here to stay unless some governmental action is taken to stop it, and, while this has been discussed, it appears unlikely. There is nothing inherently dishonest about the practice, but if both parties are not equally informed and of equal bargaining ability, inequities can result.

In a bargaining situation, the general proposition is one of a trade-off between pencil shrink and price. An equal trade-off occurs where the price adjustment exactly offsets the pencil shrink. This can be illustrated as follows: suppose a rancher wants to realize $60.00 per cwt for his 450-lb steer calves. What price would he need if he agreed to a 4% pencil shrink? In the absence of pencil shrink each calf would be worth $\frac{450}{100} \times \$60.00 = \$270.00$. A pencil shrink of 4% would amount to a deduction of 18 lb per calf or a pay weight of 432 lb. Thus, the trade-off price would need to be $\$270.00 \div \frac{432}{100} = \$62.50$ per cwt. The same result can be obtained by direct calculation as follows:

$$\frac{\$60.00 \div 96}{100} = 62.50$$

Here the 96 is the difference between 100% and the 4% pencil shrink. If the rancher did not bargain for a higher price to offset the effect of the 4% pencil shrink, but instead settled for $60.00 per cwt *and* also granted the shrink, he actually would have realized $\frac{\$60.00\ (432)}{450} = \$57.60$ per cwt for his calves. Again, this result can be obtained more simply as follows: $\frac{(\$60.00)\ (96)}{100} = \$57.60$. The same general principles apply to any class of livestock. Differences in value received can amount to substantial sums of money. Tables have been prepared showing the trade-off price for various levels of pencil shrink allowance and cattle price level. Two such examples are reproduced in Tables 15.8 and 15.9. These tables are derived by straight arithmetic as illustrated in the examples above. A glance at Table 15.9 shows that with $50.00 calves each 1% shrink amounts to 50¢ per cwt. Few professional marketers carry tables. They have enough experience to make accurate mental estimates. The nonprofessional, occasional trader would do well either to resort to a table or to make the necessary calculations.

One aspect of the shrinkage-bargaining issue that needs special mention is the combination of pencil shrink and weighing conditions. A set of yearling steers sold with, say, a 3% pencil shrink after an overnight stand and a 10-mile haul before weighing, are likely to sustain at least 5–6% actual shrinkage in addition to the pencil shrink (i.e., a total of 8–9%).

TABLE 15.8.   SHRINKAGE TABLE (OFFER) REALIZED PRICES: DOLLARS PER HUNDREDWEIGHT SHRINKAGE DEDUCTED

OFFER ($)	2%	3%	4%	6%	8%	OFFER ($)	2%	3%	4%	6%	8%
50.00	49.00	48.50	48.00	47.00	46.00						
49.75	48.75	48.26	47.76	46.77	45.77	34.75	34.05	33.70	33.36	32.66	31.97
49.50	48.51	48.02	47.52	46.53	45.54	34.50	33.81	33.46	33.12	32.43	31.74
49.25	48.27	47.77	47.28	46.29	45.31	34.25	33.56	33.22	32.88	32.20	31.51
49.00	48.02	47.53	47.04	46.06	45.08	34.00	33.32	32.98	32.64	31.96	31.28
48.75	47.77	47.29	46.80	45.83	44.85	33.75	33.08	32.74	32.40	31.72	31.05
48.50	47.53	47.04	46.56	45.59	44.62	33.50	32.83	32.50	32.16	31.49	30.82
48.25	47.29	46.80	46.32	45.35	44.39	33.25	32.58	32.25	31.92	31.26	30.59
48.00	47.04	46.56	46.08	45.12	44.16	33.00	32.34	32.01	31.68	31.02	30.36
47.75	46.79	46.32	45.84	44.89	43.93	32.75	32.10	31.77	31.44	30.79	30.13
47.50	46.55	46.08	45.60	44.65	43.70	32.50	31.85	31.51	31.20	30.55	29.90
47.25	46.31	45.83	45.36	44.41	43.47	32.25	31.60	31.28	30.96	30.32	29.67
47.00	46.06	45.59	45.12	44.18	43.24	32.00	31.36	31.04	30.72	30.08	29.44
46.75	45.81	45.35	44.88	43.95	43.01	31.75	31.12	30.80	30.48	29.84	29.21
46.50	45.57	45.10	44.64	43.71	42.78	31.50	30.87	30.56	30.24	29.61	28.98
46.25	45.33	44.86	44.40	43.47	42.55	31.25	30.62	30.31	30.00	29.38	28.75
46.00	45.08	44.62	44.16	43.24	42.32	31.00	30.38	30.07	29.76	29.14	28.52
45.75	44.83	44.38	43.92	43.01	42.09	30.75	30.14	29.83	29.52	28.90	28.29
45.50	44.59	44.14	43.68	42.77	41.86	30.50	29.89	29.58	29.28	28.67	28.06
45.25	44.35	43.89	43.44	42.53	41.63	30.25	29.64	29.34	29.04	28.44	27.83
45.00	44.10	43.65	43.20	42.30	41.40	30.00	29.40	29.10	28.80	28.20	27.60
44.75	43.85	43.41	42.96	42.07	41.17	29.75	29.16	28.86	28.56	27.96	27.37
44.50	43.61	43.16	42.72	41.83	40.94	29.50	28.91	28.62	28.32	27.73	27.14
44.25	43.37	42.92	42.48	41.59	40.71	29.25	28.66	28.37	28.08	27.50	26.91
44.00	43.12	42.68	42.24	41.36	40.48	29.00	28.42	28.13	27.84	27.26	26.68
43.75	42.87	42.44	42.00	41.13	40.25	28.75	28.18	27.89	27.60	27.02	26.45
43.50	42.63	42.20	41.76	40.89	40.02	28.50	27.93	27.64	27.36	26.79	26.22
43.25	42.39	41.95	41.52	40.65	39.79	28.25	27.68	27.40	27.12	26.56	25.99
43.00	42.14	41.71	41.28	40.42	39.56	28.00	27.44	27.16	26.88	26.32	25.76
42.75	41.89	41.47	41.04	40.19	39.33	27.75	27.20	26.92	26.64	26.08	25.53
42.50	41.65	41.22	40.80	39.95	39.10	27.50	26.95	26.68	26.40	25.85	25.30
42.25	41.41	40.98	40.56	39.71	38.87	27.25	26.70	26.43	26.16	25.62	25.07
42.00	41.16	40.74	40.32	39.48	38.64	27.00	26.46	26.19	25.92	25.38	24.84
41.75	40.91	40.50	40.08	39.25	38.41	26.75	26.22	25.95	25.68	25.14	24.61
41.50	40.67	40.26	39.84	39.01	38.18	26.50	25.97	25.71	25.44	24.91	24.38
41.25	40.43	40.01	39.60	38.77	37.95	26.25	25.72	25.46	25.20	24.68	24.15
41.00	40.18	39.77	39.36	38.54	37.72	26.00	25.48	25.22	24.96	24.44	23.92
40.75	39.93	39.53	39.12	38.31	37.49	25.75	25.24	24.98	24.72	24.20	23.69
40.50	39.69	39.28	38.88	38.07	37.26	25.50	24.99	24.74	24.48	23.97	23.46
40.25	39.45	39.04	38.64	37.83	37.03	25.25	24.74	24.49	24.24	23.74	23.23
40.00	39.20	38.80	38.40	37.60	36.80	25.00	24.50	24.25	24.00	23.50	23.00
39.75	38.96	38.56	38.16	37.36	36.57	24.75	24.26	24.01	23.76	23.26	22.77
39.50	38.71	38.32	37.92	37.13	36.34	24.50	24.01	23.76	23.52	23.03	22.54
39.25	38.46	38.07	37.68	36.90	36.11	24.25	23.76	23.52	23.28	22.80	22.31
39.00	38.22	37.83	37.44	36.66	35.88	24.00	23.52	23.28	23.04	22.56	22.08
38.75	37.98	37.59	37.20	36.42	35.65	23.75	23.28	23.04	22.80	22.32	21.85
38.50	37.73	37.34	36.96	36.19	35.42	23.50	23.03	22.80	22.56	22.09	21.62
38.25	37.48	37.10	36.72	35.96	35.19	23.25	22.78	22.55	22.32	21.86	21.39
38.00	37.24	36.86	36.48	35.72	34.96	23.00	22.54	22.31	22.08	21.62	21.16

TABLE 15.8.   (Continued)

OFFER ($)	2%	3%	4%	6%	8%	OFFER ($)	2%	3%	4%	6%	8%
37.75	37.00	36.62	36.24	35.48	34.73	22.75	22.30	22.07	21.84	21.38	20.93
37.50	36.75	36.38	36.00	35.25	34.50	22.50	22.05	21.82	21.60	21.15	20.70
37.25	36.50	36.13	35.76	35.02	34.27	22.25	21.80	21.58	21.36	20.92	20.47
37.00	36.26	35.89	35.52	34.78	34.04	22.00	21.56	21.34	21.12	20.68	20.24
36.75	36.02	35.65	35.28	34.50	33.81	21.75	21.32	21.10	20.88	20.44	20.01
36.50	35.77	35.41	35.04	34.31	33.58	21.50	21,07	20.85	20.64	20.21	19.78
36.25	35.53	35.16	34.80	34.08	33.35	21.25	20.82	20.61	20.40	19.98	19.55
36.00	35.28	34.92	34.56	33.84	33.12	21.00	20.58	20.37	20.16	19.74	19.32
35.75	35.04	34.68	34.32	33.60	32.89	20.75	20.34	20.13	19.92	19.50	19.09
35.50	34.79	34.44	34.08	33.37	32.66	20.50	20.09	19.88	19.68	19.27	18.86
35.25	34.54	34.19	33.84	33.14	32.43	20.25	19.84	19.64	19.44	19.04	18.63
35.00	34.30	33.95	33.60	32.90	32.20	20.00	19.60	19.40	19.20	18.80	18.40

Source: Wellman (undated).

TABLE 15.9.   SHRINKAGE TABLE (ASKING) PRICES: DOLLARS PER HUNDRED-WEIGHT NEEDED TO COMPENSATE FOR SHRINKAGE

ASKING ($)	2%	3%	4%	6%	8%	ASKING ($)	2%	3%	4%	6%	8%
50.00	51.02	51.54	52.08	53.19	54.36						
49.75	50.77	51.29	51.83	52.93	54.10	34.75	35.46	35.82	36.20	36.97	37.78
49.50	50.51	51.03	51.56	52.65	53.81	34.50	35.20	35.57	35.94	36.70	37.50
49.25	50.26	50.78	51.31	52.39	53.55	34.25	34.95	35.31	35.68	36.44	37.22
49.00	50.00	50.51	51.04	52.12	53.27	34.00	34.69	35.05	35.42	36.17	36.96
48.75	49.75	50.26	50.79	51.86	53.00	33.75	34.44	34.79	35.16	35.90	36.68
48.50	49.49	50.00	50.52	51.59	52.72	33.50	34.18	34.54	34.90	35.64	36.41
48.25	49.24	49.75	50.27	51.33	52.43	33.25	33.93	34.28	34.64	35.37	36.14
48.00	48.98	49.48	50.00	51.05	52.17	33.00	33.67	34.02	34.37	35.11	35.87
47.75	48.73	49.23	49.74	50.79	51.90	32.75	33.42	33.76	34.11	34.84	35.60
47.50	48.47	48.97	49.48	50.52	51.62	32.50	33.16	33.51	33.85	34.57	35.33
47.25	48.22	48.72	49.23	50.26	51.36	32.25	32.91	33.25	33.59	34.31	35.05
47.00	47.96	48.45	48.96	49.99	51.08	32.00	32.66	32.99	33.33	34.04	34.78
46.75	47.71	48.20	48.70	49.72	50.81	31.75	32.40	32.73	33.07	33.78	34.51
46.50	47.45	47.94	48.44	49.46	50.54	31.50	32.14	32.47	32.81	33.51	34.24
46.25	47.20	47.69	48.19	49.20	50.28	31.25	31.89	32.22	32.55	33.24	33.97
46.00	46.94	47.42	47.92	48.93	50.00	31.00	31.63	31.96	32.29	32.98	33.70
45.75	46.69	47.17	47.66	48.66	49.72	30.75	31.38	31.70	32.03	32.71	33.42
45.50	46.43	46.91	47.40	48.40	49.46	30.50	31.12	31.44	31.77	32.45	33.15
45.25	46.18	46.66	47.15	48.14	49.19	30.25	30.87	31.19	31.51	32.18	32.88
45.00	45.92	46.39	46.88	47.87	48.92	30.00	30.61	30.93	31.25	31.91	32.61
44.75	45.67	46.14	46.62	47.60	48.64	29.75	30.36	30.67	30.99	31.65	32.34
44.50	45.41	45.88	46.36	47.34	48.38	29.50	30.10	30.41	30.73	31.38	32.07
44.25	45.16	45.63	46.11	47.08	48.11	29.25	29.85	30.15	30.47	31.12	31.79
44.00	44.90	45.36	45.84	46.81	47.83	29.00	29.59	29.88	30.21	30.85	31.52

TABLE 15.9.   *(Continued)*

ASKING ($)	2%	3%	4%	6%	8%	ASKING ($)	2%	3%	4%	6%	8%
43.75	44.65	45.11	45.58	46.54	47.56	28.75	29.34	29.64	29.95	30.59	31.25
43.50	44.39	44.85	45.32	46.28	47.29	28.50	29.08	29.38	29.69	30.32	30.98
43.25	44.14	44.60	45.07	46.02	47.03	28.25	28.83	29.12	29.43	30.05	30.71
43.00	43.88	44.33	44.80	45.75	46.75	28.00	28.57	28.87	29.17	29.79	30.43
42.75	43.63	44.08	44.54	45.48	46.48	27.75	28.32	28.61	28.91	29.52	30.16
42.50	43.37	43.82	44.28	45.21	46.20	27.50	28.06	28.35	28.65	29.26	29.89
42.25	43.12	43.57	44.03	44.96	45.94	27.25	27.81	28.09	28.39	28.99	29.62
42.00	42.86	43.30	43.76	44.68	45.66	27.00	27.55	27.84	28.12	28.72	29.35
41.75	42.61	43.05	43.50	44.42	45.39	26.75	27.30	27.58	27.86	28.46	29.08
41.50	42.35	42.79	43.24	44.15	45.12	26.50	27.04	27.32	27.60	28.19	28.80
41.25	42.10	42.54	42.99	43.90	44.86	26.25	26.79	27.06	27.34	27.93	28.53
41.00	41.84	42.27	42.71	43.61	44.57	26.00	26.53	26.80	27.08	27.66	28.26
40.75	41.59	42.02	42.46	43.36	44.31	25.75	26.28	26.55	26.82	27.39	27.99
40.50	41.33	41.76	42.20	43.09	44.03	25.50	26.02	26.29	26.56	27.13	27.72
40.25	41.08	41.51	41.95	42.84	43.78	25.25	25.77	26.03	26.30	26.86	27.45
40.00	40.82	41.24	41.67	42.55	43.48	25.00	25.51	25.77	26.04	26.60	27.17
39.75	40.56	40.98	41.41	42.29	43.21	24.75	25.25	25.52	25.78	26.33	26.90
39.50	40.31	40.72	41.15	42.02	42.93	24.50	25.00	25.26	25.52	26.06	26.63
39.25	40.05	40.46	40.89	41.76	42.66	24.25	24.74	25.00	25.26	25.80	26.36
39.00	39.80	40.21	40.63	41.49	42.39	24.00	24.49	24.74	25.00	25.53	26.09
38.75	39.54	39.95	40.36	41.22	42.12	23.75	24.23	24.48	24.74	25.27	25.82
38.50	39.29	39.69	40.10	40.96	41.85	23.50	23.98	24.23	24.48	25.00	25.54
38.25	39.03	39.43	39.84	40.69	41.58	23.25	23.72	23.97	24.22	24.73	25.27
38.00	38.78	39.18	39.58	40.43	41.30	23.00	23.47	23.71	23.96	24.47	25.00
37.75	38.52	38.92	39.32	40.16	41.03	22.75	23.21	23.45	23.70	24.20	24.73
37.50	38.27	38.66	39.06	39.89	40.76	22.50	22.96	23.20	23.44	23.94	24.46
37.25	38.01	38.40	38.80	39.63	40.49	22.25	22.70	22.94	23.18	23.67	24.18
37.00	37.76	38.14	38.54	39.36	40.22	22.00	22.45	22.68	22.92	23.40	23.91
36.75	37.50	37.89	38.28	39.10	39.95	21.75	22.19	22.42	22.66	23.14	23.64
36.50	37.24	37.63	38.02	38.83	39.67	21.50	21.94	22.16	22.40	22.87	23.37
36.25	36.99	37.37	37.69	38.56	39.40	21.25	21.68	21.91	22.14	22.61	23.10
36.00	36.73	37.11	37.50	38.30	39.13	21.00	21.43	21.65	21.87	22.34	22.83
35.75	36.48	36.86	37.24	38.03	38.86	20.75	21.17	21.39	21.61	22.07	22.55
35.50	36.22	36.60	36.98	37.77	38.59	20.50	20.92	21.13	21.35	21.81	22.28
35.25	35.97	36.34	36.72	37.50	38.32	20.25	20.66	20.88	21.09	21.54	22.01
35.00	35.71	36.08	36.46	37.23	38.04	20.00	20.41	20.62	20.83	21.28	21.74

Source: Wellman (undated).

This can be offset by adequate consideration in the price. But, as in any other transaction in a competitive market, the "consideration" will depend upon demand and supply conditions for livestock at the particular time and the astuteness and bargaining ability of the participants.

One overriding consideration in shrinkage is the desirability of having an accurate set of scales at the farm, ranch, or feedlot. Shrinkage is so variable that results of average, or unknown, conditions from other sources are at best only guidelines or approximations. With a good set of scales, an operator can establish shrinkage with a reasonable degree of accuracy for his particular situation. The availability of scales also serves the extremely important function of permitting an operator to check rates of gain and feed consumption for his particular, and often unique, conditions.

Bruising, Crippling, and Death Loss

Bruised carcasses are costly. In the first place, the bruised area, normally, is condemned and must be trimmed away; and secondly, the remainder of the carcass is docked due to a less appealing appearance. Cripples are docked severely on a live basis as some must be condemned, and others may be partially condemned. Dead animals, of course, have virtually no salvage value. Livestock Conservation, Inc., a Chicago-based organization which has collected data and carried out educational programs for many years to reduce such losses, estimates that millions of dollars are lost each year. The irony of the situation is that most bruises occur on the higher priced cuts and much of the loss could be prevented by simple and sensible precautions.

Among major causes of bruising are: overcrowding; trampling; striking the animals with clubs, canes, and whips; kicking, prodding, and horning; fork and nail punctures; slipping; lifting sheep and lambs by the wool; and poor loading facilities.

Optimum density of loading depends somewhat on weather conditions, but too few or too many animals in a truck or rail car will increase the incidence of bruises. There is little excuse for striking animals to the extent of bruising (in addition to bruise loss, the associated agitation and excitement will increase shrinkage). Inadequate and poorly designed pen and loading facilities can be corrected. Narrow gateways, protruding boards, nails, and sharp edges are frequent causes of bruising and can be corrected.

Many of the factors mentioned above also contribute to crippling and death. In addition, inadequate provision for temperature control is a major cause of death. Improper loading density and lack of adequate partitioning together with faulty loading and unloading facilities are major causes of crippling. Most of these things are controllable to a large degree.

Bruising, crippling, and death loss probably cannot be eliminated. Accidents will happen. However, this loss certainly could be reduced by proper precautions.

BIBLIOGRAPHY

ABBENHAUS, C. R., and PENNY, R. C. 1951. Shrink characteristics of fat cattle transported by truck. Chicago Union Stockyard and Transit Co., Chicago.

AGNEW, D. B. 1973. Cost of marketing U.S. livestock through dealers and public agencies. USDA Marketing Res. Report No. 998.

AGNEW, D. B. 1975. Trends in prices and marketing spreads for beef and pork. USDA Econ. Res. Serv., ERS 556 (Revised).

BERMETTLER, E. R. 1964. Interstate transportation of Nevada cattle. Univ. Wyoming, Max C. Fleischmann Coll. Agr. Bull. *234*.

BJORKA, K. 1938. Shrinkage and dressing yields of hogs. USDA Tech. Bull. *621*.

BOLES, P. P. 1976. Operations of for-hire livestock trucking firms. USDA Econ. Res. Serv., Agr. Econ. Report No. 342.

BRANDOW, G. E. 1966. Implications for consumers in the work of the National Commission on Food Marketing. Proc. 44th Ann. Agr. Outlook Conf., Washington, D.C., Nov. 16.

CAPENER, W. N. *et al.* 1969. Transportation of cattle in the west. Wyoming Agr. Expt. Sta. Res. J. *25*.

CASAVANT, K. L., and NELSON, D. C. 1967. An economic analysis of the cost of operating livestock trucking firms in North Dakota. N. Dakota Agr. Expt. Sta. Agr. Econ. Rept. *55*.

DUEWER, L. A. 1969. Effects of specials on composite meat prices. USDA Agr. Econ. Res., Econ. Res. Serv. *21*, No. 3, 70–77.

DUEWER, L. A. 1970. Price spreads for beef and pork revised series, 1949–69. USDA Econ. Res. Serv. Misc. Publ. *1174*.

DUEWER, L. A. 1978A. Personal correspondence. Agricultural Economist, Commodity Economics Division, Econ., Stat., and Co-op. Serv., USDA.

DUEWER, L. A. 1978B. Changes in price spread measurements for beef and pork. USDA Econ., Stat., and Co-op. Serv. LMS-222, 33.

HARSTON, C. R. 1959A. Cattle shrinkage is important. Montana Agr. Expt. Sta. Circ. *220*.

HARSTON, C. R. 1959B. Cattle shrinkage depends on where, when and what you market. Montana Agr. Expt. Sta. Circ. *221*.

HARSTON, C. R. 1959C. Cattle shrinkage depends on how you market. Montana Agr. Expt. Sta. Circ. *222*.

HARSTON, C. R., and RICHARDS, J. 1965. Montana Livestock transportation. Montana Agr. Expt. Sta. Bull. *592*.

HENNING, G. F., and THOMAS, P. R. 1962. Factors influencing the shrinkage of livestock from farm to first market. Ohio Agr. Expt. Sta. Bull. *925*.

HULBERT, ARCHER B. 1920. The Paths of Inland Commerce, Vol. 21. Yale Chronicles of America Series. By permission of United States Publishers Assoc., New Rochelle, N.Y.

MADSEN, A. G. 1965. Calf shrinkage under auction market conditions. USDA Consumer Marketing Serv. Res. Rept. *718*.

MARSH, J. M. 1977. Effects of marketing costs on livestock and meat prices for beef and pork. Montana Agr. Expt. Sta. Bull. *697*.

MOSER, D. E. 1970. Changes in transportation and their implications for the livestock and meat industry. *In* Long-Run Adjustments in the Livestock and Meat Industry: Implications and Alternatives, T. T. Stout (Editor). Ohio Agr. Res. Develop. Center Res. Bull. *1037*. Also, North Central Regional Publ. *199*.

NATIONAL COMMISSION ON FOOD MARKETING. 1966A. Organization and competition in the livestock and meat industry. Natl. Comm. Food Marketing Tech. Study *1*, U.S. Govt. Printing Office, Washington, D.C.

NATIONAL COMMISSION ON FOOD MARKETING. 1966B. Organization and competition in food retailing. Natl. Comm. Food Marketing Tech. Study *7*. U.S. Govt. Printing Office, Washington, D.C.

RAIKES, R. and TILLEY, D. S. 1975. Weight loss of fed steers during marketing. Amer. J. Agr. Econ. Vol. 57, No. 1 83–89.

SCHULTE, W. (undated) Beef marketing margins and costs. S. Dakota Coop. Ext. Serv. *FS-209*.

SPERLING, CELIA. 1957. The agricultural exemption in instrastate trucking, a legislative and judicial history. USDA Agr. Marketing Serv., Marketing Res. Rept. *188*.

STOUT, T. T., and ARMSTRONG, J. H. 1960. What happens when hogs are fed at market? Purdue Univ. Dept. Agr. Econ., Econ. Marketing Inform. Indiana Farmers Mar. 30.

STOUT, T. T. and COX, C. B. 1959. Farm-to-market hog shrinkage. Indiana Agr. Expt. Sta. Res. Bull. *685*.

USDA. 1975A. Price spreads and industry margins are not the same. USDA Econ. Res. Serv., ERS-607.

USDA. 1975B. Facts of farm-retail price spreads for beef and pork. USDA Econ. Res. Serv., ERS 597.

USDA. 1977. 1977 Handbook of agricultural charts. USDA Handbook No. 524.

USDA. 1978. National food review. USDA Econ., Stat., and Co-op. Serv. NFR-1.

WELLMAN, A. C. (undated) Dressed, live price and shrinkage Tables. Nebraska Coop. Ext. Serv. *EC 70*–839.

WILEY, J. R., and COX, C. B. 1955. Hog shrinkage—farm to market. Purdue Univ. Dept. Agr. Econ., Econ. Marketing Inform. Indiana Farmers Feb. 26.

16

Meat Substitutes and Synthetics

A rapid rise in meat prices during the early 1970's gave impetus to a development which was already well underway—the production of meat substitutes and synthetics. Meat substitutes are not new. People of certain religions and cultures have survived on meat substitutes for centuries. The U.S. livestock and meat industry, however, became interested and concerned especially during the decade of the 1970's. Not only did the general consumer begin to balk for economic reasons, but a vegetarian sub-culture also became more prevalent, and reduced meat consumption (particularly fatty meats) became a more common prescription by the medical profession for certain cardiovascular and other diseases. Concurrent shortfalls of food and feed crops in many countries of the world added further concern about the feasibility of feeding scarce grain to livestock for a source of protein. A great deal of governmental and industrial effort, in the U.S. and elsewhere, was expended in technological research on meat substitutes and synthetics. Subsequent improvement in world-wide crop production and increased meat production (associated with the liquidation phase of the cattle cycle in major cattle producing countries) led to extreme declines in meat prices and grain prices. Concern abated over the economic urgency for meat substitutes. However, that concern surfaced again in the late 1970's when meat production dropped and prices rose rapidly—a predictable reflection of the cattle cycle. The basic circumstances which precipitated livestock and meat concern are still very much in evidence. Problems have not developed to the magnitude that many feared, but the enormous potential market—about 40 billion pounds of red meat—provides a great incentive to industry for the development of substitutes and synthetics.

CLARIFICATION OF TERMS

Use of the terms "substitute" and "synthetic" in connection with agricultural products has not been entirely consistent; but, generally, substitutes are considered to be products used in place of conventional natural-form agricultural products. The natural-form agricultural product may, or may not, have undergone processing. Synthetics are products derived from raw materials of nonagricultural origin and used in place of agricultural products. Thus, substitution may involve one agricultural product for another, or it may involve a synthetic for an agricultural product. As used here, then, synthetics are substitutes, but not all substitutes are synthetic.

ASSOCIATED FACTORS

The forces which stimulate development of substitutes, both agricultural and synthetic, may be classified as economic, social, technological, and institutional. In a competitive economy, businessmen constantly are searching for ways of luring customers from competitors with better, different, or less expensive products. The overriding incentive, from the producer's standpoint, is profit. Consumer demand changes rather slowly, but nevertheless it does change. Personal disposable incomes are increasing, giving people more and more discretionary spending power. Tastes change with affluence, fads, increases in knowledge, and changes in occupational status. An almost continuous stream of technological developments provides sources of know-how for product and market development. Shifts in the urban-rural distribution of population and intracity shifts to suburbia, or vice versa, are related to social changes which impinge upon the preferences of people for agricultural and other products. Institutional factors include governmental concern and action about the health and welfare of the general public. Recent years have witnessed an increase in governmental attention to consumer problems. While the government still recognizes the interests of separate sectors, agriculture's influence has declined and will decline further in the future.

The impact of substitute and synthetic products on agriculture's economic position is a complex phenomenon. Each product must be viewed separately since some agricultural sectors produce substitutes and other sectors utilize them in their productive processes, both of which presumably benefit those sectors. Other sectors find the demand for their products displaced by substitutes and suffer an adverse effect. A substitute which draws its raw materials from agriculture may have little effect on agriculture as a whole, but it may alter intra-agricultural relationships. For example, oleomargarine had an adverse effect on the dairy sector because of a reduction in demand for butterfat, but enhanced the demand for vegetable fats, primarily soybean oil. Presumably, in this case one agricultural sector's loss was another's gain. But expansion in use of soybean oil for oleomargarine affected its availability and price for other uses. Soybean meal is a joint product which also is affected. Urea, a synthetic product used to replace part of the agriculturally-based protein supplements in feeds for ruminant animals, has adversely affected the soybean and cotton sectors by reducing the demand for the respective meals. Presumably, the feeders of urea (the feeding sector) derive some benefit. And it is likely that some soybean growers also feed urea so that adverse effects to one enterprise may be at least partially offset by beneficial effects to another on the same farm. Significant increases in the use of synthetic fibers have severely cut into the demand for cotton and wool, with no offsetting benefits to agriculture, and the same has happened to citrus growers with expansion in use of synthetic citrus drinks. Other examples could be cited, but these illustrate some of the complexities and conflicts of interest within agriculture, and some between agricultural and nonagricultural industries.

RAW MATERIAL SOURCES

Vegetable Protein Substitutes

Major efforts to date in the development of meat substitutes have centered in the use of vegetable proteins. Soybeans are the major source of raw materials, but cottonseed, peanuts, sunflower, and safflower are potential alternative sources. Soy proteins are used in two general ways: (1) as a dilutant or extender in processed meat items and in comminuted meats, and (2) as the major ingredient in a meatless meat product designed to be a total substitute with characteristics of meat—referred to as a meat analog.

Soy proteins currently are used in four forms: (1) flour and grits, (2) concentrates, (3) isolates, and (4) textured items. Flour and grits are lowest in protein content with 40–55%, while concentrates are 65–70%, isolates 90–97%, and textured products range from 50% to more than 90% (Manley and Gallimore 1971). The textured products may be "extruded" or "spun." Extruded soy items are near the low end of the protein range, while spun proteins are near the upper end of the range. The various forms differ also in physical and chemical properties, use, and price. The spinning process, which is similar to that used in manufacturing rayon and nylon, is the most complex and costly. Extruded and spun proteins can be textured into a fibrous structure with chewability characteristics (varying degree of tenderness, toughness, and mouth-feel) of meat. They also can be colored, flavored, and molded into characteristic form to simulate meat. Although much improvement has been made in the meat-like characteristics of analogs, discriminating consumers generally prefer the real product. In addition to simulating the appearance, color, texture, and flavor of meat, analogs are capable of manipulation with respect to caloric content, cholesterol content, protein content, and additives of various sorts. In other words, meat analogs can be manufactured to desired specifications with complete uniformity of product and without seasonal or cyclical variation in supply. Another appealing factor is no cooking loss or shrink during meal preparation. As will be recalled from earlier chapters, large retail chain and HRI buyers are insisting on a year-round supply of uniform quality product. Meat analogs can satisfy these requirements.

Synthetic Substitutes

Among the leading developments in synthetic substitutes is the use of the single cell protein (SCP)—a microorganism (yeast, bacteria or fungi) capable of transforming organic carbon compounds into protein. The process involves cultivation (or "feeding") of the organisms in a hydrocarbon medium, then processing the cells to reclaim the protein. Numerous sources of hydrocarbons are available, but petroleum probably is one of the most abundant. A number of petroleum firms in the United States, several western European countries, and the Soviet Union reportedly are carrying on extensive experiments in attempts to develop a practical process for commercial production. It has been pointed out that waste gas which is "flared" in just one Middle Eastern country would produce the protein equivalent of about one-half the soybean acreage of the U.S. The emerging recognition of eventual world-wide depletion of petroleum and natural gas, and associated price increases of the 1970's, cast doubt on the feasibility of this source of protein. However, there are other substantial sources of hydrocarbons, some of which are waste industrial and animal products. Investigations leave no doubt that protein compounds can be produced by SCP. Limited use already is reported in Europe in the production of animal feeds. While, at this time, no known commercial application is being made in human food, technologically it is possible and may be expected if or when economic conditions warrant it. Findlen (1974) in a discussion of SCP as a source of feed protein concluded that under current costs and prices the likelihood of factory-grown protein replacing natural crop-grown protein was remote in the near future. In referring to the longer-run he stated, "Looking ahead, however, the new protein sources could become a replacement for conventional proteins in times of crop shortfalls and high meal prices, if demand for high protein supplements in animal rations outpaces available supplies."

COMPETITIVE CONSIDERATIONS

Meat substitutes already are on grocery store shelves in a variety of forms—imitation bacon bits or crumbles, bacon strips, products resembling beef and ham—and incorporated in processed and comminuted meats. Relatively high meat prices in 1973 induced widespread merchandising of soy-beef blend hamburger. Generally this was a blend of 75% beef and 25% soy protein—textured soy flour or concentrate. Gallimore (1976) analyzed the market penetration of soy-beef blend hamburger in a study of three retail chains with approximately 1500 stores operating in 21 major markets. Over a 46 week study period the blends market share was about 24% of all ground beef sold. In the same study Gallimore found, "The demand for blend was highly elastic as the quantity of soy-beef blend sales increased on the average 1.6 to 1.8 percent for each 1 percent decrease in price. As the price of regular ground increased, more blend was sold, with the cross-elasticity ranging from 1.1 to 1.6, indicating the blend was considered a close substitute for regular ground.

"The subsequent decline in soy-beef blend sales as beef prices decreased suggests that the sale of soy proteins as meat extenders directly to consumers will be more cyclic than the sale of soy products to institutional markets."

Economics is only one of several factors, but for the majority, price is critical. "The influence of selling price was shown with beef-soy blends that sold well as long as beef sold 15 to 20 cents a pound higher than the blends; when the price differential decreased, blend sales also dropped. The drop in sales of beef-soy blends with declining beef prices indicates that consumers consider beef-soy blends less satisfactory than all beef" (Wolf 1976).

Institutions are reported to be relatively heavy users in prepared dishes. The extent to which substitutes have replaced meat has not been precisely measured, but at this stage it is relatively minor. What happens in the future will depend upon economic considerations, consumer tastes and preferences, and the impact of groups in position to recommend or influence their consumption.

The economic considerations concern producers, consumers, and retail store operators. Retailers are interested in the total contribution of meat operation to net profit. They will not necessarily push a cheaper product if it contributes a smaller margin to net profit. Not enough experience has been gained yet to know the effect of meat substitutes on retail margins.

From the consumers' standpoint, the price of substitutes compared to meat is one of the controlling factors, but also important is the degree to which substitutes satisfy tastes and preferences. The quantity of substitutes in labeled meat products is relatively small and many consumers probably are unaware of its presence. According to Moede et al. (1969), federal standards of identity permit soy protein to be used in meat products up to a 3.5% level. However, regulations do not cover all uses. Manley and Gallimore (1971) report that food served in restaurants is not subject to the same labeling and identification requirements as food sold directly to consumers.

Soy proteins can be readily incorporated into items such as stews, soups, chili, stroganoffs, etc. The least expensive of the substitutes— flour and grits—is generally used. Here the soy proteins can compete with meat and probably will be used to the limit allowed by law, even though the meat they replace in most of these products is from trimmings and the less expensive cuts. Table 16.1 shows the relative cost of net utilizable protein (NPU) from several food sources. "From the information on relative costs, it is obvious why users (at this time decision-makers in institutional kitchens) are interested in substituting soy proteins for more expensive animal proteins. If proteins from soy cost 31¢ per lb and proteins from beef cost $3.26 per lb, there is a strong incentive to substitute soy proteins in uses for which the two products are interchangeable. Examples: pizza, sausage, frankfurters, meat loaf, sandwich salami, etc. Add to this the functional advantages of soy proteins (water and fat retention, improvement in keeping quality, browning effects, etc.) and soy proteins appear to have a bright future. These cost and functional advantages will no doubt do much to overcome present deterrents to acceptance identified with consumer prejudices and governmental regulations. Both these barriers are toppling much faster than most people have imagined possible" (Manley and Gallimore 1971).

The barriers did not topple to the extent implied in that study, due to a

TABLE 16.1.   RELATIVE COSTS OF NET UTILIZABLE PROTEIN COMING FROM SELECTED FOOD SOURCES

Protein Source	Price of the Food[1] ($/Lb)	Cost of the Net Utilizable Protein[2] ($/Lb)
Beef	0.49	3.26
Chicken	0.33	2.47
Fish	0.45	3.07
Whey (dry)	0.09	0.84
Milk	0.07	2.34
Skim milk (dry)	0.22	0.79
Eggs	0.25	2.09
Dry beans	0.07	0.65
Soybean flour	0.08	0.31
Wheat	0.03	0.41
Cottonseed flour	0.35	1.57
Rice	0.09	1.71

Source: Manley and Gallimore (1971).
[1] These prices are for wholesale, lots FOB, point of manufacture.
[2] Crude protein values from *Composition of Foods* USDA Agr. Handbook 8. Net utilizable protein (NPU) is the proportion of nitrogen intake that is retained in the human body. The NPU values used to construct this table from *Amino Acid Content of Food and Biological Data on Proteins*, FAO Nutritional Studies Rep. 24.

large extent to a change in price relationships associated with increased supplies of meat. However, on the basis of cost of net utilizable protein, the advantage is still with soy protein. Cyclical change to relative meat scarcity and relatively high prices will bring this comparison into the limelight again periodically.

Attempts to substitute meat analogs for the higher priced cuts involve several considerations, as follows: (1) textured items and isolates, under the present technology, are more costly to produce than soy flour or soy concentrate, and (2) consumers are more discriminating where the substitute is an imitation of the total product. A study based on 1967–1968 prices reported that "the textured meat-type products made from soy protein isolate currently available would probably need to be priced at retail near the upper range of prices for red meat. These products are reported to have a relatively high ingredient and production cost, and adding usual marketing markups would result in retail prices that might range as high as those for steaks and other cuts from sirloins" (Moede *et al*. 1969). This relationship is a tenuous one, for if the price of meat rises or the cost of meat analogs decreases, the substitutes could be competitive from a price standpoint.

Certain consumers for religious and philosophical reasons, and others in response to transitory fads, welcome imitation meats. The extent to which these groups may influence the total demand for meat is unknown. However, they probably will not have a significant effect, for it is likely most of them have been using some other substitute in the past and, if so, their acceptance of "meatless meat" will not affect the demand for meat.

A group which may already have had some effect and could have considerably more in the future are those with dietary restrictions. Some members of the medical profession recommend that patients with certain cardiovascular problems reduce the intake of animal fat as a measure of control over blood cholesterol levels. Others recommend a restricted fat diet for weight control, or other reasons. Engineering control over the specifications of substitute meat products indicates they are made to order for these situations—and cost may not be the controlling factor.

Thus, it appears that pending further cost-reducing technological developments in substitutes, the major use of soy proteins is in processed meats and prepared dishes in the hotel, restaurant and institutional trade. Institutions provide a growing market for products such as bacon bits which can be incorporated in scrambled eggs, soups, stews, etc. The elimination of preparatory steps and the ease of storage, along with competitive prices, give soy proteins an advantage in these markets.

To date, only minor inroads have been made in the market for higher priced cuts, but increased use is expected for certain dietary purposes and if cost-reducing technology is developed along with improved meat-like characteristics, the general demand for meat will be vulnerable. The magnitude of the potential market is so extensive that continued research and development in substitutes is to be expected.

One of the few reported tests in the household market sector was on a bacon substitute (Corkern and Dwoskin 1970). The test product was an analog designed to simulate the form, texture, color, and taste of bacon. This was in the form of bacon strips—not chips or crumbles—and was sold in a frozen state. The experiment was conducted for 6 months in Fort Wayne, Indiana, utilizing 40 supermarkets in national, regional, and local chains. The market test was conducted by a private research firm under contract to the manufacturer of the product, and results were published by the USDA. During the first three months (Phase I) an intensive promotional and advertising program was carried out, consisting of in-store promotion, as well as newspaper and TV advertising. During the latter three months (Phase II) the advertising was similar to that ordinarily carried on for long-established products.

Bacon analog sales amounted to 4% of bacon sales during Phase I and declined to 1.3% of bacon sales as promotion and advertising dropped off in Phase II. Bacon sales also declined during the period of Phase II, but only ⅛ as much as the drop in bacon analog sales. A survey of purchasers showed that, "In general, users expressed a high level of satisfaction. The product's strongest attributes were ease and speed of preparation and good cooking qualities. A large proportion of the users found nothing they disliked about the product" (Corken and Dwoskin 1970).

There was no clear evidence that bacon analog sales replaced bacon sales during the test period. However, this would hardly be expected in a short, introductory period. The survey showed that many consumers were attracted by the advertised low caloric and cholesterol content, and relatively low price. On an as-served basis, the bacon analog was advertised to cost ½ as much as bacon.

Results of the test showed no adverse reaction to the product in the frozen state. "All in all, test market results, though limited, indicate a good chance for further commercial success of the bacon analog" (Corkern and Dwoskin 1970).

In a consumer panel study of approximately 600 randomly selected urban, southern households Mize (1972) found no adverse effect on palatability from the addition of 2% soybits to ground beef. Weimer (1976) conducted taste tests under controlled laboratory conditions using regular hamburger and three ground beef products containing textured vegetable protein (tvp). In one test participants were told which products contained tvp. In another test that information was withheld. In the test when knowledge of the contents was unknown the participants indicated no difference in preference among the products, but when the products containing tvp were identified, regular hamburger was significantly preferred. From a merchandising standpoint this indicates problems with preconceived bias and suggests the possibility of image enhancement by proper choice of label name, e.g., avoidance of use of the word *soy*.

IMPLICATIONS FOR FOOD SUPPLIES

As a matter of self interest, the livestock and meat industries are rightfully concerned about competition from substitute and synthetic meats. A discussion of the subject, however, would be incomplete without recognition of food supply problems—domestic and world-wide. People in the United States as a whole are the best fed in the world, but recent studies have shown serious inadequacies in the diets of a substantial number. And on a comparative basis the problem is infinitely more serious in less developed countries around the world.

Demographers estimate that the world population will almost double by about the year 2000. This would mean that within a scant 20 yr, world food production will almost need to double just to maintain the present inadequate level of supplies. The problem is one not only of total quantity but also—and especially serious—of a lack of high protein foods. There is little doubt that the United States has the capacity to produce sufficient total food supplies for itself and, with proper economic incentives, to continue to produce sufficient animal protein for domestic needs. However, without some considerable shifts in income distribution, substantial numbers of people probably will not have the purchasing power to buy adequate supplies of animal proteins. It is certain on the world scene that the need will exist for more, and less expensive, proteins than can be furnished from animal sources.

The pressure on food supplies points up the need to consider the efficiency of protein production from alternative sources. On this score, "The production potential of single-cell proteins is fantastic. Production increases exponentially so that a 1 lb seeded culture would multiply to 2 tons of edible food (½ protein, ½ carbohydrates and fats) in 24 hr. The SCP yield from 1000 lb of petroleum is approximately 1000 lb of edible product which compares with 500 lb of catfish, 250 lb of dressed poultry, and 75 lb of dressed beef per 1000 pounds of feed" (Swackhamer 1969). On a conversion ratio of 7 lb of grain to produce 1 lb of beef, it would require about 39 lb of grain to produce 1 lb of animal protein.[1] Soybeans average about 37.9% protein (Morrison 1956). While this 37.9% is not all recoverable in the form of protein concentrates, insofar as technical production

---

[1]This is based on a beef protein content of 17.9% (Moede *et al.* 1969).

efficiency is an issue, it is apparent that protein production by way of animal agriculture is less efficient than by either vegetable or synthetic production. Table 16.1 shows that in 1970 the cost of 1 lb of protein from soybean flour was about $1/10$ that of protein from beef. That ratio changes continually as relative prices of the basic commodities change, but except in unusual circumstances the cost advantage will be with vegetable protein. From the standpoint of palatability and preferences, it nevertheless remains to be shown that people, even in protein deficient areas, can be induced to consume a new food if that new food constitutes a significant change from traditional consumption patterns.

Several commercial U.S. firms have promoted with some success, vegetable protein-fortified bottled and canned drinks in less developed countries. Any gains made in vegetable or synthetic protein consumption in less developed countries would be little threat to U.S. or international meat industries. Most of the people in such countries consume less than 10 lb of meat per capita. Vegetable or synthetic proteins would not be a substitute, but an addition to their present meat consumption.

ALTERNATIVE LIVESTOCK-MEAT INDUSTRY REACTION

The intrusion of substitute and synthetic products into agricultural markets has prompted various reactions. When oleomargarine began to make inroads into the butter market, dairy farmers attempted to induce legal action prohibiting its manufacture and sale. When this proved ineffective, the approach was switched to taxation by individual states and attempts to prohibit its being colored yellow prior to sale. Some states adopted such measures, but all have now faded away. Agriculture has less political power now than in the days of the oleo battle, and even if farmers and ranchers had the power to induce legal restrictions on meat substitutes, the oleo experience indicates its futility.

The cotton and wool industries have given up a considerable share of their markets to substitute and synthetic fibers. After a somewhat belated start, these industries instituted concerted research efforts to develop processing technology for improved characteristics of their products in line with consumer preferences. Improvements have been made in such things as altering the shrink characteristics of wool and cotton fabrics, improving wash-and-wear characteristics, permapress, etc. Technical improvements along with concurrent marketing and merchandising innovations have been effective. While these industries have not regained their former position in the fabric market, the market share for cotton has improved and the long-time drop in wool production has tended to obscure an improvement in consumer preference for wool. The point to be emphasized is that the approach here has been more effective than that used with oleo.

In a protein deficient world, the livestock and meat industries cannot make an effective case for opposing improved technology for the production of vegetable and synthetic protein foods. Public opinion would not support such a stand, and as pointed out above, it probably would be ineffective anyway. However, the following aspects warrant consideration: (1) The competitive structure necessitates continual efforts to improve the acceptance of meat and meat products through improvements in production, processing, and merchandising. The most effective competitive device is a product which satisfies consumers' tastes and preferences—backed up by an effective merchandising program. (2) The livestock and meat industry also has a responsibility in inducing governmental vigilence with respect to truthfulness in the labeling of substitute and synthetic products so the public will be fully aware of the ingredients and their health-related characteristics. It cannot be assumed that all such products will equal the nutritive qualities of meat.

BIBLIOGRAPHY

ANON. 1968. Food from petroleum. Standard Oil Company, SPAN *8*, No. 3, Fall.

CORKERN, R. S., and DWOSKIN, P. B. 1970. Consumer acceptance of a new bacon substitute. USDA Econ. Res. Serv. *454*.

CORKERN, R. S., and POATS, F. J. 1968. Synthetics and substitutes in food and non-food markets. USDA Marketing Transportation Situation, Econ. Res. Serv. *MTS-171*.

ETHRIDGE, M. D. 1975. Competitive potential of synthetic meat products: some efficiency implications of nutritional composition. Western Agr. Econ. Assn., Proceedings 48th Annual Conference, Reno, Nevada, July, 191–195.

FINDLEN, P. J. 1974. Factory-grown feed protein is no match for soybean meal. USDA Foreign Agr. Serv., Foreign Agr., November, 6–16.

GALLIMORE, W. W. 1972. Synthetics and substitutes for agricultural products: projections for 1980. USDA Econ. Res. Serv., Marketing Res. Rept. No. 947.

GALLIMORE, W. W. 1976. Estimated sale and impact of soy-beef blends in grocery stores. USDA Econ. Res. Serv., National Food Situation, NFS-155, 37–44.

LUBLIN, J. S. 1976. Soybean saga, revival is attempted for meat substitutes that flopped after '73. Wall Street Journal. Oct. 26, 1.

MANLEY, W. T., and GALLIMORE, W. W. 1971. Emerging product inroads into agriculture: synthetics and substitutes. Proc. 1971 Natl. Agr. Outlook Conf., Washington, D.C., Feb. 24–25.

MINER, B. D. and GALLIMORE, W. W. 1977. Soy protein use can increase 71% by 1985. USDA Farmer Co-op. Serv. Farm Cooperatives, July, 4–6.

MINER, B. D. 1976. Edible soy protein: operational aspects of producing and marketing. USDA Farmer Co-op. Serv., FCS Res. Rept. 33.

MIZE, J. J. 1972. Factors affecting meat purchases and consumer acceptance of ground beef at three levels with and without soya-bits. Georgia Agr. Expt. Sta., Southern Cooperative Series Bull. 173.

MOEDE, H. H. et al. 1969. Meat and poultry substitutes. In Synthetics and Substitutes for Agricultural Products, a Compendium. USDA Econ. Res. Serv. Misc. Publ. 1141.

MORRISON, F. B. 1956. Feeds and Feeding, 22nd Edition. Morrison Publishing Co., Clinton, Iowa.

SWACKHAMER, G. L. 1969. Synthetics and substitutes: Challenge to agriculture. Federal Reserve Bank of Kansas City Monthly Rev. Mar., 3–12.

USDA. 1967. Proceedings of International Conference on Soybean Protein Foods. USDA Agr. Res. Serv. 71–35.

WEIMER, J. 1976. Taste preference for hamburger containing textured vegetable protein. USDA Econ. Res. Serv., National Food Situation, NFS-155, 45–46.

WOLF, W. J. 1976. Edible soy protein, operational aspects of producing and marketing—market growth. USDA Farmer Co-op. Serv. FCS Res. Report 33, 40–45.

Index

Other AVI Books

COMMERCIAL FRUIT PROCESSING
Woodroof and Luh

COMMERCIAL VEGETABLE PROCESSING
Luh and Woodroof

COOKIE AND CRACKER TECHNOLOGY
2nd Edition   *Matz and Matz*

ELEMENTS OF FOOD TECHNOLOGY
Desrosier

ENCYCLOPEDIA OF FOOD SCIENCE
Peterson and Johnson

ENCYCLOPEDIA OF FOOD TECHNOLOGY
Johnson and Peterson

FLAVOR TECHNOLOGY: PROFILES, PRODUCTS,
APPLICATIONS   *Heath*

FOOD PRODUCTS FORMULARY
Vol. 3   *Tressler and Woodroof*

FUNDAMENTALS OF ENTOMOLOGY AND PLANT PATHOLOGY
Pyenson

HANDLING, TRANSPORTATION, AND STORAGE OF FRUITS AND
VEGETABLES
Vol. 1   *Ryall and Lipton*
Vol. 2   *Ryall and Pentzer*

LABORATORY MANUAL FOR ENTOMOLOGY AND PLANT
PATHOLOGY   *Pyenson and Barké*

MODERN PASTRY CHEF
Vol. 1 and 2   *Sultan*

PEANUTS: PRODUCTION, PROCESSING, PRODUCTS
2nd Edition   *Woodroof*

PLANT PHYSIOLOGY IN RELATION TO HORTICULTURE
American Edition   *Bleasdale*

POSTHARVEST BIOLOGY AND HANDLING OF FRUITS AND
VEGETABLES   *Haard and Salunkhe*

POSTHARVEST PHYSIOLOGY, HANDLING AND UTILIZATION OF
TROPICAL AND SUBTROPICAL FRUITS AND VEGETABLES
Pantastico

SOURCE BOOK FOR FOOD SCIENTISTS
Ockerman

TREE NUTS: PRODUCTION, PROCESSING, PRODUCTS
2nd Edition   *Woodroof*